T0418830

CEMENTED CARBIDES

CEMENTED CARBIDES

IGOR KONYASHIN
Element Six GmbH, Burghaun, Germany

BERND RIES
Element Six GmbH, Burghaun, Germany

Elsevier
Radarweg 29, PO Box 211, 1000 AE Amsterdam, Netherlands
The Boulevard, Langford Lane, Kidlington, Oxford OX5 1GB, United Kingdom
50 Hampshire Street, 5th Floor, Cambridge, MA 02139, United States

Notices
Knowledge and best practice in this field are constantly changing. As new research and experience broaden our understanding, changes in research methods, professional practices, or medical treatment may become necessary.

Practitioners and researchers must always rely on their own experience and knowledge in evaluating and using any information, methods, compounds, or experiments described herein. In using such information or methods they should be mindful of their own safety and the safety of others, including parties for whom they have a professional responsibility.

To the fullest extent of the law, neither the Publisher nor the authors, contributors, or editors, assume any liability for any injury and/or damage to persons or property as a matter of products liability, negligence or otherwise, or from any use or operation of any methods, products, instructions, or ideas contained in the material herein.

Library of Congress Cataloging-in-Publication Data
A catalog record for this book is available from the Library of Congress

British Library Cataloguing-in-Publication Data
A catalogue record for this book is available from the British Library

ISBN: 978-0-12-822820-3

For information on all Elsevier publications visit our website at https://www.elsevier.com/books-and-journals

Publisher: Matthew Deans
Acquisitions Editor: Christina Gifford
Editorial Project Manager: Andrea Dulberger
Production Project Manager: Poulouse Joseph
Cover Designer: Mark Rogers

Typeset by TNQ Technologies

Contents

Preface

This book provides a general overview of fundamental and technological aspects of manufacturing cemented carbides, also known as "hardmetals," as well as their microstructures, properties, examination techniques, applications, and mechanisms of wear and degradation in various areas of use.

The book contains a summary on modern cemented carbide grades for different applications and is thought to be useful for newcomers in the field as well as for application engineers dealing with the employment of cemented carbides and carbide tools in a great number of applications varying from metal-cutting to mining and construction. Basic phase diagrams of WC-based cemented carbides are presented and analyzed with an emphasis on modern trends regarding the further development, improvement, and optimization of up-to-date carbide grades for numerous applications. The book also describes all aspects related to the fabrication and examination of cemented carbides starting from the production of raw materials to ending with final operations of brazing, surface finishing, and deposition of thin wear-resistant coatings. Technological processes and equipment employed on different stages of the cemented carbide manufacture including milling, granulation, pressing, sintering, surface finishing, etc., are described in detail. Modern techniques and instruments used for the examination and control of the microstructure and properties of cemented carbide are reported. Conventional modern cemented carbide grades and advanced uncommon industrial grades, which were developed and implemented in industry in the last decade, are described. Performance and operations of cemented carbides and carbide tools in a great number of applications varying from machining tools to high-pressure high-temperature components employed for the diamond and cubic boron nitride synthesis are reported with an emphasis on mechanisms of cemented carbides' wear and degradation in various areas of use.

The major objective of this book is not to summarize all the numerous publications and patents in the field of cemented carbides, so each chapter comprises a limited number of only the major references related to the concrete chapter topic. The major idea of the book is to give a general but still comprehensive picture of the present state of the art with respect to materials science and engineering of cemented carbides as well as their applications, performance, operations, wear and failure mechanisms, etc. The book comprises basic but still wide-ranging overview of modern theoretical approaches employed for simulations and predictions of the cemented carbide structures and properties on the macro-, micro-, nano-, and atomic-level. The book also includes a comprehensive overview of modern experimental techniques employed for the examination of such structures by use of up-to-date methods (transmission electron microscopy, in situ examination of deformation processes in transmission electron microscopes, micromechanical testing, etc.). The final part of the book outlines recent trends in the research and development of cemented carbides.

In spite of the fact that cemented carbides are unique materials essential for a great number of applications and one cannot imagine modern industry without their large-scale employment, surprisingly, English language monographies on cemented carbides published in the last 50 years are limited by the only one book of Upadhyaya Gopal S. entitled "*Cemented Tungsten Carbides: Production, Properties, and Testing*" (1998). This book covers some aspects of manufacturing and properties of cemented carbides. However, it appears to be presently out of date, as the information about up-to-date technological processes, examination techniques (for example, transmission electron microscopy), and modern uncommon industrial carbide grades, that did not exist 25 years ago, is missing in this book. Also, modern approaches to ab initio simulations and predictions of the cemented carbide structures on the micro-, nano-, and atomic-level as well as up-to-date experimental techniques for examinations of cemented carbides are not reported in this book. Other books on cemented carbides, for example, the English translation of the classical book of Swarzkopf and Kiefer entitled "*Cemented Carbides*" (1960), were published a long time ago, so the cemented carbide industry has dramatically changed since that time. The other monographies on cemented carbides, for example, the classical book of H. Kolaska entitled "*Powder Metallurgy of Hardmetals*," 1992 (in German) and the comprehensive monography of V.I.Tretyakov entitled "*Bases of materials science and production technology of sintered cemented carbides*" (in Russian), are excellent books; however, besides the language barrier, they can be assessed as outdated when taking into account the significant scientific and technological advances developed in the last 25 years. The most recently published assembly of review-like articles on cemented carbides entitled "*Comprehensive Hard Materials*" (2014) does not cover all the different aspects of materials science and engineering of cemented carbides, as it is not structured as a "monography." Different chapters of this book were written by various authors resulting in the fact that some topics on cemented carbides are reported in excessive details, but other important topics are not reflected in the book at all, although the Editors did their best to get the "maximum coherence" of the book from this point.

The major audience of the present book is primarily experts from both industry and academia, for whom the field of cemented carbides is new and who want to learn about the fundamental and applied aspects of the cemented carbides' production and applications, as well as to obtain an in depth knowledge of the field. The book can be useful for MS and PhD students studying materials science and engineering, young engineers, and researchers dealing with the development, fabrication, and examination of cemented carbides and other hard and superhard materials. The book can also be used as a handbook for engineers dealing with the employment of cemented carbide tools and articles in metal-cutting, machine-building, mining, construction, oil-and-gas drilling, etc. The book is thought to be comprehensive, especially with respect to presenting and analyzing phase diagrams, manufacturing processes and modern examination techniques, which can make it interesting also for experts who have worked in the field for quite a long time. Those who want to become less of an expert in the field, in particular, application engineers, can use it as a reference book. Some chapters of the book can be useful as reference literature also for the more general materials science readership, and certain chapters (e.g., those on the powder fabrication, powder processing, sintering, microstructure analysis, etc.) are believed to be applicable also to students, researchers, and engineers working in other areas of materials science and engineering.

The authors would like to thank various organizations for their permissions to reproduce illustrations and photos of their industrial equipment, in particular, Wolfram Bergbau und Hütten AG (Austria), Saar-Harmetall und Werkzeuge GmbH (Germany), Joint Stock Company "Kirovgrad Hard Alloys Plant" (Russia), Tikomet Oy (Finland), and Tisoma Anlagen und Vorrichtungen GmbH (Germany). The authors would like to thank colleagues from these companies for fruitful discussions and assistance, particularly, A. Bicherl, H. Westermann D.A. Pelts, and A. Eckert. The authors gratefully acknowledge the assistance of their colleagues from Element Six with respect to the book preparation and fruitful discussions as well as the Element Six Management for the permissions to publish illustrative materials with special thanks to K.-G. Hildebrand, M. Wingenfeld, J. Sauer, K. Binder, L. Elter, A. Loch, H. Hinners, S. Farag, D. Müller, P. Balog, S. Lu, R. Möller, M. Reinhard, E. Weinbach, S. Hlawatschek,

R. Hlawatschek, S. Harmatzy-Simon, G. Richter, W. Talke, S. Thurnwald, G. Rollison, and G. Sciarrone. The authors also appreciate assistance of colleagues from Chalmers University of Technology (Sweden), Universitat Politecnica de Catalunya (Spain), and National Physical Laboratory (Great Britain) with special thanks to K. Mingard and H. Jones.

CHAPTER 1

The history of the invention and early development of cemented carbides

WC-based cemented carbides are the best and unique materials for applications characterized by intensive wear and abrasion, high impact loads, thermal shocks, high bending and compressive loads, elevated temperatures, and severe fatigue. Therefore, their discovery in the beginning of the 1920s resulted in revolutionary changes in different branches of industry including metal-cutting, mining, construction, wire-drawing, etc.

Tungsten carbide, the ceramic hard phase of cemented carbides, was first synthesized by French chemist Henri Moissan with the aid an electric arc furnace that he invented. In his book "*The Electric Furnace*" published in 1897, he described syntheses of different new refractory compounds including tungsten carbides by use of this furnace. It is likely that Moissan obtained a mixture of WC and W_2C, as the precise phase identification by XRD was not developed at that time.

Just tungsten carbide was presumably chosen as a hard phase of cemented carbides among different hard refractory compounds due to the fact the development of cemented carbides started at companies producing tungsten powders for fabrication of tungsten wires for filaments for electrical lightning.

At such companies, the initial tungsten metal powder and equipment for its carburization and sintering trials (hydrogen tube furnaces) were available for large-scale research with the objective to develop novel hard materials suitable to substitute very expensive diamond dies needed for drawing tungsten wires.

Intensive studies on searching such hard materials started at the company entitled "Auer Gesellschaft" in Berlin. Fig. 1.1 shows a photo of Franz Skaupy, who was the head of the group entitled "Research group for electrical lighting" at the "Auer Gesellschaft" in Berlin [1,2] and initiated research on developing new WC-based hard materials for drawing dies.

Cemented Carbides
ISBN 978-0-12-822820-3
https://doi.org/10.1016/B978-0-12-822820-3.00008-2

Figure 1.1 Photo of Prof. Franz Skaupy. *(Redrawn from Ref. [1].)*

Meanwhile, Karl Schröter, who invented WC—Co cemented carbides in the beginning of the 20s, started working at the electrical department of the German gas lamp company DGA in 1908. He was involved in experiments on obtaining WC-based hard materials by use of different approaches including the infiltration of porous carbide bodies by Fe-group metals and casting mixtures of carbides and transitions metals [3—5]. In 1919, a new company OSRAM formed as a result of merging DGA, Siemens Lamp, and AEG, which were two other German companies dealing with the electrical bulbs' production, and Franz Skaupy was appointed as director for research and development at this new company [6]. Under his supervision, Karl Schröter and coworkers started experiments on the fabrication of WC powders by mixing tungsten powders with graphite and heating in tube furnaces in the early 20s. The first drawing dies were produced by casting a mixture of tungsten carbides, but they were too brittle. Other approaches including sintering a mixture of WC and Mo_2C were tried, but the resulting materials were also too brittle for any potential application. Afterward, it was decided to mix WC powders with iron and nickel powders, and samples of such mixtures were sintered at temperatures of up to 1500°C in a protective atmosphere.

Surprisingly, the samples obtained in such a way were found to have a high density and relatively high hardness; however, the first results on sintering the WC—Fe and WC—Ni samples were controversial. Nowadays, we can explain this controversy due to understanding the W—C—Fe and W—C—Ni phase diagrams (see Chapters 4 and 6). In the W—C—Fe system, the two-phase region not comprising free carbon plus cementite and η-phase is extremely narrow, so that the microstructure of the first WC—Fe samples obtained by Karl Schröter presumably belonged to the three-phase regions of the phase diagram containing either graphite plus cementite or η-phase. The two-phase region of the W—C—Ni phase diagram is noticeably shifted toward lower carbon contents in comparison to the carbon content corresponding to stoichiometric WC, so that the WC—Ni samples obtained from stoichiometric WC were likely to contain inclusions of free carbon.

Afterward, Karl Schröter tried to mix WC and Co powders and sintered samples of such a mixture in a protective atmosphere, which leaded to unexpectedly good results. After several experiments on sintering such WC—Co mixtures, the sintered samples were tried as components employed for the fabrication of drawing dies. Amazingly, the results were excellent, so that, afterward, the WC—Co material was successfully employed to produce dies for drawing of relatively coarse tungsten wires. As a result, K. Schröter filed a patent application in 1923, and the corresponding patents, which became afterward famous, were granted in different countries [7,8] (Fig. 1.2).

At that time, the importance of this invention was not yet understood, so the patent application was only successfully filed in Germany, England, and the United States of America, whereas for all other countries, the timely priority claim was missed.

OSRAM offered the patent to the Krupp company, which started a small-scale production of the new hard material. Patent attorneys and technical people at Krupp designated the new material by the trade name "WIDIA," which was a combination of the German words "wie" and "Diamant," meaning "like diamond."

The production of the WIDIA cemented carbides according to the "*WIDIA Handbook*" [9] is shown in Fig. 1.3. The production at Krupp had to be up-scaled from the pilot-batch scale to the large-scale manufacture,

DEUTSCHES REICH

AUSGEGEBEN AM
30. OKTOBER 1925

REICHSPATENTAMT

PATENTSCHRIFT

— Nr. 420689 —

KLASSE 40b GRUPPE 1

(P 46003 VI/40b)

Patent-Treuhand-Gesellschaft für elektrische Glühlampen m. b. H. in Berlin*).

Gesinterte harte Metallegierung und Verfahren zu ihrer Herstellung.

Patentiert im Deutschen Reiche vom 30. März 1923 ab.

Figure 1.2 The first page of Schröter's patent on WC—Co cemented carbides granted in Germany.

which was conducted at a Krupp plant in Essen. The quality testing of the WC—Co cemented carbides was initially very simple: the toughness test was performed by hitting a carbide sample by a hammer. The hardness test was performed by scratching of glass. It is reported that after some time no more window glass without scratches was available for this purpose in the quality testing laboratories. The microstructure of cemented carbides was examined by optical microscopy of fracture surfaces especially with respect to the residual porosity. The initially used coal gas as sintering atmosphere was substituted by hydrogen. The application of hydrogen was dangerous and a catastrophic hydrogen explosion was reported within a gas purification unit which caused heavy damage but fortunately no casualties [10]. The sintered carbide tips shown in Fig. 1.4 were then soldered onto steel with a brass solder and subsequently ground. Fig. 1.5 shows the cover page of the "*WIDIA Handbook*" published in six languages in 1930.

The first presentation of the new material with the brand name WIDIA—"Widia-N"—by Friedrich Krupp AG was carried out at the Spring Leipzig Fair in 1927 and caused a sensation (Fig. 1.6). The first cemented carbides were extremely expensive in the beginning of the 1930s, so 1 g of cemented carbide cost roughly 1 USD, which was comparable with the price of 1 g gold at that time. On that ground, the demand on such an expensive material and consequently the total production volume of cemented carbides in the beginning of the 1930s were limited. Nevertheless, the prices on cemented carbides articles decreased before the

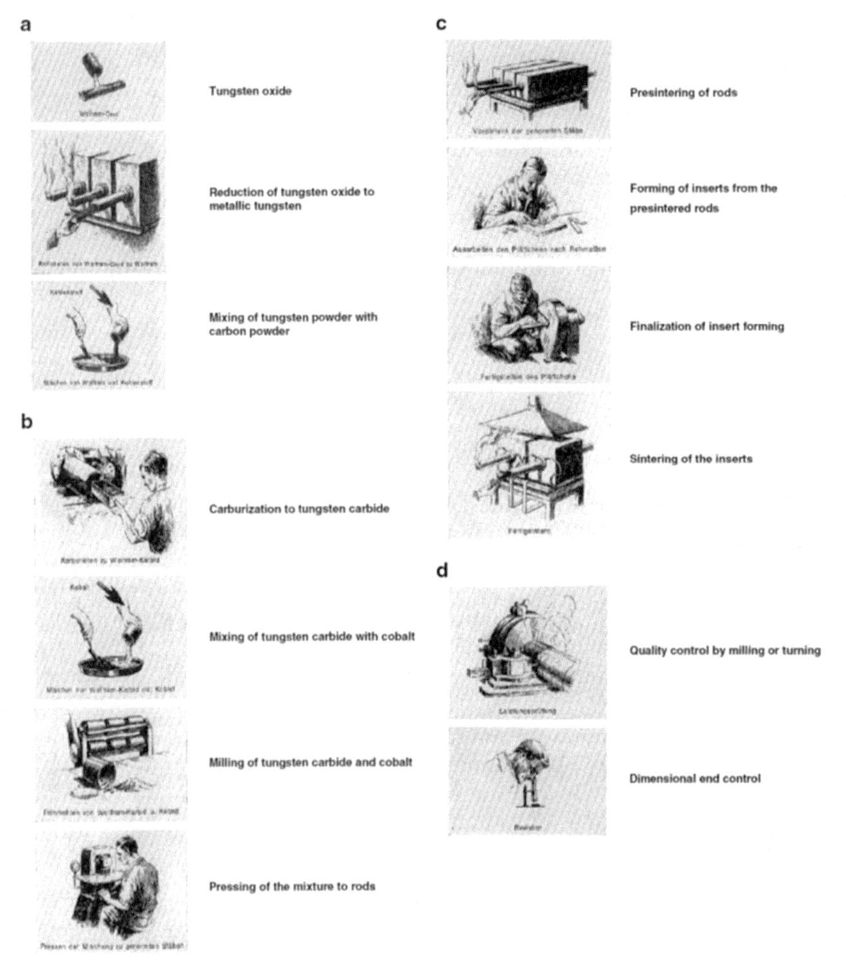

Figure 1.3 The production of hardmetal "WIDIA" (from the "*WIDIA-Handbook*", 1936). Titles to the pictures of the workflow: (a) Tungsten oxide – Reduction of tungsten oxide to metallic tungsten – Mixing of tungsten with carbon; (b) Carburization to tungsten carbide – Mixing of tungsten carbide with cobalt – Milling of tungsten carbide with cobalt – Pressing of green articles; (c) Presintering – Forming of inserts from the presintered green articles – Finalization of inserts' forming – Sintering of the inserts; (d) Quality control by milling or turning – Dimensional end control. *(Redrawn from Ref. [1].)*

beginning of the Second World War and during the war: in 1937 – 0,44 \$/g, in 1941 – 0,22 \$/g, and in 1944 – 0,15 \$/g. In the mid-1930s, the production of cemented carbides by Krupp WIDIA was roughly 40 t per year.

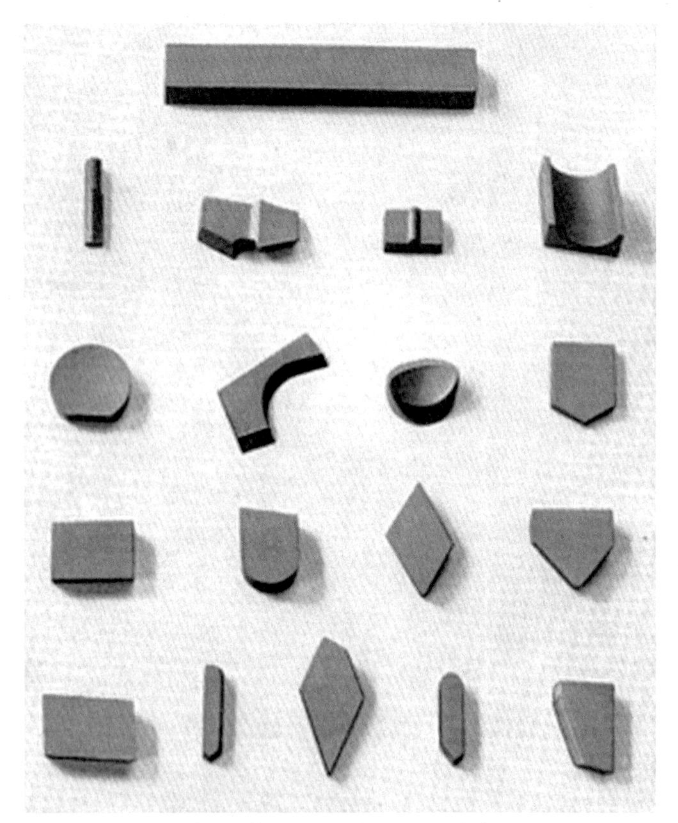

Figure 1.4 Carbide articles for different applications of the WIDIA cemented carbide (from the "*WIDIA Handbook*"). *(Redrawn from Ref. [1].)*

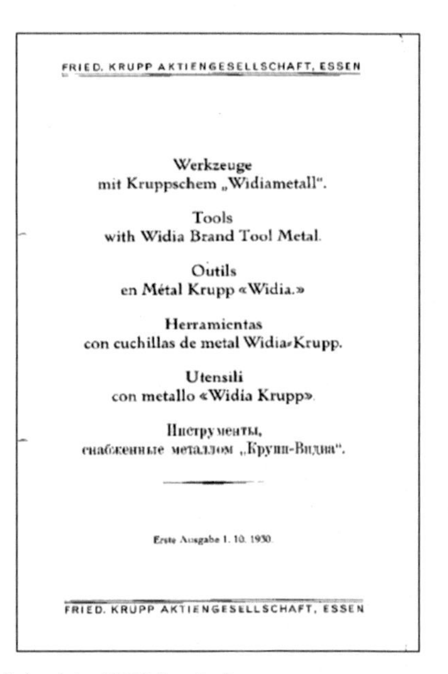

FRIED. KRUPP AKTIENGESELLSCHAFT, ESSEN

Werkzeuge
mit Kruppschem „Widiametall".

Tools
with Widia Brand Tool Metal.

Outils
en Métal Krupp «Widia.»

Herramientas
con cuchillas de metal Widia-Krupp.

Utensili
con metallo «Widia Krupp».

Инструменты,
снабженные металлом „Крупп-Видиа".

Erste Ausgabe 1. 10. 1930.

FRIED. KRUPP AKTIENGESELLSCHAFT, ESSEN

Figure 1.5 The *WIDIA Handbook* published in 1930 in six languages.

Figure 1.6 The first presentation of the WIDIA cemented carbide at the Leipzig Spring Fair in 1927. *(Redrawn from Ref. [2].)*

Interestingly, as early as in the end of the 1930s, the assortment of carbide grades produced by Krupp WIDIA was quite advanced including TiC-containing grades for steel-cutting, which were invented in 1929, and fine-grain grades containing VC and TaC as grain growth inhibitors, which were developed in 1930–31. These grades were described in the report entitled "*The German Hard Metal Industry*" written by a group of British experts as a result of their visit to Germany in 1945 (Figs. 1.7 and 1.8). The production plant of Krupp WIDIA in Essen was heavily bombed during the Second World War, so the carbide production was stopped in 1944. Nevertheless, directly after the occupation of Essen by the American Troops in April 1945, intensive reconstruction works started and the carbide fabrication was recovered soon afterward.

Several companies producing cemented carbides were founded in the Western countries in the decade following the discovery of WC–Co materials, namely: Carboloy Corp. by General Electric, USA (1928); Sumitomo, Japan (1929); Plansee Titanit, Austria (1930); Wimet, Great Britain (1931); Mitsubishi, Japan (1931), Fagersta Bruks AB, Sweden (nowadays, Seco Tools) (1932), Kennametal Inc., USA (1938); Sandvik, Sweden (1942). Interestingly, cemented carbide factories in different countries, particularly in Germany, were located in small towns and villages, as in this case, they did not require high expenses for transport and logistics. Fig. 1.9 shows covers of brochures of some companies producing cemented carbides in the 1930s and 1940s of the last century.

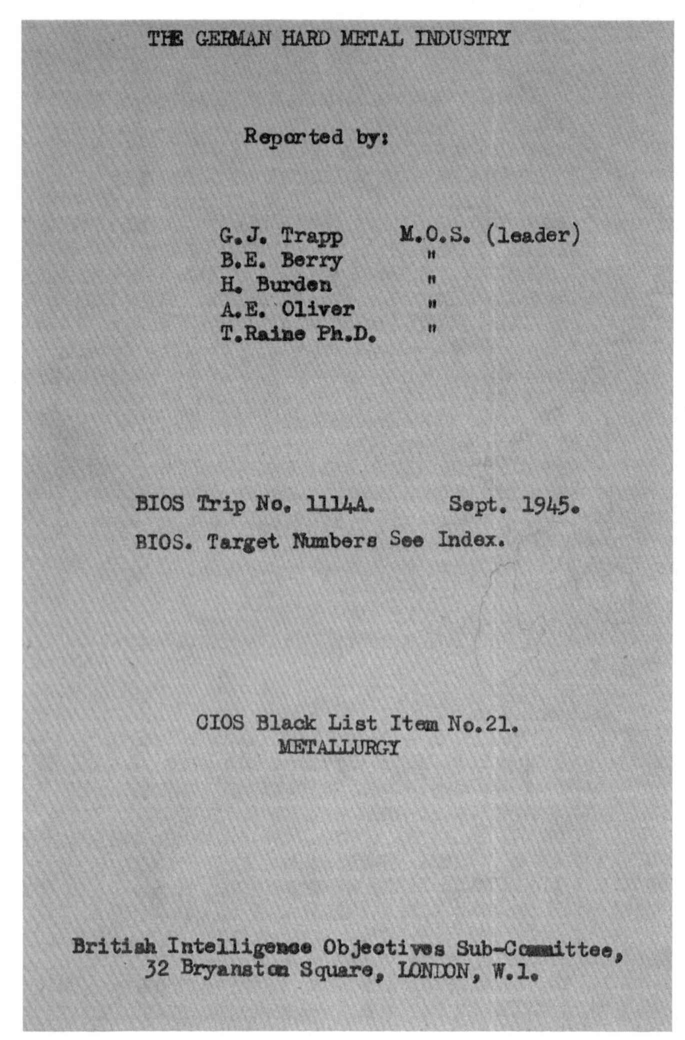

THE GERMAN HARD METAL INDUSTRY

Reported by:

G.J. Trapp M.O.S. (leader)
B.E. Berry "
H. Burden "
A.E. Oliver "
T.Raine Ph.D. "

BIOS Trip No. 1114A. Sept. 1945.
BIOS. Target Numbers See Index.

CIOS Black List Item No.21.
METALLURGY

British Intelligence Objectives Sub-Committee,
32 Bryanston Square, LONDON, W.1.

Figure 1.7 The cover page of the report on the cemented carbide industry in Germany written by a group of British experts as a result of their visit of Germany in 1945.

Gewerkschaft Wallram, a German company, filed a patent application on carbide nozzles for sand blasting in 1930 and started the carbide production in Essen in the beginning of the 1930s of the last century. Wallram took over the Bayer and Müller tool plant in Burghaun, Germany, and partially moved the carbide production to this plant in 1972; this plant is now the major cemented carbide plant of the Element Six Carbide Division. Another plant for producing cemented carbides and mining tools was

Table 5

PROPERTIES OF KRUPPS HARD METAL TOOL TIPS

Grade	Composition	Sintering Temperature °C	Sintering Time in Min. for Tip Thickness 2mm-15mm	Transverse Rupture Strength Kg/sq.mm.	Co-efficient of Thermal Expansion 10^{-6} cm/cm/°C	Heat Conduction Cal./Sec/°C	Specific Heat Cal./gm/°C	Electric Resistance Ohms/Meter 1 sq.mm section
S1	78 W C 16 Ti C 6 Co	1600	20 - 100	125	6	0.09	0.06	0.43
S2	78 W C 14 Ti C 8 Co	1550	20 - 100	140	6.2	0.08	-	0.44
S3	88 W C 5 Ti C 7 Co	1500	20 - 100	150	5.5	0.15	0.05	0.25
G1	94 W C 6 Co	1420	20 - 100	160	5	0.19	0.05	0.2
G2	89 W C 11 Co	1400	20 - 100	180	5.5	0.16	0.05	0.18
G3	85 W C 15 Co	1380	17 - 60	205	-	-	-	-
H1	94 W C 6 Co	1420	17 - 60	160	5	0.19	0.05	0.21
H2	91.5 W C 0.5 V C 1 Ta C 7 Co	1500	66 - 220	115	5	-	-	0.25
F1	69 W C 25 Ti C 6 Co	1550	66 - 220	110	7	0.05	-	0.65
F2	34.5 W C 60 Ti C 5.5 Co	1700	66 - 200	80	-	-	-	-

Figure 1.8 Carbide grades fabricated by Krupp WIDIA during the Second World War according to the report on the cemented carbide industry in Germany shown in Fig. 1.7.

built by Boart in South Africa, Johannesburg, in 1949 and a plant for producing synthetic diamonds was founded by De Beers in the neighborhood of this carbide plant in South Africa in 1952. The carbide plant in Germany and the carbide plant in South Africa as well as a carbide department of the synthetic diamond plant in South Africa merged together later and became the basis of the Element Six Carbide Division, a subsidiary of Element Six. Element Six, part of the De Beers Group, is a global leader in the design, development, and production of synthetic diamond and cemented carbide supermaterials.

As it was mentioned above, Schrötter's patents were applied only in Germany, Great Britain, and the United States of America and the priority date in other countries was missed, so that there was no need to purchase a license from Krupp WIDIA to produce WC—Co cemented carbides in these countries. This was presumably one of the major reasons explaining the fact that Sweden became soon one of the major players on the cemented carbide market, so in the 1940s, cemented carbides became known in many countries worldwide as "Swedish steel." Although the

Figure 1.9 Cover pages of brochures of some companies producing cemented carbides in the first half of the 20th century. *(Redrawn from Ref. [1].)*

carbide production started at Sandvik, one of the biggest carbide companies nowadays, relatively lately in comparison with other companies (in 1942), its Coromant brand name became very popular soon. In 1969, Sandvik launched onto the market the first indexable cutting inserts with TiC wear-resistant thin coatings, which was a revolutionary discovery in the field of cemented carbides for metal-cutting.

The production of cemented carbides in the Soviet Union started in 1929 and was based on the tungsten metal manufacture at Moscow Electro-Plant, which is shown in Fig. 1.10 [11]. In 1929, Prof. G.A. Meerson, the father of the Soviet carbide industry, with a colleague developed the first Soviet WC−10%Co carbide grade at this plant and gave it the brand name "POBEDIT," the word originating from the Russian word POBEDA—victory. The brand name POBEDIT soon became a common noun and even now many people use it to designate cemented carbides in Russia instead of the technically correct terms "hard alloy" or "hardmetal." Interestingly, the same happened in Germany, where the first brand name of cemented carbide "WIDIA" also became a common noun, so that many people still employ it when they talk about cemented carbides. The Soviet Union did not need a license from Friedrich Krupp AG to fabricate WC−Co materials, as Schröter's patent was not granted in the Soviet Union. Nearly 16 tons of cemented carbides were fabricated in the USSR

Figure 1.10 Moscow Electro-Plant in the 1920s, where the first Soviet cemented carbide grade with the brand name POBEDIT was produced in 1929. *(Redrawn from Ref. [11].)*

in 1931, which was more than the total production of cemented carbides in the Western countries in the early 1930s estimated in Ref. [2]. According to the report of the British experts "*The German Hard Metal Industry*" (see Fig. 1.7), the carbide output by Krupp WIDIA was some 9 tons in 1929, dropped during 1931 and 1932, but rose again to nearly 24 tons in 1935.

To illustrate how the first cemented carbide products were developed and fabricated in the late 1920s of the last century, Fig. 1.11 shows ball mills employed for the fabrication of WC–Co mixtures and hydrogen graphite tube furnaces used for sintering of carbide articles [11].

Figure 1.11 Equipment employed to produce the first carbide grade POBEDIT in the Soviet Union in 1929: above - ball mills, and below - hydrogen furnaces. *(Redrawn from Ref. [11].)*

References

[1] H.M. Ortner, P. Ettmayer, H. Kolaska, The history of the technological progress of hardmetals, Int. J. Refractory Met. Hard Mater. 44 (2014) 148–159.
[2] H. Kolaska, The dawn of the hardmetal age, Powder Met. Int. 24/5 (1992) 311–314.
[3] Z. Yao, J.J. Stiglich, T.S. Sudarshan, WC-Co enjoys proud history and bright future, Met. Pow. Rep. 53 (1998) 32–36.
[4] G.G. Goetzel, Treatise on Powder Metallurgy, vol. III, Interscience, New York, NY, 1952.
[5] W.W. Engle, in: J. Wulff (Ed.), Cemented Carbides, Powder Metallurgy, American Society for Metals, Cleveland OH, 1942, pp. 436–453.
[6] K.J.A. Brookes, Half a century of hardmetas, Met. Powder. Rep. 50 (1995) 22–28.
[7] K. Schröter, DRP 420.689: Sintered Hardmetal Alloy and Procedure for its Fabrication, 1923.
[8] K. Schroter, W. Jenssen, Tool and Die, 1925. U.S. Patent 1,551,333.
[9] WIDIA Handbook, Friedrich Krupp AG (Ausgabe B), Essen, 1936.
[10] H. Kolaska, Hardmetal — yesterday, today and tomorrow, Metall 12 (2007) 825–832 (in German).
[11] I. Konyashin, L. Klyachko, History of cemented carbides in the Soviet union, Int. J. Refractory Met. Hard Mater. 49 (2015) 9–26.

CHAPTER 2

Classification and general characteristics of cemented carbides

Cemented carbides belong to a special class of metal—ceramic composite materials, the microstructure of which consists of ceramic grains, WC grains, embedded in a metallic binder, a solid solution of tungsten and carbon in cobalt, or other iron group metals in the amount of 3%—30% by mass (Fig. 2.1). In some cases, the binder phase is alloyed with different chemical elements (V, Cr, Ta, etc.), which are known as grain growth inhibitors, as they suppress the WC grain growth during liquid phase sintering. In cemented carbide grades designed for steel-cutting, the microstructure comprises a second phase of mixed cubic carbides comprising Ti, Ta, W, and, in many cases, Nb in the amount of 8%—20% by mass (Fig. 2.2).

According to the ISO standard (ISO 4499-2:2:2010), WC—Co cemented carbides are divided into the following groups, in accordance with the WC mean grain size:

- nano - 0.2 μm;
- ultrafine - 0.2—0.5 μm;
- submicron - 0.5—0.8 μm;
- fine-grain - 0.8—1.3 μm;
- medium-grain - 1.3—2.5 μm;
- coarse-grain - 2.5—6.0 μm;
- ultracoarse - 6.0 μm and coarser.

It is well known that the main physical and mechanical properties of WC—Co cemented carbides, affecting their performance properties, are hardness, fracture toughness, compression strength, and transverse rupture strength (TRS), which are in turn dependent mainly on the Co content and WC mean grain size for conventional WC—Co grades.

Fig. 2.3 shows curves indicating the dependencies of the major mechanical properties of cemented carbides of different classes according to

Cemented Carbides
ISBN 978-0-12-822820-3
https://doi.org/10.1016/B978-0-12-822820-3.00004-5

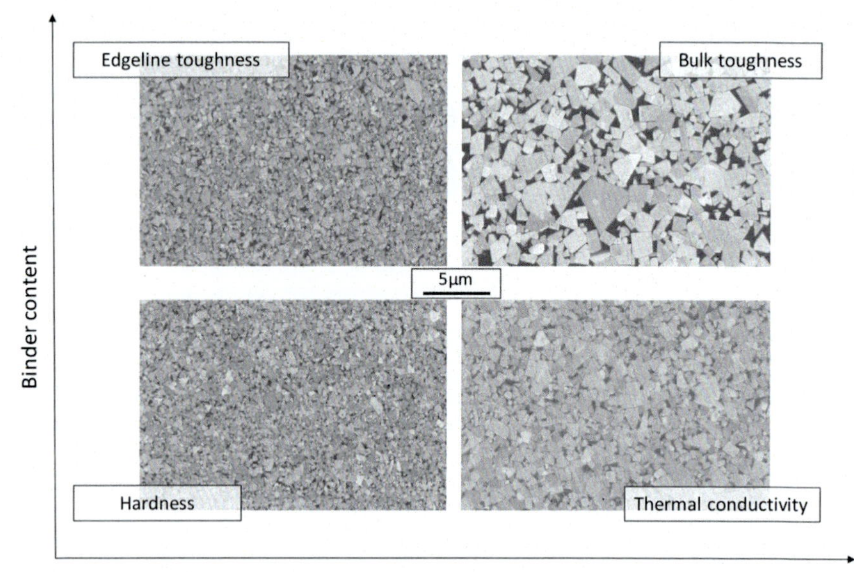

Figure 2.1 Microstructures of WC—Co cemented carbides with different Co contents and WC mean grain sizes, varying from ultrafine and submicron (left) to medium- and coarse-grain (right). *(According to Ref. [1].)*

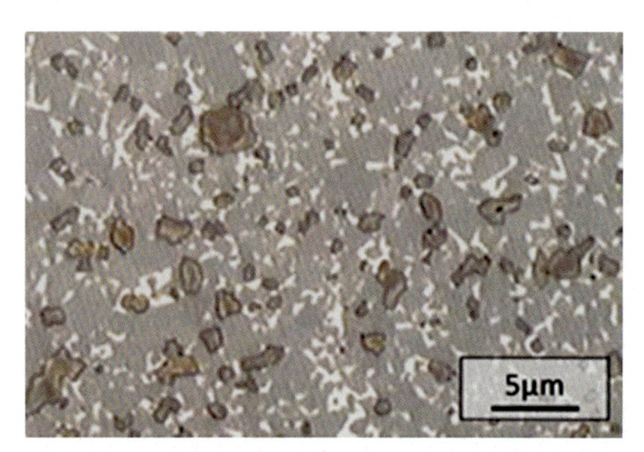

Figure 2.2 Microstructure of the WC-(Ta,Nb,W)C—Co cemented carbide showing the WC grains (*gray phase*), Co binder phase (*white*) and cubic (Ti,Ta,Nb,W)C γ-phase (*brown*) (light microscopy image). *(According to Ref. [1].)*

their WC mean grain sizes on the Co content. As it can be seen in Fig. 2.3, the hardness decreases, and fracture toughness increases when increasing the Co content, and this also affects the wear resistance. Although there are some up-to-date approaches allowing to simultaneously increase hardness,

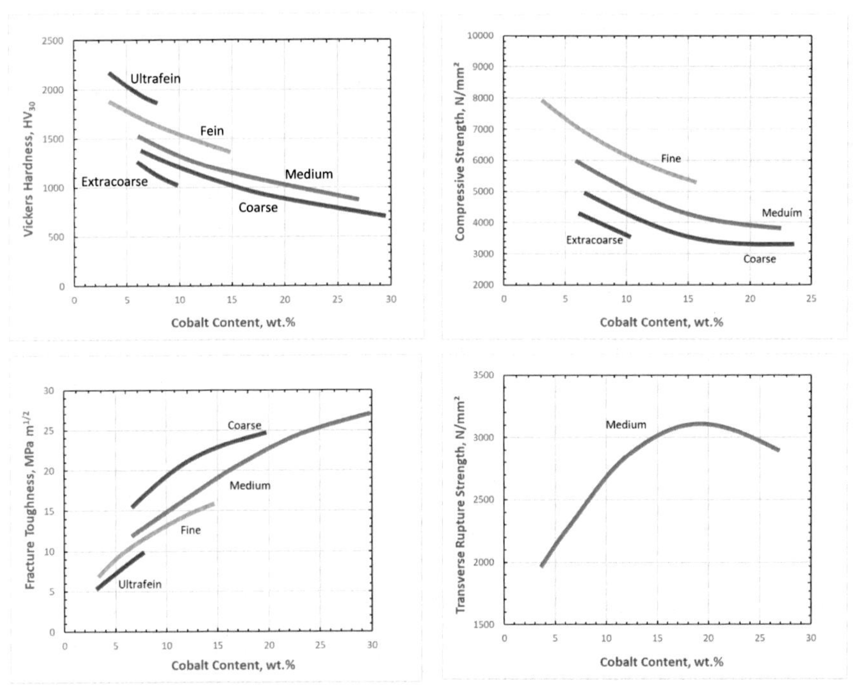

Figure 2.3 Hardness, fracture toughness, compressive strength, and transverse rupture strength of WC—Co cemented carbides depending on Co contents for different groups of cemented carbides in accordance with the WC mean grain size.

wear resistance, and fracture toughness of WC—Co materials, described in detail in Chapter 12, for conventional WC—Co grades hardness and consequently wear resistance can be increased only at the expense of fracture toughness. The effect of the Co content and WC mean grain size on the compression strength and TRS for medium- and coarse-grain cemented carbides is complex; this is considered in detail in Chapter 10. Nevertheless, in general, the compressive strength decreases when increasing the Co content above nearly 3 to 6 wt.% Co, as it is shown in Fig. 2.3. Thermal conductivity of WC—Co cemented carbides can be increased by increasing the WC grain size and WC content.

Fig. 2.4 shows major properties of WC—Co cemented carbides in comparison with ceramics, stellite, cast iron, and steel. As one can see, WC—Co materials have a unique combination of different properties, particularly hardness and compressive strength, distinguishing them from other ceramic and metal materials. This ensures their employment in a great number of different applications, including metal-cutting, mining,

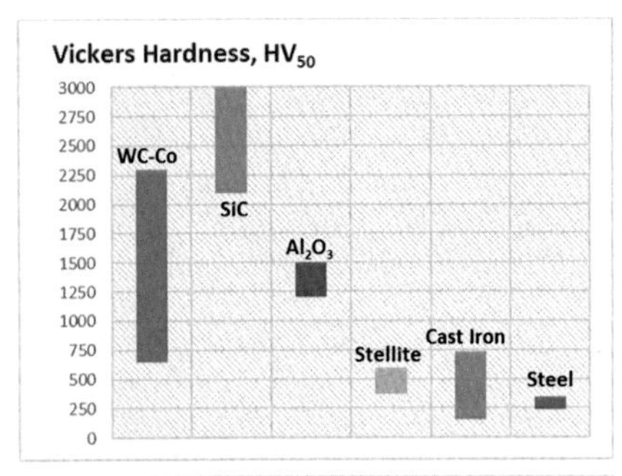

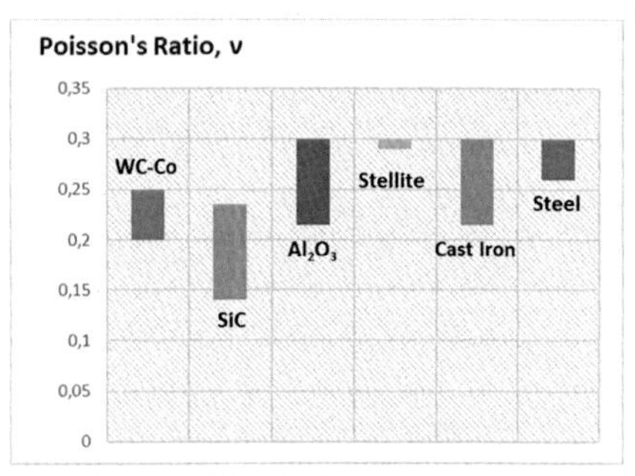

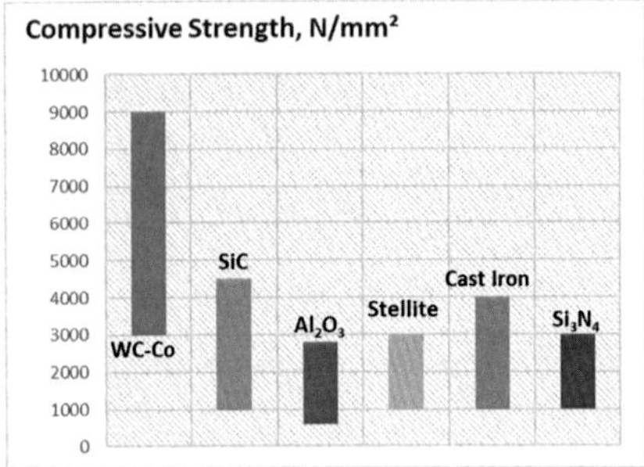

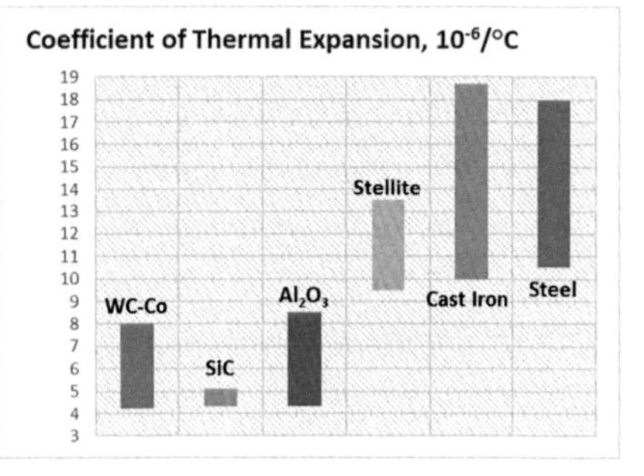

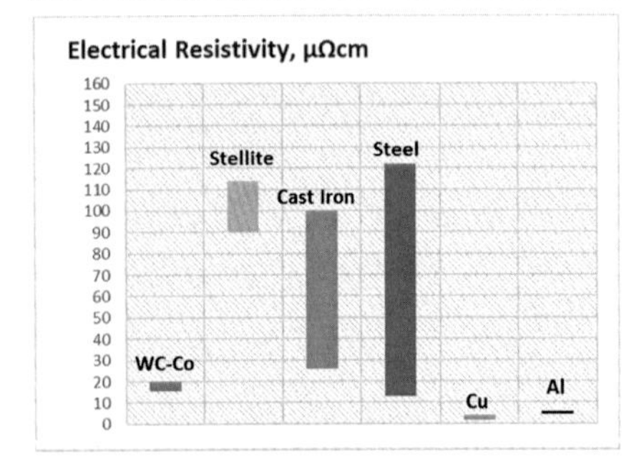

Figure 2.4 Vickers hardness, compressive strength, Poisson's ratio, coefficient of thermal expansion, and electrical resistivity of WC—Co cemented carbides in comparison with ceramics, stellite, cast iron, and steel.

construction, woodworking, rolls for hot rolling, cold forming tools, dies for wire drawing, etc., which is illustrated in Fig. 2.5. Application ranges of WC—Co cemented carbides grades are determined by their physical and mechanical properties, which are in turn dependent on their WC mean grain size and binder content (Fig. 2.5 and Fig. 2.6).

According to Ref. [2], currently, nearly 50,000—60,000 tons of cemented carbides are produced worldwide. The market share of cemented carbides in Europe is about 28%, in China - 39%, the BRICS countries - 13%, Japan - 10%, and 10% for the rest of the world. Global sales in

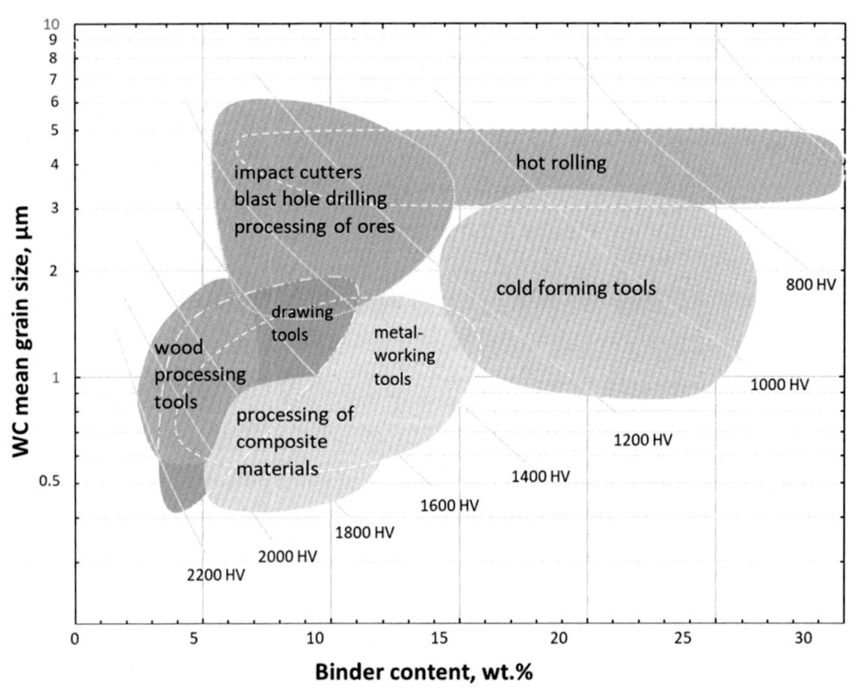

Figure 2.5 Application ranges of WC–Co cemented carbides according to their Vickers hardness, WC mean grain size, and binder content.

the cemented carbides' market exceeded €22 billion in 2011. The share of cemented carbides by sales for various applications is estimated in Ref. [2] to be 65% for metal processing, 10% for wood and plastics processing, 10% for wear parts, 10% for mining, and 5% for cold forming tools. The share of cemented carbides for various applications by mass is estimated in Ref. [2] as the following: metal processing - 22%, wood and plastics processing - 26%, production of wear parts - 17%, mining tools - 26%, and cold forming tools - 9%.

According to Ref. [4], the greatest share of the total tungsten production in 2016 was consumed by the cemented carbides industry (Fig. 2.6). The total input for the production of intermediates was 108,500 t from which 37,500 t were scrap (new and old scrap), leading to a recycling input rate of nearly 35%.

Fig. 2.7 shows different cemented carbide articles for various applications according to Ref. [3].

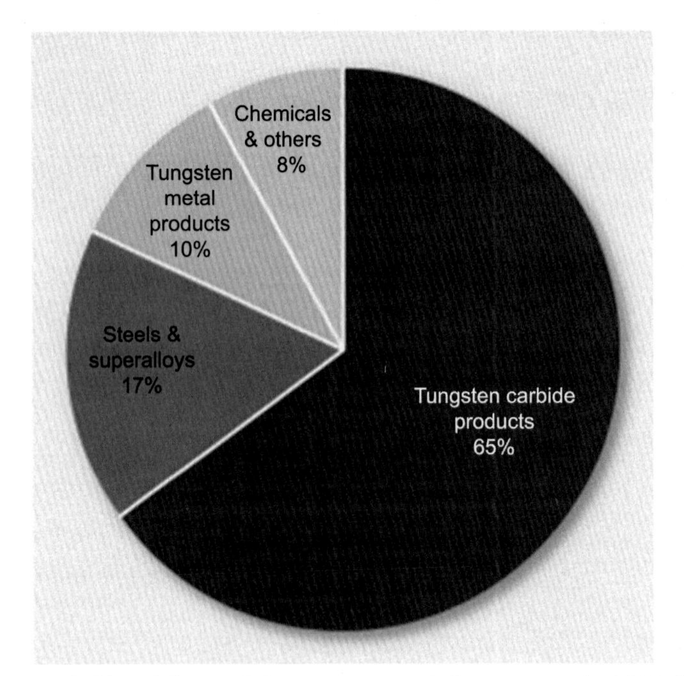

Figure 2.6 Global breakdown of first-use segments for tungsten in 2016. *(Redrawn from Ref. [4].)*

Figure 2.7 Cemented carbide articles for different applications. *(According to Ref. [3]. With the permission of Element Six.)*

References

[1] J. García, V. Collado, C.A. Blomqvist, B. Kaplan, Cemented carbide microstructures: a review, Int. J. Refract. Met. Hard Mater. 80 (2019) 40–68.

[2] B. Williams, Powder metallurgy: a global market review, in: International Powder Metallurgy, fifteenth ed., Innovar Communications Ltd., UK, 2012.
[3] Internet site, 2021. www.e6.com.
[4] B. Zeiler, A. Bartl, W.-D. Schubert, Recycling of tungsten: current share, economic limitations, technologies and future potential, Int. J. Refract. Metals Hard Mater. 98 (2001) 105546.

Further reading

[1] L. Prakash, in: V. Sarin (Ed.), Fundamentals and General Applications of Hardmetals, Comprehensive Hard Materials, Elsevier Science and Technology, 2014, pp. 29–89.

CHAPTER 3

Materials science of cemented carbides

3.1 General overview of the W–C–Co and related phase diagrams

The W–C–Co phase diagram was examined in detail in numerous publications summarized in the monographs [1–4]; therefore, this chapter contains only major references related to this diagram.

The carbide microstructure forms at relatively high temperatures of liquid-phase sintering, which makes it possible to consider the conditions of the microstructure formation as the equilibrium or near-equilibrium conditions. Phase diagrams of the W–C–Me systems, where Me is Co, Ni, or Fe, as well as of the W–C–Ti–Ta(Nb)–Co system, allow one to understand the processes of sintering and formation of the microstructure of cemented carbides as well as to explain their properties depending on the composition and fabrication parameters.

The first study of the W–C–Co system was conducted by Wyman and Kelley in 1931, but the W–C–Co phase diagram was not fully understood at that time. The W–C–Co system was examined in detail by different methods in the works of Takeda [5] and Rautala and Norton [6], who established a series of important regions and vertical sections of the W–C–Co phase diagram. The approach to examinations of phase diagrams initially was exceptionally experimental, but later on, thermodynamic calculations were also employed, although, in many cases, there is a distinct difference between the experimental results and those obtained by thermodynamic calculations. The phase diagrams presented in this chapter were obtained exceptionally experimentally and can be compared to the corresponding vertical sections of the WC–Co, WC–Ni, WC–Fe, and WC–Fe–Ni diagrams obtained by thermodynamic calculations reported in Chapter 5. It should be noted that the experimentally obtained phase diagrams describe real cemented carbides of different compositions

Cemented Carbides
ISBN 978-0-12-822820-3
https://doi.org/10.1016/B978-0-12-822820-3.00001-X

more precisely than the calculated ones with respect to temperatures corresponding to different phase transformations. However, the precise determination of border lines between different regions of the experimentally obtained phase diagrams with respect to carbon contents can be not accurate enough, as the position of such border lines depends not only on the initial total carbon content but also on the carbon potential of the gas atmosphere in sintering furnaces. The carbon potential depends on the composition of the gas atmosphere during sintering, which is usually either slightly carburizing or decarburizing thus resulting in deviations of the total carbon content in the sintered carbide articles from the initial carbon content in WC−Co powders.

Fig. 3.1 shows a simplified schematic horizontal section of the W−C−Me phase diagram (where Me is Co, Ni, or Fe) indicating the equilibrium state in cemented carbides after liquid phase sintering [7]. Generally, such a section comprises three major regions: (1) the region of carbon precipitates corresponding to microstructures comprising inclusions of free carbon, (2) the two-phase region, in which the microstructure comprises only WC grains and a binder phase consisting of a solid solution of W and C in Me, and (3) the region corresponding to microstructures consisting of

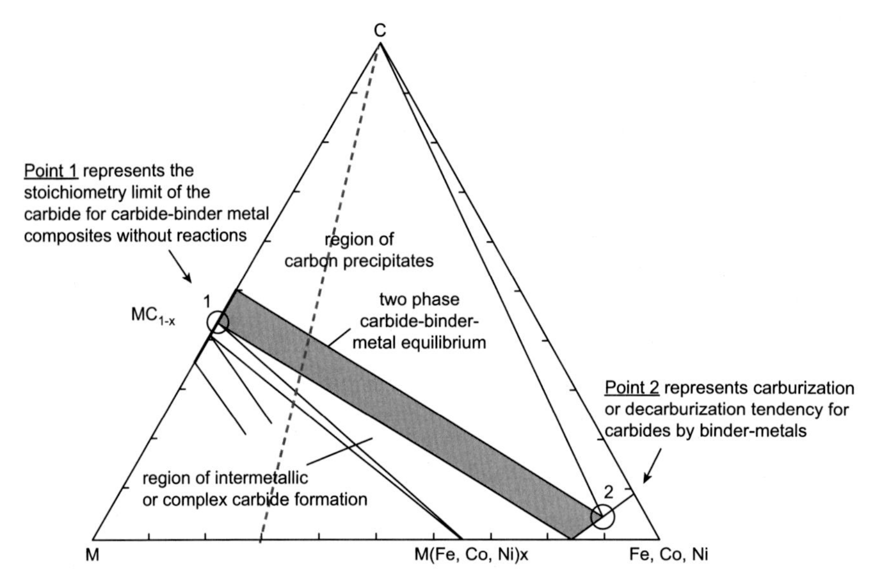

Figure 3.1 Simplified schematic diagram indicating the equilibrium in cemented carbides, where M = W (the *dashed line* indicates a vertical section through the carbon angle). *(According to Ref. [7].)*

at least three phases: the carbide phase (WC or W_2C), binder phase, and either W—Me—C phases of the η-phase type or similar triple W—Me—C compounds or intermetallic phases of the W—Me system forming at a very low carbon content. The simplified schematic horizontal section of the W—C—Me phase diagram shown in Fig. 3.1 allows one to better understand several horizontal sections of different W—C—Me phase diagrams reported below. The most important vertical section of this simplified schematic horizontal section of the phase diagram marked by the dashed line is a section through the carbon angle. As it can be seen in Fig. 3.1, this vertical section crosses the three major regions of the phase diagram mentioned above and provides information on the carbon contents corresponding to the border lines among the three major regions mentioned above.

Another important simplified section (vertical section) of the phase diagram is a vertical section between stoichiometric WC and iron group metals shown in Fig. 3.2. As it is shown below, in some cases, such a vertical section is significantly more complicated than that presented in Fig. 3.2, which has an impact on the microstructure of cemented carbides obtained by liquid-phase sintering followed by very rapid cooling. Nevertheless, the simplified vertical section illustrates the major phenomenon allowing dense WC—Me materials to be sintered at relatively low temperatures, namely, the formation of the low-melting point eutectics between WC and Fe group metals.

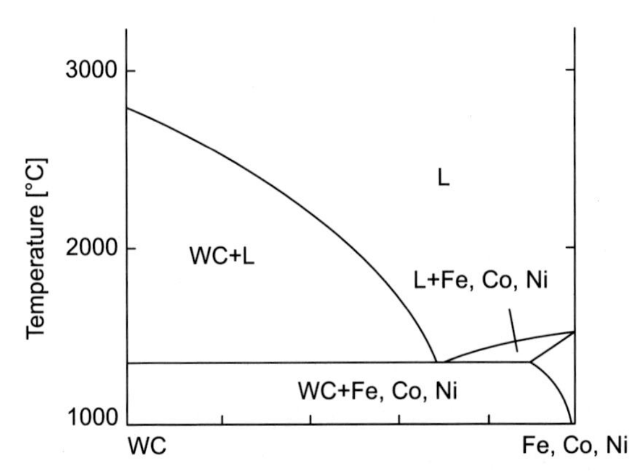

Figure 3.2 Simplified schematic pseudobinary vertical section of the W—C—Co phase diagram between stoichiometric WC and iron group metals. *(According to Ref. [7].)*

The horizontal section of the W–C–Co phase diagram shown in Fig. 3.3 indicates the different regions of the W–C–Co diagram and the border lines corresponding to the formation of various double and triple compounds. When taking into account this horizontal section, the following major features of the W–C–Co system can be outlined:

1. The single-phase area of the Co-based binder phase (β-phase) located close to the Co angle (which is indicated by black color) is a solid solution of tungsten and carbon in cobalt. It should be noted that the composition of the β-phase strongly depends on the total carbon content of the cemented carbide. In particular, the solubility of tungsten significantly increases when decreasing the total carbon content.
2. The Co–WC triangle region indicated by blue color (gray in print) shows the two-phase region (Fig. 3.3), in which only the WC phase and Co-based binder (β-phase) are present in the microstructure. This region is quite wide at high Co contents, but becomes narrower when decreasing the Co content.
3. The microstructure of cemented carbides in the region above the Co–WC triangle region comprises inclusions of free carbon.
4. The microstructure of cemented carbides in the areas below the Co–WC triangle region comprises inclusions of four triple compounds of cobalt, tungsten, and carbon, three of which are characterized by the face-centered cubic crystal lattice (η_1 – Co_3W_3C, η_2 – Co_6W_6,C, and θ – Co_2W_4C) and one of which has a hexagonal crystal lattice

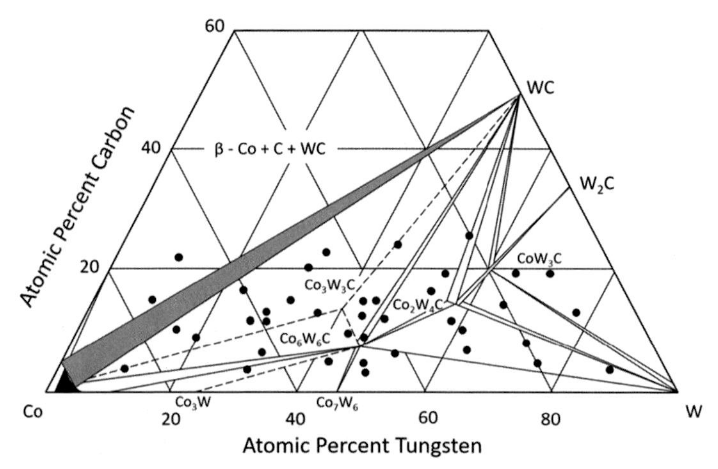

Figure 3.3 Horizontal section of the W–C–Co phase diagram at 1000°C [3]. The *blue triangle* (gray in print) indicates the two-phase region not comprising free carbon and η-phase.

(æ - CoW_3C). Furthermore, two cubic phases $Co_3W_6C_2$ and Co_4W_4C as well as three hexagonal phases $Co_2W_8C_3$, $Co_3W_{10}C_4$, and Co_3W9C_4 can form in the region below the Co—WC triangle area according to Refs. [2,6]. Intermetallic W—Co phases form at very low carbon contents (close to the compositions corresponding to the Co—W line).

5. In the two-phase triangle region (β-phase + WC), the line corresponding to stoichiometric WC is located so that it is close to the high-carbon boundary of this region. The width of this two-phase region indicates acceptable fluctuations in the carbon content of WC—Co cemented carbides that can occur without a risk of the formation of other phases: graphite at high carbon contents or η-type phases at reduced carbon contents.

3.2 Some important sections of the W—C—Co phase diagram

Fig. 3.4 shows detailed vertical section of the W—C—Co phase diagram at the WC—Co line suggested by Takeda. As one can see, it is significantly more complicated than the simplified schematic pseudobinary vertical section of the W—C—Co phase diagram (Fig. 3.2), as this diagram cannot explain the presence of inclusions of η-phase or η-phase + free carbon in cemented carbide microstructures after rapid cooling from sintering temperatures (the metastable crystallization mode) at medium carbon contents.

The microstructure of carbide samples having the eutectic composition of the WC—Co pseudobinary vertical section of the W—C—Co phase diagram (nearly WC—65% wt.% Co) comprises thin WC needles or platelets after slow cooling (Fig. 3.5b). If liquid-phase sintering of WC—Co is followed by very rapid cooling (for example, quenching in water), the microstructure of carbide samples with medium carbon contents can contain both the WC phase and η phase, or inclusions of free carbon and η-phase, which is illustrated in Fig. 3.5a. This phenomenon was explained in Ref. [5], where it was suggested that some η-phase may remain in equilibrium with WC and liquid phase even at medium carbon contents. While η-phase completely decomposes resulting in the formation of WC + β at low and medium-low cooling rates, its inclusions can remain in the microstructure at very high cooling rates as a result of the metastable mode of crystallization according to the diagram shown in Fig. 3.5, as there is a region, where liquid + β-phase + η-phase coexist (Fig. 3.4, the region

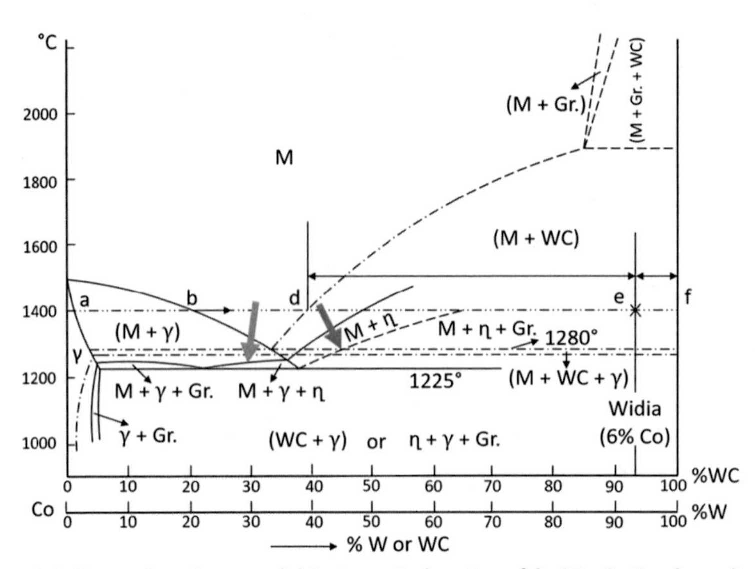

Figure 3.4 Comprehensive pseudobinary vertical section of the W—C—Co phase diagram between stoichiometric WC and cobalt indicating the phase composition regions occurring at the metastable crystallization mode (*solid lines*) and stable crystallization mode (*dashed lines*) [5]. Note that the β-phase is designated as "ϒ-phase," liquid phase is designated as "M" (melt), and free carbon is designated as "Gr," which correspond to the original designations of these phases in Ref. [5]. Note also, that "Widia" indicates the composition of the first carbide grade developed and implemented in Germany. The blue and red arrows (light gray arrow and dark gray arrows in print, respectively) indicate the regions of the phase diagram, in which both free carbon and η-phase simultaneously form at high cooling rates (the metastable crystallization mode, see Fig. 3.5a).

is marked by a blue arrow [light gray in print]). During the metastable crystallization process occurring at very high cooling rates, the simultaneous formation of η-phase and graphite described in Ref. [8] and illustrated in Fig. 3.5a can also take place. This phenomenon can be explained by the metastable phase diagram shown in Fig. 3.4, as it comprises a region, where liquid + η-phase + graphite coexist in a certain temperature range (in Fig. 3.4, this region is marked by a red arrow [dark gray arrow in print]).

Thus, the comprehensive phase diagram suggested by Takeda (Fig. 3.4) predicts that as a result of extremely rapid cooling from the sintering temperatures (for example, quenching in water), the primary crystals of η-phase, binary eutectics (η + graphite), and ternary eutectics (β + η + graphite) can crystallize from the melt having a eutectic or nearly eutectic composition according to the metastable crystallization mode. As a result, the microstructure of carbide samples can be very unusual comprising η-phase and free carbon inclusions embedded in the Co-based matrix, which are shown in

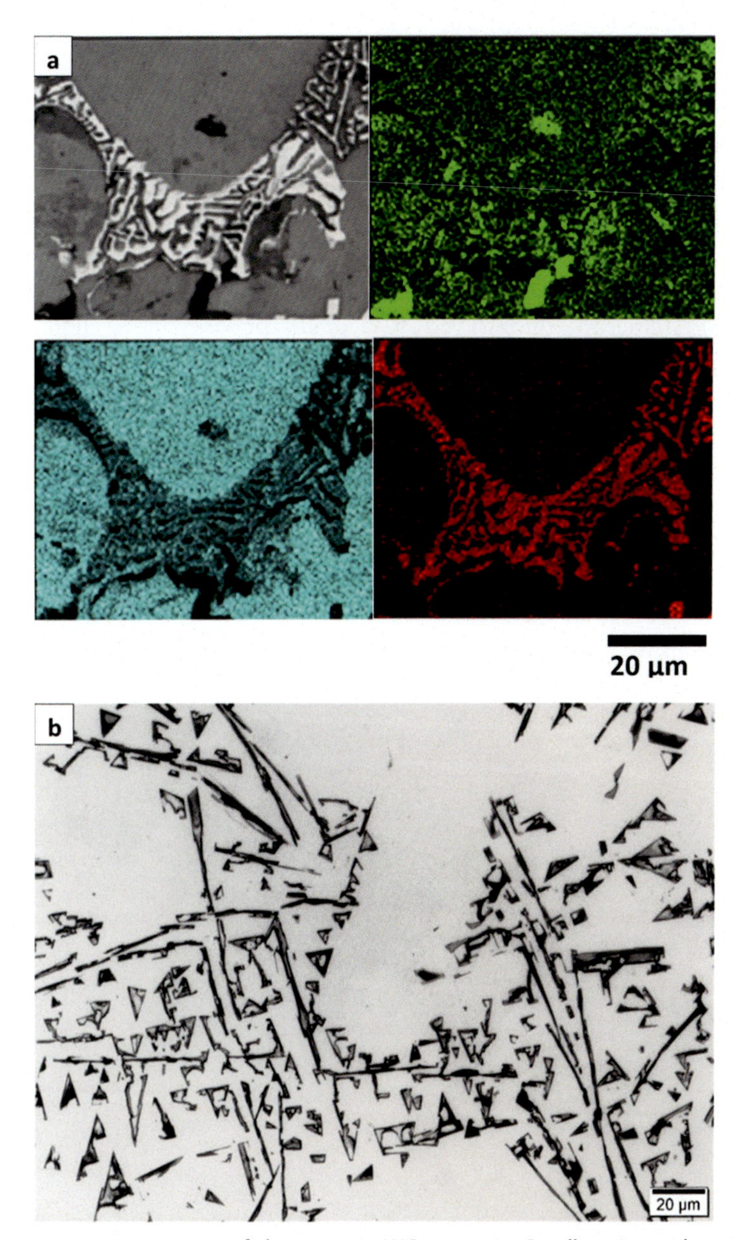

Figure 3.5 Microstructures of the eutectic WC–35 wt.% Co alloy sintered at 1400°C followed by cooling at different rates: (A) - very rapid cooling corresponding to the metastable crystallization mode (above, left – SEM image; above, right – carbon EDX map in green color; below, left – cobalt EDX map in blue color; and below right – tungsten EDX map in red color); note that the microstructure comprises dendritic grains of η-phase and rounded inclusions of free carbon embedded in the Co-based matrix; and (B) - slow cooling corresponding to the stable crystallization mode (the microstructure comprises dark-grey needles of WC embedded in the Co-based matrix; light microscopy image after etching in the Murakami reagent). *(With the permission of Element Six.)*

Fig. 3.5A. Meantime, at the same cobalt and carbon contents, the crystallization of the WC crystals shown in Fig. 3.5B embedded in the Co-based matrix occurs at the stable crystallization mode, when the melt is cooled down at moderate or low cooling rates. In this case, primary WC crystals form from the eutectic or nearly eutectic melt first acting as nuclei for the final decomposition of residual amounts of liquid phase corresponding to the binary WC/Co eutectics. In industrial WC–Co grades containing significantly less Co than the Co content corresponding to the WC/Co eutectics, there are numerous undissolved WC grains present during liquid-phase sintering, so that when the eutectics decomposes, W and C atoms contained in it precipitate on the existing WC grains leading to their coarsening.

A simplified pseudobinary vertical section through the ternary phase diagram suggested by Rautala and Norton [6] is shown in Fig. 3.6. This vertical section is very close to a quasibinary diagram and can be employed in practice. However, it illustrates a special case according to the authors of ref. [6] because experimental data are lacking. On that ground, it must be considered only as a relatively coarse approximation in comparison with the comprehensive pseudobinary vertical section of the WC–Co phase diagram shown in Fig. 3.4.

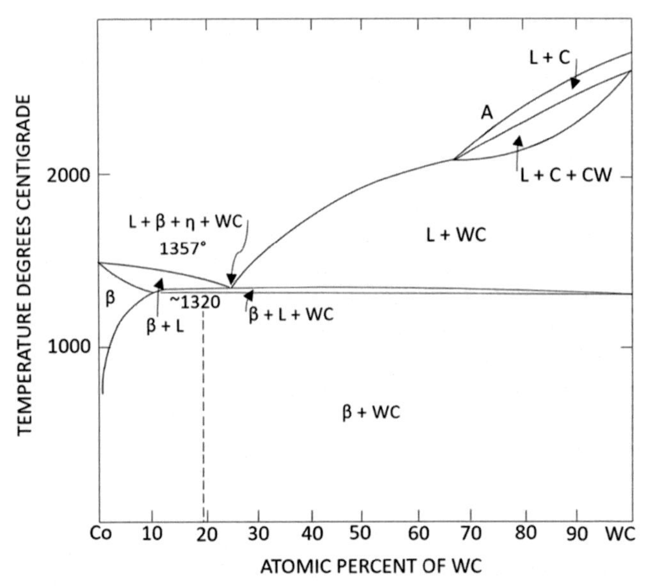

Figure 3.6 Simplified pseudobinary vertical section of the WC–Co phase diagram between stoichiometric WC and cobalt suggested by Rautala and Norton [6]. According to the authors, because experimental data are lacking, only a special case is illustrated by this vertical section.

The vertical section of the W—C—Co phase diagram through the carbon angle is very important from the practical point. The practice of the carbide industry shows that the melting point of WC—Co materials varies from nearly 1300°C to 1360—1370°C depending on the carbon content, which is clearly seen in Fig. 3.7. It is very important that according to the vertical section shown in Fig. 3.7, sintering of stoichiometric WC with 16 wt.% Co allows obtaining two-phase cemented carbides not containing free carbon and η-phase. For this composition, the width of the two-phase region regarding the carbon content is nearly 0.13 wt.% with respect to that in WC. According to the horizontal section of the W—C—Co phase diagram shown in Fig. 3.3, the two-phase region becomes narrower when reducing the Co content. This occurs due to shifting the border line between the two-phase region and the region corresponding to the η-phase formation. Nevertheless, even for cemented carbide grades containing 6 wt.%Co, which are produced on a large scale in the carbide industry, the width of the two-phase region regarding the carbon content with respect to that in WC is relatively broad (about 0.1 wt.%), which ensures obtaining two-phase WC—Co materials relatively easily. It is possible nowadays to

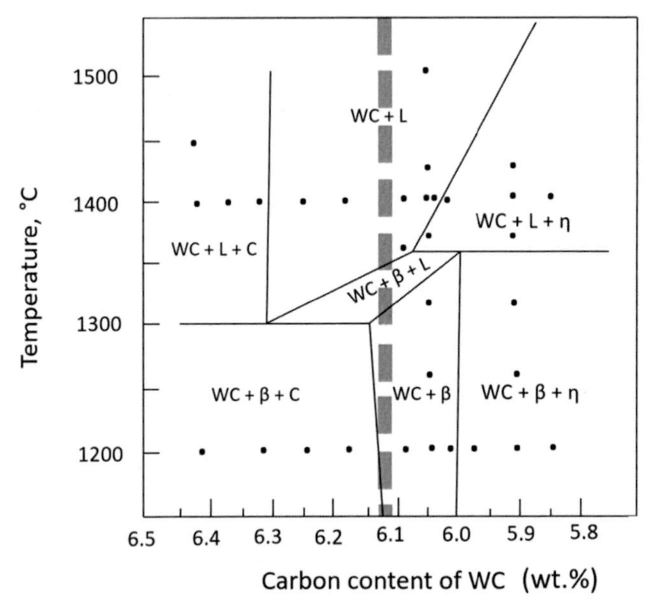

Figure 3.7 Vertical section of the W—C—Co phase diagram through the carbon angle for the WC—16 wt.%Co cemented carbide [9]; the *dashed line* indicates the stoichiometric carbon content in WC.

precisely control and regulate the carbon content in industrial carbide grades with medium and medium low Co contents by the use of modern production equipment.

The melting points of different carbide grades depend on the total carbon content. According to the results of ref. [10], the melting intervals of WC-Co model binder alloys containing 65 wt.% Co strongly depend on the carbon content. They lie between 1299°C and 1335°C for alloys with high carbon contents containing free carbon, between 1332°C and 1389°C for alloys with medium-high carbon contents, between 1350°C and 1395°C for alloys with medium-low carbon contents, and between 1364°C and 1400°C for alloys with low carbon contents comprising η-phase.

Nowadays, when taking into account the data of different authors summarized in Ref. [4], it can be considered that the solubility of tungsten in cobalt is close to nearly 10 wt.% for the carbon content in WC—Co materials containing stochiometric tungsten carbide. The tungsten solubility increases up to roughly 20—22 wt.% at a low carbon content close to that corresponding to the η-phase formation and decreases down to nearly 2—3 wt.% at a high carbon content close to that corresponding to the free carbon formation.

The difference in the tungsten solubility in the Co-based binder mentioned above is considered without taking into account the existence of two allotropic modifications of cobalt with the fcc and hcp crystal lattices. The dissolution of tungsten in cobalt stabilizes the cubic modification; therefore, if the binder phase contains much dissolved tungsten at low carbon contents, it consists almost exceptionally of fcc Co at room temperature, although this cobalt modification is a metastable phase of cobalt at ambient conditions.

3.3 General overview of the W—C—Ni phase diagram

Although the binder phase of most of the carbide grades consists of Co, Ni is also employed in the carbide industry for the fabrication of some special grades characterized by high corrosion and oxidation resistance.

First results on the W—C—Ni phase diagram were obtained by Takeda [6]. The experimental sections of the W—C—Ni phase diagram shown in Figs. 3.8 and 3.9 were examined in detail in Ref. [11] and several unpublished works and were, afterward, summarized in the monographs [2,3].

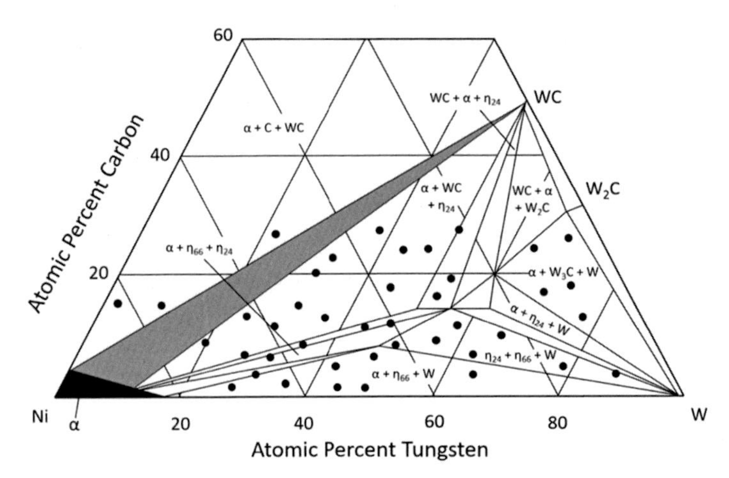

Figure 3.8 Horizontal section of the W—C—Ni phase diagram at 1300°C [3]. The *blue triangle* (gray in print) indicates the two-phase region not comprising free carbon and η-phase.

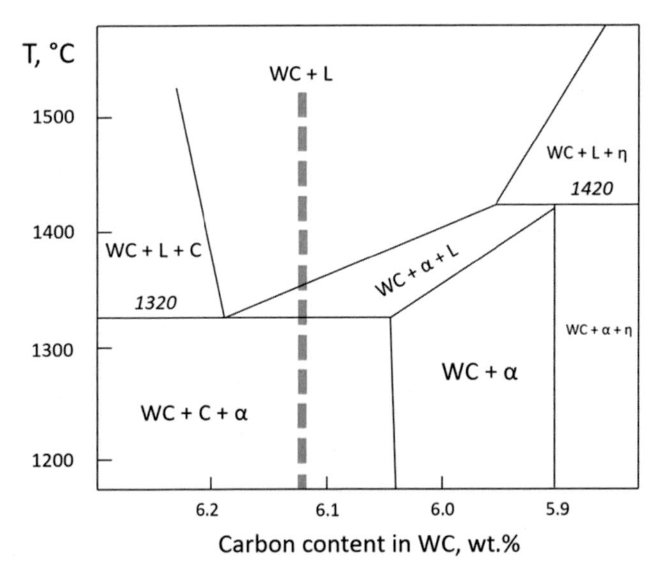

Figure 3.9 Vertical section of the W—C—Ni phase diagram through the carbon angle at 15 wt.% Ni [2]; the *dashed line* indicates the stoichiometric carbon content in WC.

The W–C–Ni system is characterized by many common features in comparison with the W–C–Co system; nevertheless, there are some differences between these two-phase diagrams.

1. The single-phase area of the Ni-based binder phase (α-phase) located close to the Ni angle (which is indicated by black color in Fig. 3.8) is a solid solution of tungsten and carbon in nickel. The solubility of tungsten and carbon in the carbon-rich Ni-based solid solution is dependent on the total carbon content similar to the W–C–Co system; however, it is significantly higher (by nearly 50%) in the Ni-based binders in comparison with that in the Co-based binders according to ref. [2].
2. The two-phase Ni–WC triangle region indicated by blue color (gray in print) in Fig. 3.8 is slightly wider with respect to the carbon content in comparison with that of the W–C–Co system. As in the W–C–Co system, this region becomes narrower when decreasing the Ni content.
3. The microstructure of cemented carbides in the regions above the Ni–WC triangle area indicated by blue color (gray in print) in Fig. 3.8 comprises inclusions of free carbon as in the W–C–Co system. However, the microstructure of WC–Ni alloys containing stoichiometric WC consists of WC + α-phase + C.
4. The microstructure of cemented carbides in the regions below the Co–WC triangle region comprises inclusions of triple compounds of nickel, tungsten, and carbon (η_1 and η_2 phases, which are designated in Fig. 3.8 as "η_{24} and η_{66},", η_3-type phase, and æ-type phase); the compositions and crystal lattices of these phases are close to those of the corresponding Co–W–C triple compounds. Intermetallic W–Ni phases form at very low carbon contents.
5. In the horizontal section of the W–C–Ni phase diagram (Fig. 3.8), the two-phase triangle region (α-phase + WC) is shifted toward lower carbon contents in comparison with the W–C–Co system. Therefore, when fabricating WC–Ni grades from stoichiometric WC powders, one must employ additions of tungsten metal to reduce the total carbon content for obtaining the carbide microstructure not containing inclusions of free carbon.
6. The temperatures of the liquid phase formation both at a low and a high carbon content of the W–C–Ni phase diagram are noticeably higher than those of the W–C–Co diagram, so that WC–Ni cemented carbides must be sintered at higher temperatures compared to WC–Co alloys.

3.4 General overview of the W−C−Fe−Ni phase diagram

Fig. 3.10 shows the horizontal section of the W−C−Fe phase diagram at 1000°C. It is well known that the two-phase region, in which inclusions of free carbon, cementite, and η-type phases are absent, is very narrow in the W−C−Fe system. Due to extreme difficulties in regulation of the microstructure of WC−Fe cemented carbides with respect to their carbon content, cemented carbides with Fe binders are not fabricated in the carbide industry on an industrial scale.

Nevertheless, it was established that additions of nickel to iron makes it possible to significantly broaden the two-phase region [11,12], so that WC−Fe−Ni cemented carbides containing only the WC phase and binder phase (ϒ-phase) after sintering can potentially be produced with minor difficulties.

Fig. 3.11 shows the horizontal section of the W−C−Fe−Ni phase diagram at 1200°C and the ratio of Fe to Ni of 3:1 by weight. Fig. 3.12 shows the vertical section of the W−C−Fe−Ni phase diagram through the carbon angle for the WC - 20% (Fe, Ni) at Fe:Ni = 85:15 by weight. As one can see, there are some common features and differences between the W−C−Co and W−C−Ni phase diagrams on the one hand and the WC−C−Fe−Ni phase diagram on the other hand.

1. The relatively narrow two-phase region is located between the three-phase region with high carbon contents comprising free carbon and that with low carbon contents comprising η-type phases. The presence

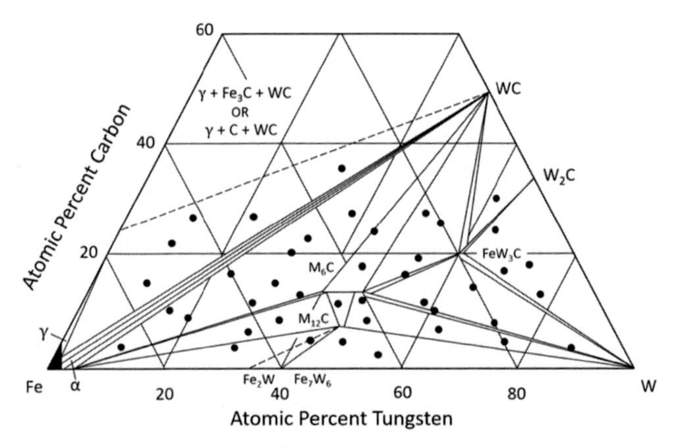

Figure 3.10 Horizontal section of the W−C−Fe phase diagram at 1000°C [3].

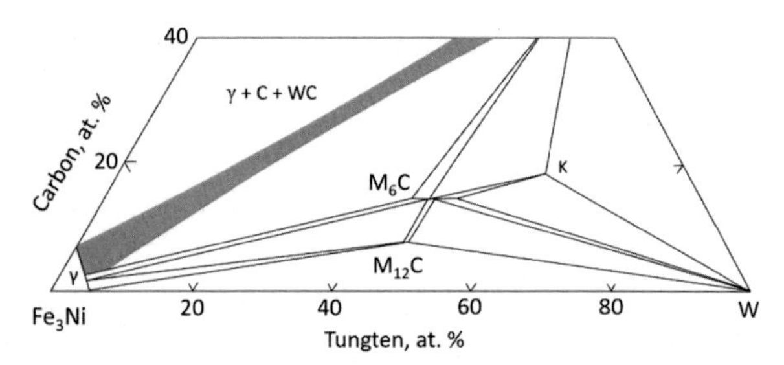

Figure 3.11 Horizontal section of the W—C—Fe—Ni phase diagram at 1200°C and the ratio of Fe:Ni = 3:1 [3]. The *blue area* (gray in print) indicates the two-phase region not comprising free carbon and η-phase.

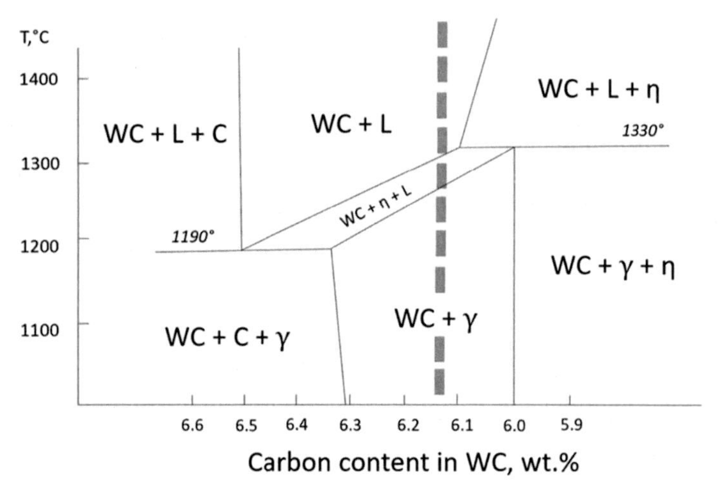

Figure 3.12 Vertical section of the W—C—Fe—Ni phase diagram through the carbon angle for the WC - 20% (Fe, Ni) grade at the ratio of Fe:Ni = 85:15 [12]; the *dashed line* indicates the stoichiometric carbon content in WC.

of Ni in the binder suppresses the formation of cementite in the microstructure of cemented carbides with high carbon contents.

2. The line corresponding to stoichiometric WC (Fig. 3.12) lies nearly in the middle of the two-phase region of the W—C—Fe—Ni phase diagram at the ratio of Fe:Ni = 85:15.
3. The two-phase region of the phase diagram for the alloy containing 20% (Fe + Ni) at the Fe to Ni ratio of 85:15 is quite broad regarding the carbon content (about 0.3 wt.% C) with respect to the carbon content in WC. It appears therefore that WC—Fe—Ni cemented carbides can be relatively easily fabricated on an industrial scale

4. The melting points of WC–Fe–Ni alloys at low and high carbon contents are significantly lower than those of the W–C–Co and W–C–Ni alloys(Fig. 3.12).
5. Properties of the WC–Fe–Ni cemented carbides can be regulated and improved by final heat treatments, as, in fact, the binder phase in this case consists of alloyed Fe–Ni–W steel.

Although the WC–Fe–Ni cemented carbides are not produced in the carbide industry on a large scale, their potential could presumably be realized in future, as their mechanical and performance properties are comparable with those of conventional WC–Co cemented carbides according to Ref. [2].

3.5 General overview of the W–C–Ti–Ta(Nb)–Co phase diagrams

Additions of mixed (Ti,Ta(Nb),W)C carbides are widely employed for the fabrication of metal-cutting grades; therefore, it is important to estimate the influence of such additions on the phase composition of WC–Co materials. In literature, there is information on the (Ti,Ta,W)C–Co system; however, one can expect that the partial substitution of Ta atoms by Nb atoms in the mixed carbide should not noticeably affect the phase equilibrium, as Ta and Nb are very similar in this respect. Carbides of titanium and tantalum (as well as niobium) form a continuous series of solid solutions. The solubility of WC in these solid solutions decreases with increasing the TaC content[2].

Fig. 3.13 shows the horizontal section of the (Ti,Ta,W)C–Co phase diagram at 1000°C [2]. The mixed carbide phase (Ti,Ta,W)C is slightly dissolved in cobalt and the solubility increases when increasing the WC content in the TiC–TaC–WC solid solution; it also depends on the total carbon content. For example, for the solid solution with a ratio of (TiC,TaC):WC = 1:3, the solubility of the mixed carbide in cobalt decreases from nearly 5.0 wt.% at a low carbon content down to 0.2 wt.% at a high carbon content [2]. The solubility of solid solutions TiC–TaC–WC in the Co-based binder does not depend on the content of tantalum carbide (and presumably on the NbC content). The melting point of the triple eutectics Co + (Ti,Ta,W)C + C is nearly 1260–1270°C and that of the double eutectics Co + (Ti,Ta,W)C is roughly 1360–1380°C [2].

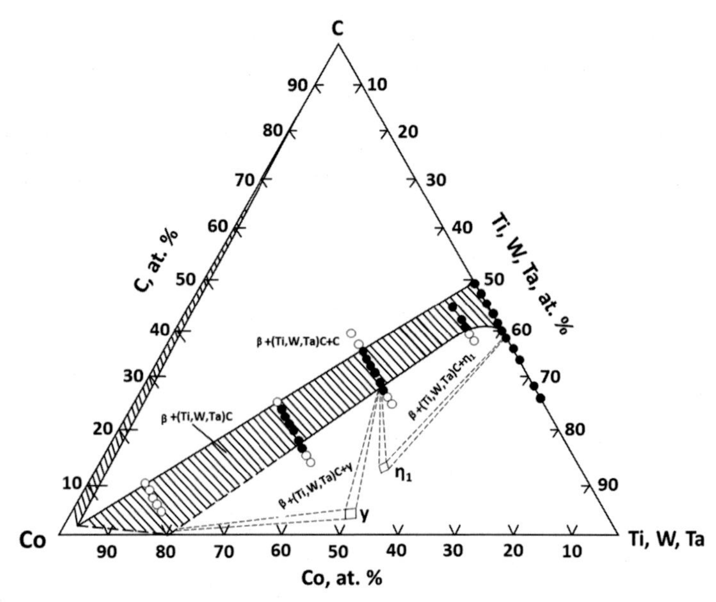

Figure 3.13 Horizontal section of the W—C—Ti—Ta—C—Co phase diagram at 1000°C [2].

In general, the horizontal section of the (Ti,Ta,W)C—Co phase diagram is to some extent similar to that of the WC—Co phase diagram. The two-phase region β + (Ti,Ta,W)C is significantly broader with respect to the total carbon content compared to that of the WC—Co and WC—Ni phase diagrams (by more than one order of magnitude). The microstructure of cemented carbides corresponding to the region located above the two-phase region comprises free carbon. The microstructure of carbide samples corresponding to the area located below the two-phase region comprises inclusions of the η_1 phase and y-phase, the composition and crystal structure of which are assumed to be close to those of the η-type phases. According to the results of EDX [2], its composition roughly corresponds to the formula W_4CoC_x or, presumably, $(W,Ta)_4CoC_x$.

The two-phase region of the (Ti,Ta,W)C—Co phase diagram with respect to the carbon content is quite broad and equal to nearly 3.3 at.% at 6 wt.% Co and 6.3 at.% at 30 wt.% Co according to Ref. [1]. When taking into account the broad two-phase region of the (Ti,Ta,W)C—Co phase diagram, adding some amounts of cubic (Ti,Ta,W)C carbides to the WC—Co graded powders should not significantly affect the fabrication process of carbide grades for steel-cutting.

References

[1] I. Chaporova, K. Chernyavsky, Structure of Sintered Cemented Carbides, Metallurgiya, Moscow, 1978.
[2] V.I. Tretyakov, Bases of Materials Science and Production Technology of Sintered Cemented Carbides, Metallurgiya, Moscow, 1976.
[3] G.S. Upadhyaya, Cemented Tungsten Carbides: Production, Properties, and Testing, Noyes Publications, New Jersey, USA, 1998.
[4] V.S. Panov, I.Yu. Konyashin, E.A. Levashov, A.A. Zaitsev, Hardmetals, MISIS, Moscow, 2019.
[5] S. Takeda, Metallographic study of the action of the cementing materials for cemented tungsten carbide. Outline of equilibrium diagrams of cobalt-tungsten-carbon and nickel-tungsten-carbon systems, in: Science Reports of the Tohoku Imperial University, Series 1: Mathematics, Physics, Chemistry, 1936, pp. 864–881.
[6] P. Rautala, J.T. Norton, Tungsten-cobalt-carbon system, J. Med. Eng. Technol. 4 (1952) 1045–1050. Does not exist any more.
[7] C.M. Fernandes, A.M.R. Senos, Cemented carbide phase diagrams: a review, Int. J. Refract. Met. Hard Mater. 29 (2011) 405–418.
[8] I. Konyashin, A. Zaitsev, A. Meledin, J. Mayer, Interfaces between model Co-W-C alloys with various carbon contents and tungsten carbide, Materials 11 (3) (2018) 404, https://doi.org/10.3390/ma11030404.
[9] J. Gurland, A study of the effect of carbon content on the structure and properties of sintered tungsten carbide-cobalt alloys, J. Metal. AIME Trans. 6 (1954) (1954) 285–290.
[10] I. Konyashin, A.A. Zaitsev, et al., Wettability of tungsten carbide by liquid binders in WC–Co cemented carbides: is it complete for all carbon contents? Int. J. Refract. Met. Hard Mater. 62 (2017) 134–148.
[11] M.-L. Fiedler, H.H. Stadelmaier, Ternary system nickel-tungsten-carbon, Z. Metallkd. 66 (7) (1975) 402–404.
[12] I.N. Chaparova, V.I. Kudryavtseva, Z.N. Sapronova, Structure and Properties of Iron-Nickel-Based Hardmetals, Cemented Carbide and Refractory Metals, Metallurgiya, Moscow, 1973, pp. 22–26.

Further reading

[1] C.B. Pollok, H.H. Stadelmaier, Eta carbides in the iron-tungsten-carbon and cobalt-tungsten-carbon systems, Met. Trans. 1 (1970) 767.

CHAPTER 4

Refractory carbides employed in the fabrication of cemented carbides

4.1 Tungsten carbides

Major properties of tungsten carbides and their structural characteristics are reported in Refs. [1–4]. Tungsten monocarbide (δ-WC) has a hexagonal crystal lattice shown in Fig. 4.1 at room temperature; periods of the crystal lattice are a = 0.2906 nm and c = 0.2839 nm. It is an intercalation compound, in which carbon atoms with a much smaller atomic radius than that of tungsten atoms are located between tungsten atoms. It should be noted that the tungsten metal is characterized by a cubic body-centered crystal lattice, which transforms into the hexagonal lattice of tungsten carbide during carburization. Tungsten monocarbide contains about 6.13 wt.% (50 at.%) carbon and even insignificant deviations from the stoichiometric carbon content lead to the formation of tungsten metal/β-W_2C or graphite, which can be seen in the W–C phase diagram shown in Fig. 4.2. β-W_2C forms three polymorphs stable at different temperatures, which are designated as β-W_2C, β-W_2C', and β-W_2C''. At temperatures above 2516°C, another modification of tungsten carbide (ϒ-WC) forms according to the W–C phase diagram; it has a cubic crystal lattice with a lattice parameter of 0.4220 nm. In the temperature range of 2516°C –2785°C, the simultaneous existence of all the three tungsten carbide allotropes is possible. Only tungsten monocarbide δ-WC is employed in the cemented carbide industry, so the designation "tungsten carbide" below and in different chapters of this book implies just the δ-WC phase. δ-WC is the only binary phase stable at room temperature, as at a lower carbon content than the stochiometric one, a mixture of δ-WC and β-W_2C should decompose to a mixture of W and δ-WC according to the phase diagram. Nevertheless, if δ-WC forming in the technological process of the tungsten metal carburization is slightly understochiometric, in other words, if there is a mixture of WC and W_2C on the final stage of carburization, this mixture

Cemented Carbides
ISBN 978-0-12-822820-3
https://doi.org/10.1016/B978-0-12-822820-3.00009-4

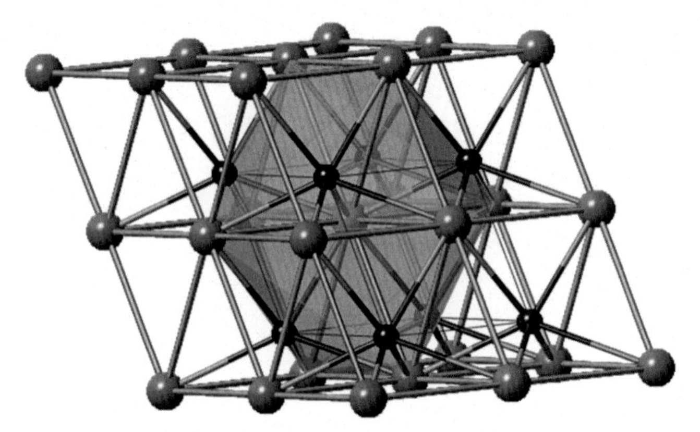

Figure 4.1 Crystal lattice of tungsten monocarbide, in which carbon atoms are black.

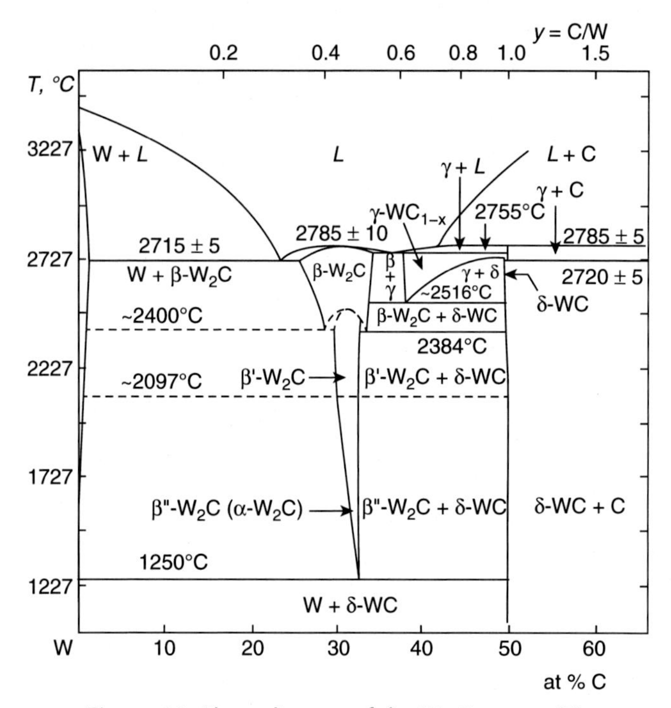

Figure 4.2 Phase diagram of the W–C system [5].

remains in the final product, as cooling rates in industrial carburization furnaces are usually too high to ensure the decomposition of W_2C.

Tungsten carbide crystals have anisotropy of hardness on various crystallographic planes of a single crystal, which has a shape of a trigonal prism.

Thus, depending on the orientation, the minimum microhardness is about 13 GPa on the side prism plane and the maximum microhardness is roughly is 22 GPa on the basal plane.

Tungsten carbide has the following major properties:
Melting point - 2870°C,
Density of 15.5—15.7 g/cm^3,
Thermal conductivity - 0.29—0.47 kal/cm/K,
Coefficient of linear thermal expansion - 4.4 K^{-1},
Young's modulus - 727 GPa,
Bulk modulus - 630—655 GPa,
Shear modulus - 274 GPa,
Compressive strength - 1.4 GPa,
Poisson's ratio - 0.31,
Specific electrical resistivity - 0.2 μΩ m.

Tungsten carbide is practically the only hard refractory compound that is characterized by some degree of plasticity before failure when applying loads at room temperature. Evidence for a noticeable rate of plasticity of WC single crystals is visible on their surface as slip strips forming as a result of dislocation gliding [6—12]. According to various authors [6—9], the slip of tungsten carbide single crystals occurs mainly on the {1010} planes of the prism but can also take place in the directions of [0001] and [1120]. The formation of the slip strips was also observed as a result of Vickers indentation of the plane {1100}.

Results of some experimental studies indicate the contribution of dislocation gliding in the predominant slip system {10-10} [11—23] with splitting by the reaction 1/3[11—23] = 1/6[11—23] + 1/6[11—23], which requires rather low stresses [10—12]. The formation of dislocation networks with bowed-out dislocation links in the WC grains of WC—15%Co samples subjected to deformation is believed to be responsible for the peculiar mechanical properties of tungsten carbide [13].

Results of refs. [14,15] appear to be very important from a practical point of view. In these works, the formation of slip strips in WC grains of cemented carbides was observed as a result of friction of WC—Co inserts against rocks and in tests on cutting Ni-based alloys. These results indicate that tungsten carbide grains in cemented carbides undergo some plastic deformation before damage (the formation of micro- and macrocracks) in mining and metal-cutting applications. Due to this unique behavior, tungsten carbide and WC-based cemented carbides have high performance properties in a great number of different applications characterized by the presence of high impact loads, thermal and mechanical fatigue, and severe abrasive wear. Numerous attempts to substitute tungsten carbide in cemented carbides by any other refractory compound in applications such as mining and construction or oil and gas drilling failed. This occurred as a result of very peculiar operation conditions in these applications requiring some residual plasticity of the carbide phase for obtaining the optimal combination of hardness/wear resistance and fracture toughness/strength.

In addition to the noticeable rate of plasticity before failure, other properties of WC, such as its high Young's modulus, relatively low electrical resistance, high values of compressive strength, and thermal conductivity, play an important role in achieving a unique combination of mechanical and performance properties of WC—Co cemented carbides in numerous applications.

An important property of WC from a practical point of view is the good wettability of tungsten carbide by liquid iron group metals, particularly cobalt, at temperatures of 1400°C—1500°C. Up to recent times, it was believed that liquid Co—based binders with any carbon content wet tungsten carbide very well; however, recently it was established that the wettability of WC by Co-based binders with high carbon contents is incomplete [16].

The wettability of tungsten carbide by melts of other metals with relatively low melting points (copper, zinc, silver, etc.) is incomplete; for example, tungsten carbide is very poorly wetted by liquid copper.

4.2 Carbides of the Ti—Ta—Nb—W—C system

WC-based cemented carbides containing grains of cubic carbides (W,Ti,Ta)C and (W,Ti,Ta,Nb)C are widely used for the manufacture of indexable cutting inserts for steel-cutting and similar metal-working operations. It is well known that in operations of steel-machining the friction coefficient between surfaces of steel chips and cemented carbides on the insert rank face become significantly lower if the cemented carbides comprise grains of cubic titanium-containing carbides. This leads to a significant reduction of cutting forces and temperatures resulting in decreased wear rates of the cemented carbides containing grains of such cubic carbides in comparison with WC—Co cemented carbides.

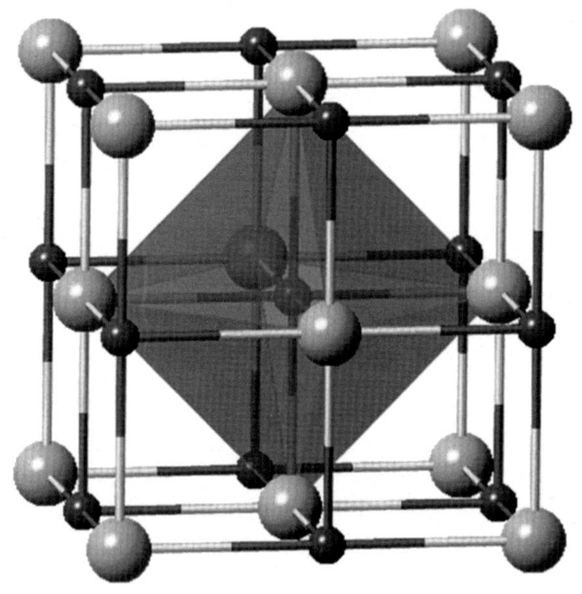

Figure 4.3 The crystal lattice of titanium carbide (a cubic lattice of the NaCl type); carbon atoms are shown by dark-blue color (dark gray in print). In mixed cubic carbides of the W–Ti–Ta–Nb–C system, atoms of W, Ta, and Nb substitute the Ti atoms in the cubic crystal lattice.

It is well known that the solubility of titanium, tantalum, and niobium in tungsten carbide is close to zero at room temperature, while the solubility of tungsten carbide in titanium carbide is quite significant, so that there is a range of TiC–WC solid solutions.

Fig. 4.3 shows the crystal lattice of titanium carbide, in which W, Ta, and Nb atoms in (W,Ti)C, (W,Ti,Ta)C, and (W,Ti,Ta,Nb)C carbides substitute Ti atoms and occupy their positions. Titanium carbide has a wide homogeneous range (from $TiC_{0.62}$ to $TiC_{0.96}$) and its microhardness varies from 19.3 GPa (for low carbon contents) to 31.7 GPa (for high carbon contents) [1].

The Ti–W–C phase diagram was studied in Ref. [17]. The vertical section of this phase diagram on the TiC–WC line shown in Fig. 4.4 is the most important one from a practical viewpoint. This section, in principle, cannot be considered as a pseudobinary one, as one of the components (WC) melts with decomposition. Nevertheless, in the production of (W,Ti) C powders, the carburization temperatures are much lower than the melting point of tungsten carbide, so that this section, in the first

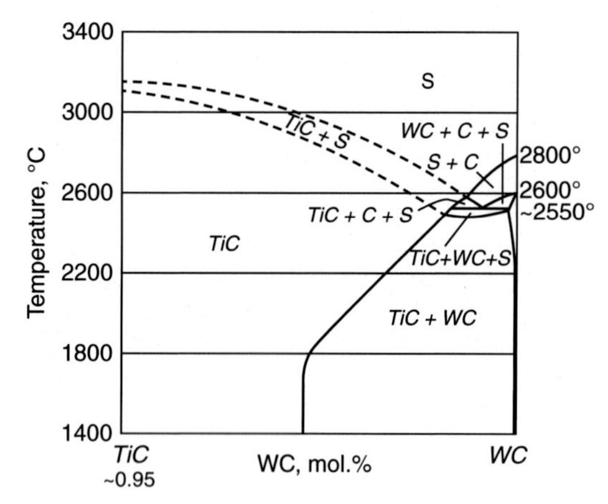

Figure 4.4 Vertical section of the Ti—W—C phase diagram on the TiC—WC line (liquid is designated by S) [17].

approximation, can be considered as pseudobinary. As one can see in Fig. 4.4, there are only two areas in this phase diagram at temperatures of up 2500°C: a single-phase solid solution based on titanium carbide and a two-phase area (the solid solution and WC phase). The solubility of tungsten carbide in titanium carbide varies from nearly 70 wt.% at 1500°C to 90 wt.% near the temperature corresponding to the formation of the liquid phase (about 2600°C).

The microhardness of the carbide solid solutions in the Ti—W—C system varies from about 20 GPa at 30 wt.% titanium carbide up to nearly 30 GPa for the compositions close to the TiC content of 80 wt.%. The microhardness decreases with decreasing the total carbon content within the homogeneity range. The Young's modulus of (W,Ti)C practically does not change in the homogeneity range and is roughly 440—460 GPa. The linear thermal expansion coefficient also almost does not change and is nearly 5.3—5.5 × 10^{-6} K^{-1}. The wetting angle of (W,Ti)C by cobalt melts in hydrogen is equal to roughly 20—26 degrees and in a vacuum is 3—7 degrees. The wetting angle of (W,Ti)C by nickel melts in hydrogen is nearly 16—17 degrees and in a vacuum is 6—10°. The (W,Ti)C phase containing nearly 70 wt.% TiC is characterized by the highest oxidation resistance.

Results on the Ta—W—C phase diagram were reported in Ref. [3] (Fig. 4.5). The solubility of tungsten carbide in tantalum carbide at 1450°C is about 6 wt.%. The microhardness of the WC—TaC solid solutions increases with an increase in tungsten carbide concentrations and varies from nearly 16 to 18 GPa.

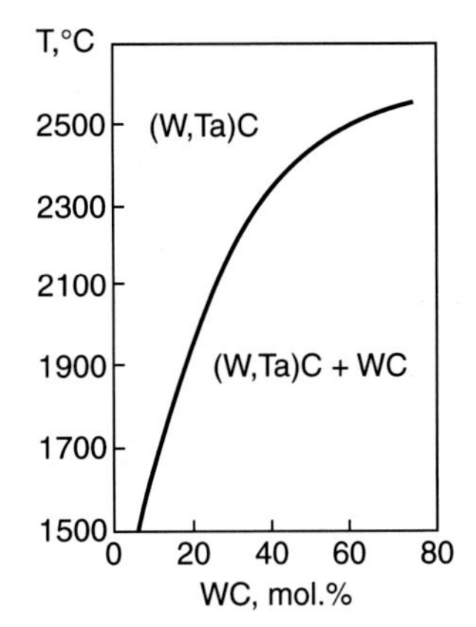

Figure 4.5 The temperature dependence of the solubility of WC in TaC [3].

It was established that TiC, TaC, and NbC form a continuous series of solid solutions; the solubility of WC in the solid solutions decreases with increasing the amount of TaC or NbC. The TiC–TaC–WC system was investigated in Ref. [18] in the temperature range between 1450°C and 2200°C. It was established that there is a continuous series of TiC-based solid solutions in this system. The solubility of WC in TiC is approximately 76 wt.% and of TaC in TiC is about 15 wt.% at 1450°C. There is, therefore, a wide range of the TiC–TaC–WC solid solutions at temperatures above 1500°C.

NbC is very close with respect to its crystal lattice to TaC and, therefore, Nb atoms can easily substitute Ta atoms in the crystal lattice of (W,Ti,Ta,Nb)C carbides. This means that the TiC–TaC–NbC–WC system (which, to our knowledge, was not reported in literature) should be similar to the TiC–TaC–WC system.

Increased concentrations of both tungsten and tantalum in (Ti,W,Ta)C lead to a decrease in the microhardness of the TiC–TaC–WC solid solutions, similar to the reduction of the (Ti,W)C microhardness as a result of its alloying with TaC.

The transverse rupture strength (TRS) of (Ti,W,Ta)C increases as a result of adding TaC to (Ti,W)C and steadily grows with further increase in the TaC content. The oxidation resistance also increases due to adding TaC to (Ti,W)C and further increasing its content up to the highest oxidation resistance obtained at 23 mol.% TaC. The fracture toughness of the (Ti,Ta,W)C solid solutions noticeably increases and their specific electrical resistivity decreases when increasing the molar percentage of TaC from 0 up to 35 mol.% [3]. The wettability of cubic carbide (Ti,W,Ta)C by liquid cobalt or nickel practically does not change with increasing the TaC content above 0; the wetting angle is nearly 5–10 degrees at 1500°C in a vacuum [3]. The solubility of (Ti,Ta,W)C in Co-based binders does not depend on the TaC content, nevertheless, it is strongly dependent on the carbon content in cemented carbide samples containing (Ti,Ta,W)C. Adding TaC to (W,Ti)C and increasing its content in the (Ti,W,Ta)C phase lead to an increase in the Young's modulus and a decrease in the linear expansion coefficient.

In ref. [3], it was concluded that the major advantages achieved as a result of adding TaC to (W,Ti)C in amounts of up to 35 mol.% are higher values of TRS of the (Ti,W,Ta)C solid solutions without sacrificing their hardness, their noticeably improved Palmqvist fracture toughness, and significantly (by up to 3 times) improved oxidation resistance.

References

[1] G.V. Samsonov, I.M. Vinitskii, Handbook of Refractory Compounds, Plenum Press, New York, N. Y., 1980.
[2] T. Kosolapova, Carbides, Moscow, Metallurgiya, 1968.
[3] V.I. Tretyakov, Bases of Materials Science and Production Technology of Sintered Cemented Carbides, Moscow, Metallurgiya, 1976.
[4] A.I. Gusev, A.S. Kurlov, Tungsten carbides and W-C phase diagram, Inorg. Mater. 42 (2) (2006) 121–127.
[5] C.M. Fernandes, A.M.R. Senos, Cemented carbide phase diagrams: a review, Int. J. Refract. Metals Hard Mater. 29 (2011) 405–418.
[6] T. Takahasi, E. Freise, Determination of the slip systems in single crystals of tungsten monocarbide, Phylosoph. Mag. 12 (115) (1965) 1–8.
[7] D. French, D. Thomas, Hardness anisotropy and slip in WC crystals, Trans. Am. Inst. Min. Metall. Petrol. Eng. 233 (5) (1965) 950–952.
[8] A. Miyoashi, A. Hara, J. Sugimoto, An investigation of cracks around Vickers hardness indentation in cemented carbides, Planseeber. für Pulvermetall. 15 (3) (1967) 187–196.
[9] S.B. Luyckx, Slip system of tungsten carbide crystals at room temperature, Acta Metall. 18 (1970) 233–236.
[10] S. Lay, HRTEM investigation of dislocation interactions in WC, Int. J. Refract. Metals Hard Mater. 41 (2013) 416–421.
[11] M.K. Hibbs, R. Sinclair, Room-temperature deformation mechanisms and the defect structure of tungsten carbide, Acta Metall. 29 (1981) 1645–1654.
[12] T. Csanádi, M. Bľanda, A. Duszová, N.Q. Chinh, P. Szommer, J. Dusza, Deformation characteristics of WC micropillars, J. Eur. Ceram. Soc. 34 (2014) 4099–4103.
[13] T. Johannesson, B. Lehtinen, On the plasticity of tungsten carbide, Phys. Status Solidi 16 (1973) 615–622.
[14] L. Ponds, in: F.W. Vahldiek, Mersol (Eds.), Anisotropy in Single Crystal Refractory Compounds, vol. 2, Plenum Press, New York, 1968.

[15] E.M. Trent, Special Report, Metallurgical Changes at the Tool-Work Interface, vol. 94, Iron and Steel Institute, London, 1967, pp. 77–87. Volume Date 1965.
[16] I. Konyashin, A.A. Zaitsev, et al., Wettability of tungsten carbide by liquid binders in WC–Co cemented carbides: is it complete for all carbon contents? Int. J. Refract. Metals Hard Mater. 62 (2017) 134–148.
[17] H. Nowotny, E. Parthe, R. Kieffer, F. Benesiwsky, The ternary system titanium-tungsten-carbon, Z. Metallkunde 45 (1954) 97–101.
[18] H. Nowotny, R. Kieffer, O. Knotek, The structure of the carbide system TiC-TaC-WC, Berg- Huettenmaennische Monatshefte der Montanistischen Hochschule Leoben 96 (1951) 6–8.

CHAPTER 5

Structure and properties of binder phases in cemented carbides

5.1 Co-based binders

Cobalt-based binders are used in most industrial grades of cemented carbides.

The major properties of pure Co are described in the monograph [1]. Cobalt exists in two allotropes having an fcc crystal lattice (above 422°C) or an hcp crystal lattice (below 422°C). Cobalt has ferromagnetic properties and is characterized by the Curie temperature of 1121°C. This property of Co is important from a practical point of view, since it makes it possible to characterize the structure and composition of WC—Co cemented carbides by measuring their magnetic properties.

The density of Co is 8.85 g/cm^3 for the hexagonal modification and 8.80 g/cm^3 for the cubic modification; the density of liquid Co is 7.8 g/cm^3 at 1850°C.

The Vickers hardness of cobalt is dependent on the thermo-mechanical treatment and varies from about 3.1 GPa for samples subjected to a process of thermo-mechanical treatment to nearly 2.1 GPa for samples annealed at a temperature of 800°C for 1 h [1].

Results on experimental studies of the phase diagrams in the systems tungsten carbide—iron group metal (Co, Ni, and Fe) are given in Chapter 3. However, different research groups studied the W—C—Co, W—C—Ni, and W—C—Fe—Ni phase diagrams at different binder contents, so that their direct comparison is not possible. Also, as it is mentioned in Chapter 3, the total carbon content in sintered carbide samples is dependent on the gas atmosphere during sintering and cannot therefore be fully controlled in experimental studies of the phase diagrams, leading to some uncertainties in the position of border lines in the vertical sections of the phase diagrams through the carbon angle.

Cemented Carbides
ISBN 978-0-12-822820-3
https://doi.org/10.1016/B978-0-12-822820-3.00005-7

Calculated phase diagrams of the system: tungsten carbide—metals of the iron group with identical contents of cobalt, nickel, and iron (20 wt.%) are reported in Ref. [2]. Based on the analysis of the vertical sections of these phase diagrams through the carbon angle, it is possible to directly compare the phase compositions of cemented carbides containing different Fe group metals depending on the total carbon content.

The vertical section of the W—C—Co phase diagram obtained as a result of thermodynamic calculations [2] is shown in Fig. 5.1.

As it can be seen in Fig. 5.1, the width of the two-phase region with respect to the carbon content in the phase diagram is of the order of 0.3 wt.% at a temperature of 1000°C. In this case, the two-phase region is shifted somewhat to the left in comparison with the carbon content corresponding to stoichiometric WC; nevertheless, cemented carbides produced from stoichiometric WC should contain only the WC and binder phase in their microstructure. Note that the melting points at low and high carbon contents in the calculated W—C—Co phase diagram are noticeably lower than those in the experimentally obtained phase diagram shown in Fig. 3.7.

Fig. 5.2 shows EBSD images of the microstructure of the WC—11%Co cemented carbide according to ref. [3]. As it can be seen from the images,

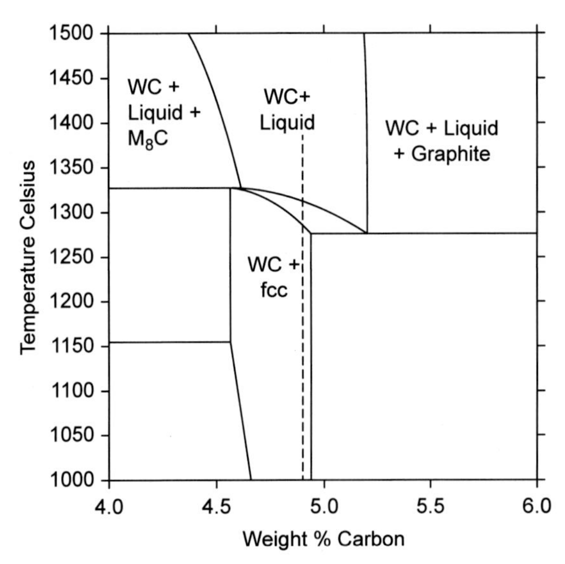

Figure 5.1 Vertical section of the calculated W—C—Co phase diagram through the carbon angle at a cobalt content of 20 wt.%. The carbon content corresponding to the stoichiometric WC is marked by a *dashed line. (Redrawn from Ref. [2].)*

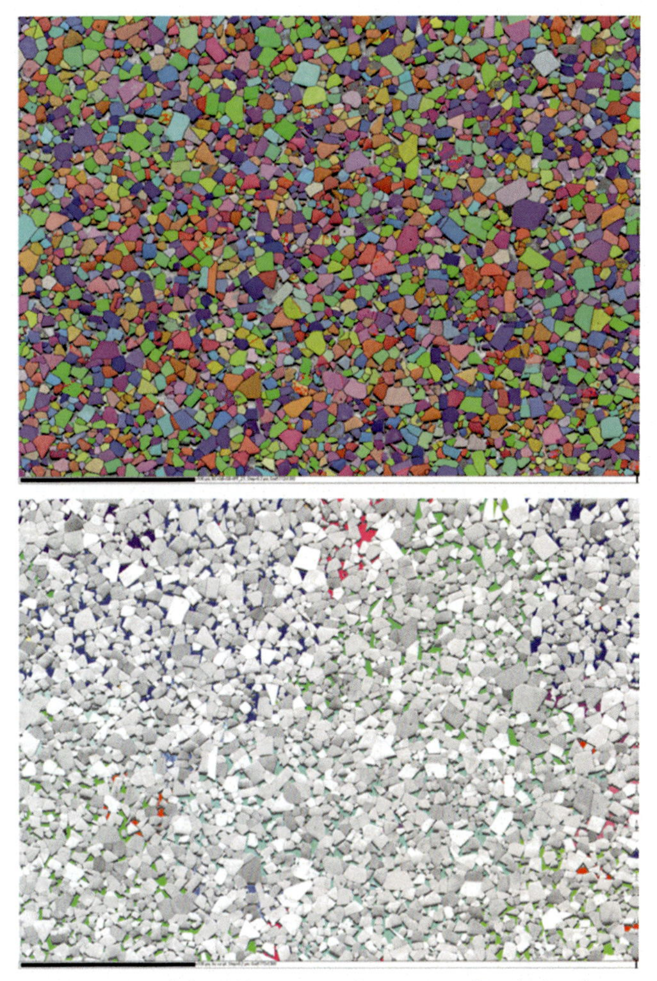

Figure 5.2 EBSD images of the WC–11 wt.%Co cemented carbide: above – grains of tungsten carbide with different crystallographic orientation are shown by different colors and Co-based binder phase is indicated by *gray color*, and below – the grains of tungsten carbide are shown by *gray* and different grains (single crystals) of the Co-based binder with various crystallographic orientations are indicated by different colors (the scale bar corresponds to 100 μm). *(Redrawn from Ref. [3].)*

the mean grain size of Co grains (single crystals) in the carbide microstructure is of the order of 100 μm or greater. In fact, as it is reported in Ref. [4], the grain size of Co grains in WC–Co cemented carbides can be as high as 0.5 mm.

As it is mentioned above, the hcp cobalt allotrope of pure Co is stable at room temperature. The Co-based binder in WC–Co cemented carbides contains relatively high amounts of dissolved tungsten, which stabilizes the high-temperature fcc cobalt allotrope. The hcp/fcc ratio in cemented carbides depends on the concentration of tungsten dissolved in the binder and consequently on the total carbon content. According to the results of ref. [3], the binder of a carbide sample at a very low carbon content, containing 7.8 wt.% tungsten dissolved in the binder phase, comprises very little hcp-Co (2.1%). The binder of a sample with a medium–high carbon content containing 3.0 wt.% tungsten dissolved in the binder phase is characterized by a relatively high proportion of hcp-Co (51.6%). In many cases, the binder phase containing significant amounts of hcp-Co consists of a mixture of fcc-Co and hcp-Co grains of the order of 100 nm up to several hundred nanometer in size, which are shown in Fig. 5.3, left.

It is well known that the Co-based binder is subjected to the partial martensitic transformation from the fcc to the hcp crystal lattice under conditions of cyclic loads or fatigue [5–7]. Nearly 3–4 decades ago, it was considered that it would be advantageous to suppress this transformation and stabilize the fcc-Co phase due to alloying the binder phase by, for example, Ni. It was believed that cracks forming as a result of fatigue would propagate through the more brittle hcp-Co phase [6,7]. The partially transformed binder comprising lamella of hcp-Co was thought to be unable to fulfill the compatibility requirements and loose its ability to impede crack propagation.

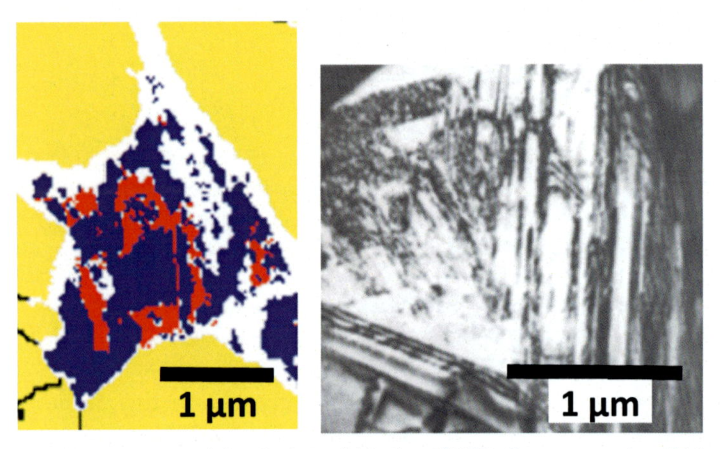

Figure 5.3 Fine structure of the Co-based binder of WC–Co cemented carbides: left - EBSD image of the WC–6 wt.%Co cemented after sintering (*yellow* – WC, *red* – fcc Co, and blue – hcp Co) [3]; and right – TEM image of the binder phase of the WC–17 wt.%Co cemented carbide after sintering followed by fatigue treatment showing lamellae of hcp-Co in a fcc-Co matrix [5].

Nevertheless, according to the generally accepted nowadays viewpoint on this issue, it is believed that the martensitic fcc-hcp transformation occurring in the binder phase subjected to cyclic loads plays a positive role with respect to performance of WC—Co materials under conditions of severe fatigue due to the highest work-hardening rates achieved in this case. It was established that additions of Fe and Ni to the Co-based binder stabilizing the fcc cobalt allotrope do not improve performance properties of WC—Co cemented carbides subjected to severe fatigue. In contrast, there is some degradation of performance properties as a result of partial substitution of Co by Fe—Ni alloys in such applications as, for example, percussive drilling. The results of ref. [5] on examining microstructural responses to monotonic and cyclic compressive loading of WC—Co and WC—Co—Ni cemented carbides shed light on this phenomenon. It was established that predominant binder deformation mechanisms shifted from the fcc-hcp martensitic transformation to slip plus twinning when increasing the Ni content in the binder phase up to 30 wt.%. In the WC—Co samples, 44.5 vol.% of the binder transformed at the highest strain level, while only 11.4% transformation occurred at the same strain in the high-Ni samples resulting in a significant change regarding the stress—strain behavior. It was concluded that the cemented carbide binders based on pure cobalt are characterized by increased work-hardening rates due to the fcc-hcp martensitic transformation occurring as a result of fatigue treatment leading to the formation of thin lamellae of hcp-Co in a fcc-Co matrix, which are shown in Fig. 5.3, right. Addition of up to 30 wt.% Ni to Co-based binders stabilizing the fcc Co modification results in decreasing the binder work-hardening rates due to suppressing or full elimination of the martensitic fcc-hcp transformation under the impact of fatigue treatments. The major reason for the different behavior of the Co binder in comparison with the Ni and Co—Ni binders under impact of severe fatigue is presumably the significantly higher stacking fault energy of Ni (125 mJ/m^2) compared to Co (20 mJ/m^2). This results in dramatically reduced rates of work hardening of the Ni-based binders compared to those of the Co binders.

In recent years, numerous attempts were made to increase the microhardness of the Co-based binder of WC—Co cemented carbides due to nanostructuring the binder phase (dispersion hardening and so-called 'nanograin reinforcement') [8—16].

The first work on the binder dispersion hardening of WC—Co cemented carbides was ref. [8]. In this work, the results of experiments on heat treatment (aging) of cemented carbides with different carbon contents were reported. It was found that as a result of aging of carbide samples with a low carbon content for a long time (up to 20 h), hard nanoparticles precipitated in the binder, which led to some increase in its microhardness accomplished by a noticeable decrease of the transverse rupture strength (TRS).

The results of experiments on the binder dispersion hardening of WC–Co cemented carbides due to their annealing are also described in Refs. [9–12]. The formation of nanoparticles consisting of η-phase or intermetallic phases of the W–Co system as well as metastable phases of the W–Co–C system was established [9,10]. The coercive force of cemented carbides was significantly increased as a result of the heat treatments [12]. The formation of nanoparticles in the Co-based binders was found to result in an increase in its microhardness accompanied by a significant decrease in fracture toughness and strength characteristics of the heat-treated carbide samples. For this reason, the heat treatments based on prolonged aging have never been employed in the cemented carbide industry.

Efforts were also made to strengthen and reinforce the binders of WC–Co cemented carbides by introducing nanoparticles of refractory compounds, diamond, and cubic boron nitride as well as carbon nanotubes, which was claimed to result in improved properties of cemented carbides (see Chapter 10). However, it is unclear how the well-known phenomenon of fast dissolution of the refractory compounds' nanoparticles, diamond, carbon fine particles, etc., in the binder melt during liquid-phase sintering can be prevented, so the claimed improved properties of cemented carbides with such binders appear to be questionable.

The first industrial cemented carbide grades with nanograin reinforced binder having the trademark MasterGrade™ were developed and implemented at Element Six GmbH (Germany); they are described in detail in Chapter 11.

5.2 Ni-based binders

Nickel is a ductile silver-white metal having a face-centered cubic crystal lattice. Nickel is chemically inactive and characterized by significantly greater corrosion resistance in aggressive media than cobalt. In its pure form, it is very plastic and can be processed by forging. It is characterized by ferromagnetic properties with the Curie temperature of 358°C.

It is well known that cemented carbides with nickel-based binders are inferior in properties compared to cemented carbides with cobalt binders. One of the main reasons for this according to Ref. [4] is a fact that the Vickers hardness and tensile strength of nickel, which are equal correspondingly to 0.8 GPa and 90–100 MPa, are significantly lower than these properties of cobalt (nearly 2–3 GPa and 480–530 MPa). Cemented carbides with Ni, Ni–Co, Ni–Cr, and Ni–Co–Cr binders are fabricated in the cemented carbide industry on a medium–large scale and used for producing wear-resistant parts and tools operating in corrosive environments and at elevated temperatures (see Chapter 10).

The vertical section of the W–C–Ni phase diagram through the carbon angle obtained as a result of thermodynamic calculations [2] is shown in Fig. 5.4. As one can see in Fig. 5.4, the two-phase region is even slightly wider with respect to the carbon content (about 0.4 mass%) in comparison with the W–C–Co phase diagram. At the same time, the two-phase region is strongly shifted towards lower carbon contents, so that WC–Ni powders employed for fabrication of WC–Ni cemented carbides must have a lower total carbon content than corresponding WC–Co powders. In addition, the temperatures of the liquid phase formation for WC–Ni

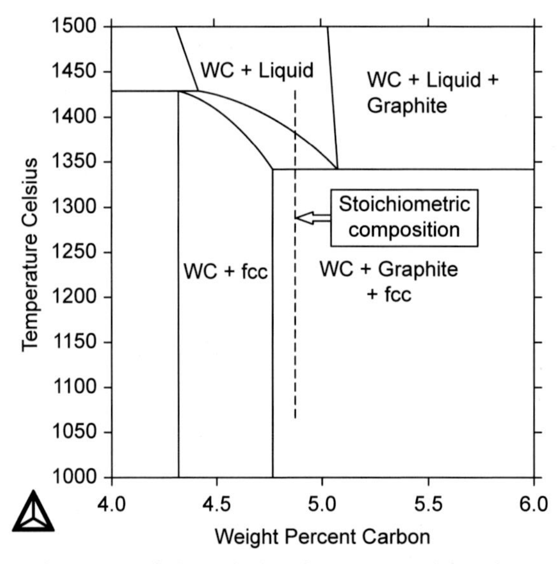

Figure 5.4 Vertical section of the calculated W–C–Ni phase diagram through the carbon angle at a cobalt content of 20 wt.%. The carbon content corresponding to the stoichiometric WC is marked by a *dashed line. (Redrawn from Ref. [2].)*

Figure 5.5 EBSD image of the WC−10 wt.%Ni cemented carbide: the grains of tungsten carbide are shown by *gray* and different grains (single crystals) of the Ni-based binder with different crystallographic orientations are indicated by different colors (the scale bar corresponds to 20 μm). *(Redrawn from Ref. [3].)*

alloys lie significantly higher in comparison with those of WC−Co alloys, so that WC−Ni cemented carbides must be sintered at noticeably higher temperatures.

As it can be seen in Fig. 5.5, grains (single crystals) of nickel in the WC−Ni cemented carbide have a substantially smaller size (10−20 μm) than the Co grains in WC−Co alloys.

Despite the fact that cemented carbide grades with nickel binders or nickel-containing binders do not have a wide application range, most cemented carbide companies produce such grades. According to Ref. [17], cemented carbides with a nickel−cobalt binder alloyed with chromium have a significantly improved corrosion−erosion resistance in comparison with similar WC−Co alloys. Examples of such cemented carbides are the grades fabricated by Element Six GmbH (Germany), which are described in detail in Chapter 10.

5.3 Fe-based binders

The vertical section of the W−C−Fe phase diagram through the carbon angle obtained as a result of thermodynamic calculations [2] is shown in Fig. 5.6. As one can see in Fig. 5.6, the two-phase region for cemented carbides with iron-based binders at 1000°C is very narrow (of the order of 0.05 wt.% or less), which is significantly lower in comparison with the

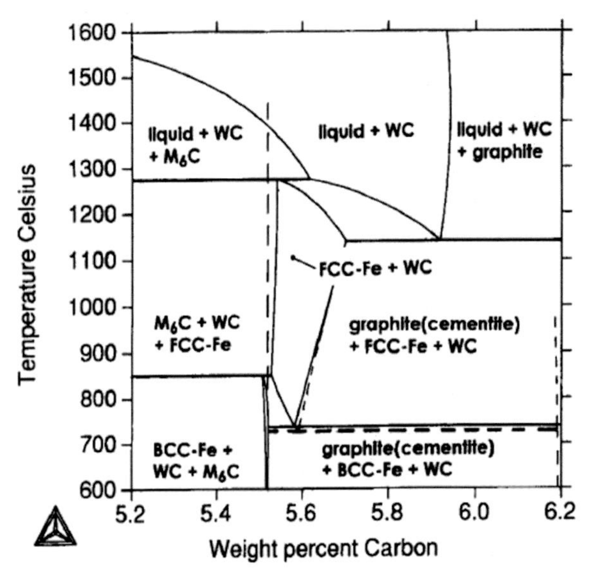

Figure 5.6 Vertical section of the calculated W—C—Fe phase diagram through the carbon angle at a cobalt content of 20 wt.%. The carbon content corresponding to the stoichiometric WC is marked by a *dashed line. (Redrawn from Ref. [2].)*

W—C—Co phase diagram. This causes difficulties in the production of cemented with iron binders, since very insignificant deviations in the carbon content in the WC—Fe powder and in the atmosphere of sintering furnaces lead to the formation of either η-phase or free carbon plus cementite in the cemented carbide microstructure. In addition, cemented carbides with iron binders must be subjected to heat treatment after sintering to obtain the desired structure and properties of the binder phase, since, in fact, the binder is a sort of tungsten-alloyed steel. In addition, the temperatures of the liquid phase formation for WC—Fe alloys lie significantly lower in comparison with WC—Co cemented carbides. On these reasons, there are no industrial carbide grades with binders containing pure iron in the cemented carbide industry; iron is employed exceptionally as an alloying component in carbide grades with mixed alternative binders.

5.4 Alternative binders

Cobalt is known to be a strategical material with a limited number of industrial mines worldwide, which causes significant fluctuations of cobalt prices from time to time. Recent data indicate that there are some health and ecological issues with respect to employing Co powders (although, up to now, there has not been clear evidence that cobalt powders are more harmful that nickel powders). In order to solve the problems mentioned above, it was suggested to substitute Co binders in cemented carbides by Fe—Ni or Fe—Co—Ni binders (see, e.g., Refs. [18,19]).

As it can be seen in Fig. 5.7 (and also in Fig. 3.12) the employment of Fe—Ni alloys instead of pure Fe allows one to significantly widen the two-phase region with respect to the carbon content in the W—Fe—Ni—C system, so the WC—Fe—Ni cemented carbides can be manufactured relatively easily in comparison with the binders consisting of pure iron. According to Ref. [20], insignificant additions of Co to the Fe—Ni binders do not noticeably change the phase equilibrium and do not increase the width of the two-phase region with respect to the carbon content. At the same time, the iron—nickel—cobalt binders with a tailored martensitic/austenitic structure can have an improved plasticity in comparison with conventional Co-based binders [20]. The brittle ordered 50Fe + 50Co binder phase was found to have inferior values of TRS, fracture toughness, and fatigue strength in comparison to the martensitic/austenitic 65Fe + 20Co + 15Ni binder that facilitates transformation-induced

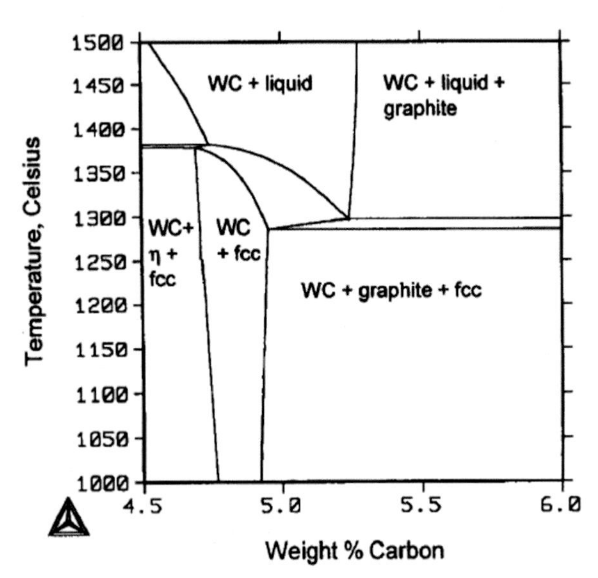

Figure 5.7 Vertical section of the calculated W—C—Fe—Ni phase diagram through the carbon angle (13% Ni + 7% Fe) through the carbon angle at a binder content of 20 wt.%. *(Redrawn from Ref. [2].)*

plasticity at high yield stresses. Cemented carbides with iron-based binders also show a better oxidation resistance than conventional WC–Co cemented carbides and this can be further improved by alloying the binder phase with Cr and Mo, so they can be employed for fabricating wear parts and tools operating at elevated temperatures.

However, as one can see in Fig. 5.8, cemented carbides with Fe–Ni–Co binders have a reduced wear resistance in percussive drilling of quartzite in comparison with the corresponding conventional WC–6 wt.%Co grade. The conventional WC–Co grade for percussive drilling (T6) was compared with two grades of the WC–Fe–Co–Ni cemented carbides either with the same weight percentage (T6-Fe–Ni–Co-1) or the same volume percentage (T6-Fe–Ni–Co-2) of the binder phase in comparison with the conventional WC–Co grade. The hardness of the WC–Fe–Co–Ni cemented carbide with the same binder weight percent was slightly lower than that of the WC–Co grade, which can explain its reduced wear resistance. However, the hardness of the WC–Fe–Co–Ni cemented carbide with the same binder volume percent was similar to that of the WC–Co grade. Therefore, the noticeably lower wear resistance of

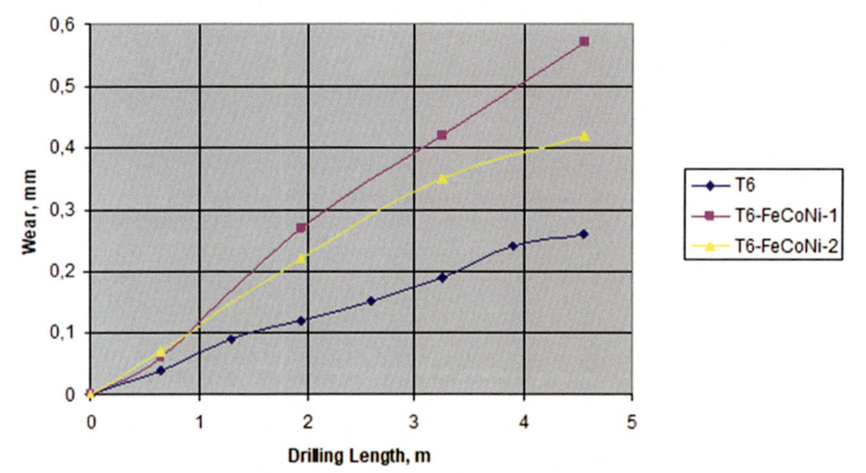

Figure 5.8 Gauge wear of inserts consisting of the standard grade WC–6wt.%Co for percussive drilling (**T6**) versus experimental grades with the alternative Fe–Ni–Co binders (**T6–Fe–Ni–Co**) in laboratory performance tests on percussion drilling of quartzite. The Fe–Ni–Co binder is characterized by the ratio Fe:Ni:Co = 40:40:20 (by weight) and either by the same weight percentage (T6-Fe–Ni–Co-1) or the same volume percentage (T6-Fe–Ni–Co-2) of the binder phase in comparison with the WC–6wt.%Co carbide grade. *(With the permission of Element Six.)*

the carbide grade with the Fe—Co—Ni binder was likely a result of its reduced performance properties (mainly fatigue resistance) in conditions of intensive abrasive wear and severe fatigue occurring in percussive drilling. This takes place presumably due to a decreased work-hardening rate of the Fe—Co—Ni binder in comparison with the cobalt binder as a result of the suppression or complete elimination of the fcc-to-hcp martensitic transformation in conditions of high fatigue loads.

Reduced performance properties of the WC—Fe—Co—Ni cemented carbides in the majority of applications characterized by high impact loads and severe fatigue as well as some technological difficulties related to the fabrication of such cemented carbides and their control by use of magnetic properties led to a fact that they are presently produced on a relatively small scale. Such cemented carbides can have advantages over WC—Co materials in applications, where high corrosion and oxidation resistance is needed. One of the few examples of such WC—Fe—Co—Ni cemented carbides is the CTU30/SNK30 grade manufactured by Ceratizit for wear applications [21].

References

[1] B.R. Breckpot, L. Habraken, J.C. Jungers, et al., Cobalt Monograph, Centre ED'Information du Cobalt, Belgium, 1958.
[2] H.P., E.P. Uhrenius, On the composition of Fe-Ni-Co-WC-based cemented carbides, Int. J. Refract. Met. Hard Mater. 15 (1997) 139—149.
[3] K.P. Mingard, B. Roebuck, J. Marshall, G. Sweetman, Some aspects of the structure of cobalt and nickel binder phases in hardmetals, Acta Mater. 59 (2011) 2277—2290.
[4] V.I. Tretyakov, Bases of Materials Science and Production Technology of Sintered Cemented Carbides, Metallurgiya, Moscow, 1976.
[5] C.H. Vasel, A.D. Krawitz, E.F. Drake, E.A. Kenik, Binder deformation in tungsten carbide-(cobalt, nickel) cemented carbide composites, Metal. Trans. A 16A (12) (1985), 2309-2307.
[6] U. Schleinkofer, H.-G. Sockel, K. Görting, W. Heinrich, Microstructural processes during subcritical crack growth in hard metals and cermets under cyclic loads, Mater. Sci. Eng. 209 (1996) 103—110.
[7] V.K. Sarin, T. Johannesson, The deformation of tungsten carbide-cobalt cemented carbides, Met. Sci. 9 (10) (1975) 472—476.
[8] H. Suzuki, H. Kubota, The influence of binder phase composition on the properties of WC-Co cemented carbides, Plnaseeberichte für Pulvermetallurgie 14 (1966) 96—109.
[9] H. Jonsson, Studies of the binder phase in WC-Co cemented carbides heat-treated at 950°C, Planseeber. für Pulvermetall. 23 (1975) 37—55.
[10] C. Wirmark, G.L. Dunlop, Phase transformation in the binder phase of Co-W-C cemented carbides, in: R.K. Viswandham, D. Rouclihle, J. Gurland (Eds.), Proc. Int. Conf. Sci. Hard Mater., Plenum, New York, 1983, pp. 311—327.
[11] D.L. Tillwick, I. Joffe, Precipitation and magnetic hardening in sintered WC-Co composite materials, J. Phys. D 6 (1973) 1585—1592.

[12] H. Grewe, J. Kolaska, Gezielte Einstellungen von Lösungszuständen in der Binderphase technischer Hartmetalle und Folgerungen daraus, Metall 7 (1981) 563–566.
[13] W. Bryant, Schneideinsatz zum Fräsen von Titan und Titanlegierungen, 1981. German Patent DE19810533A1.
[14] S. Kang, Solid-solution Powder, Method to Prepare the Solid-Solution Powder, Cermet Powder Including the Solid-Solution Powder, Method to Prepare the Cermet Powder and Method to Prepare the Cermet, 2009. US Patent Application US2009-0133534.
[15] G. Zhan et al. Polycrystalline Composites Reinforced with Elongated Nanostructures, European Patent Application EP1923475.
[16] Y. Zhang et al. Nano-reinforced WC-Co for Improved Properties. European Patent Application EP1923476.
[17] I. Konyashin, B. Ries, S. Hlawatschek, H. Hinners, Novel industrial hardmetals for mining, construction and wear applications, in: L. Sigl, H. Kestler, A. Pilz (Eds.), Proc. 19th Int. Plansee Sem., Metallwerk Plansee, Reutte, Austria, 2017. HM6/1–HM6/12.
[18] L. Prakash, B. Gries, WC hardmetals with iron based binders, in: L. Sigl, P. Rödhammer, H. Wildner (Eds.), Proc. 17th Int. Plansee Sem, vol. 2, Plansee Group, Reutte, Austria, 2009, pp. 5/1–5/14.
[19] H.-W. Heinrich, M. Wolf, D. Schmidt, U. Schleinkoffer, A Cermet Having a Binder With Improved Plasticity, a Method for Manufacture and Use Thereof, 1999. PCT Patent Application WO 99/10549.
[20] L. Prakash, Fundamentals and general applications of hardfmetals, in: V. Sarin (Ed.), Comprehensive Hard Materials, Elsevier Science and Technology, 2014, pp. 29–89.
[21] I. Konyashin, Cemented carbides for mining, construction and wear parts, in: V. Sarin (Ed.), Comprehensive Hard Materials, Elsevier Science and Technology, 2014, 425-251.

CHAPTER 6

Materials engineering of cemented carbides

6.1 Technological processes for fabrication of tungsten metal and tungsten carbide powders

6.1.1 Fabrication of tungsten metal powders

Physicochemical fundamentals of and technologies for the fabrication of tungsten metal powders for manufacturing cemented carbides are described in detail in several monographs [1−4]; therefore, this chapter reports only major features of the tungsten powder production processes.

The major starting compound employed for the fabrication of tungsten metal and tungsten products is ammonium paratungstate (APT), which is a white crystalline salt with the chemical formula $(NH_4)_{10}(H_2W_{12}O_{42})\cdot 4H_2O$. It consists of monoclinic crystals having a mean grain size in the range of 5−80 μm. Usually, APT is subjected to calcination at temperatures of above 450°C leading to the formation of either "yellow oxide" WO_3, or "blue oxide" (tungsten blue oxide, TBO), which is a mixture of WO_3 and $WO_{2.9}$ depending on the calcination conditions. Generally, in both "yellow" and "blue" oxide, the original morphology and size of the APT crystals remain nearly unchanged, nevertheless, the oxide particles are characterized by cracks and porosity due to their increased density in comparison with APT. Typically, each oxide pseudomorphic particle consists of agglomerates of very fine oxide single crystals of 0.1−0.2 μm in size. Processing APT is very important because it determines the grain size, morphology, and other properties of the tungsten oxide powders. Varying the processing conditions (temperature, duration, type, and pressure of gas atmospheres) allows obtaining different types of tungsten oxides having various morphologies, structures, and stoichiometries. In the atmosphere of air or by using special calcination conditions, one can get the "violet oxide" $WO_{2.72}$. By use of additions of hydrogen to the calcination atmosphere, "brown oxide" WO_2 can be produced.

Cemented Carbides
ISBN 978-0-12-822820-3
https://doi.org/10.1016/B978-0-12-822820-3.00016-1

Physicochemical fundamentals of the reduction of tungsten oxides by hydrogen are well understood and the influence of reduction parameters on the mean grain size and grain size distribution of tungsten powders is examined in detail.

The general reaction of the tungsten metal formation from WO_3 as a result of hydrogen reduction is described by the following equation:

$$WO_3 + 3H_2 \leftrightarrow W + 3H_2O$$

However, the existence of a range of intermediate oxides makes this reaction more complicated and it occurs in four stages in accordance with the existence of four tungsten oxides with different stoichiometries, so that the consequence of these intermediate stages can be described as follows:

$$WO_3 \leftrightarrow WO_{2.9} \leftrightarrow WO_{2.72} \leftrightarrow WO_2 \leftrightarrow W$$

The intermediate reactions at these stages comprise the formation of intermediate oxides as well as their interaction with each other, hydrogen, and, in some cases, with tungsten metal grains being formed. Tungsten oxide hydrate $WO_2(OH)_2$, a highly volatile compound, always forming as a result of the interaction of tungsten oxides with water vapor at elevated temperatures, plays a very important role in occurring the transformation stages described above by chemical vapor transport, thus significantly affecting the grain sizes of the intermediate oxides powders and tungsten metal powders.

Considering the equilibrium conditions of the process of hydrogen reduction allows one to outline the major features of this process:

1. As the temperature rises, the equilibrium of the chemical reactions mentioned above shifts toward the right-hand side.
2. The process of hydrogen reduction can occur at significant concentrations of water vapors in the bed of oxides contained in boats, which are located in hydrogen reduction furnaces.
3. By changing the temperature of hydrogen reduction or the composition of the hydrogen-based gas mixtures, one can activate or suppress the reduction process, which allows one to regulate the mean grain size of the tungsten metal powder.

Experimental results indicate that the process should be conducted at high temperatures in order to obtain a greater output of the final product. However, one must consider the noticeable volatility of different tungsten oxides as well as $WO_2(OH)_2$, which leads to significant changes in the mean grain size of the tungsten powder obtained at elevated temperatures.

In practice, the composition of the gas atmosphere (hydrogen humidity), hydrogen flow rate, and oxide bed height rather than the temperature is usually varied to regulate the mean grain size of the tungsten powder. In conditions close to the equilibrium, the rate of the hydrogen reduction is relatively low, so that a significant excess of hydrogen is needed, which is achieved due to high rates of the hydrogen flow in the reduction furnace. For the same reason, hydrogen is usually dried and purified with respect to the oxygen impurities before being fed into the furnace. Any fluctuations of the moisture content in hydrogen are undesirable due to the fact that they would strongly affect the mean grain size of the tungsten powder. The reduction temperature usually varies from nearly 750°C to 900°C for producing fine-grain and medium-grain W powders. The process temperature is typically increased up to 1200°C for the fabrication of coarse-grain and ultracoarse-grain tungsten powders.

Intermediate oxides and metallic tungsten formed during the reduction process are characterized by different crystal lattices, so that processes of the crystal lattice transformation and particles' morphology changes are believed to have a significant impact on the reduction process itself and consequently on the mean grain size of the tungsten powder. On the first stage of the reduction process, the orthorhombic crystal lattice of WO_3 transforms into the monoclinic lattice of $WO_{2.9}$, almost without any morphological changes. On the further stages of the reduction process, the morphology and structure of the oxide pseudomorphic particles noticeably change resulting in the formation of relatively fine WO_2 particles. The size of WO_2 single crystals typically determines the final grain size of the tungsten powder.

When the humidity of hydrogen and oxide bed height increase, coarse-grain tungsten powders are produced at the correspondingly optimized process temperature, gas permeability of the oxide layer in the bed (except for using rotary furnaces), and gas flow rate. Adding compounds of alkali metals, for example, sodium, to initial tungsten oxide powders leads to significant coarsening the tungsten powder due to the formation of intermediate highly volatile species in the oxide bed, so that ultracoarse tungsten powders can be produced in such a way.

Fine-grain and submicron tungsten powders, which are shown in Fig. 6.1, are produced by use of low temperatures, high flow rates of dry hydrogen with dew points of −30°C to −40°C and low heights of the oxide layer in the bed. Further refinement of such tungsten powders can be achieved using the cocurrent flow of hydrogen with the boats.

Figure 6.1 HRSEM image of a fine-grain tungsten metal powder. *(With the permission of Element Six.)*

Due to various concentrations of water vapor in the bottom and top layers of the oxide bed in the boats, tungsten powders are usually characterized by some inhomogeneity with respect to their grain size and grain size distribution. The technological approach suggested in Refs. [5,6] allows one to significantly improve the uniformity of medium-coarse and coarse-grain tungsten powders and consequently tungsten carbide powders produced on their basis as well as cemented carbides obtained from these powders. For example, a medium-coarse WC—Co grade fabricated by use of such tungsten carbide powders has a microstructure characterized by the distribution symmetry decreased by a factor of about two and the deviation of WC grain sizes from the average value reduced by about 60%.

The reduction processes are performed either in push-type furnaces, one of which is shown in Fig. 6.4, or in rotary reduction furnaces having a diameter between 300 and 750 mm and length between 3 and 8 m.

The influence of impurities and trace elements on the grain size and properties of tungsten metal and tungsten carbide powders as well as and cemented carbides fabricated on their basis is described in Chapter 9.

6.1.2 Fabrication of tungsten carbide powders

As in the case of the tungsten metal fabrication, technological parameters employed to produce WC powders are reported in detail in several monographs [1—4]; therefore, this chapter describes them only briefly.

Tungsten carbide is fabricated by the reaction between tungsten metal and carbon black at temperatures of 1400°C–2200°C; the carburization process is usually carried out in a hydrogen atmosphere.

Blending tungsten powders and carbon black is usually performed in V-type blenders because of the significant difference of densities of tungsten powders and carbon black. The carbon content in the mixtures is usually slightly higher than the stoichiometric content in WC and varies from 6.15 to 6.20 wt.% to compensate some carbon loss during the carburization process. The concrete amounts of carbon black needed to obtain stoichiometric WC are usually determined by empirical estimation of carbon loses or uptake during carburization.

The carburization process is usually performed in graphite boats located into tube furnaces in a hydrogen atmosphere. Hydrogen is typically introduced into the discharge end and burnt out at the loading end.

The process of tungsten carburization is carried out mainly through the formation of carbon-containing gaseous species as a result of interaction of fine particles of carbon black with hydrogen. The mechanism of the process of transferring carbon atoms from the carbon black particles to tungsten particles through such carbon-containing species was examined in detail and is presently well understood. One of such species should be acetylene, which is thermodynamically stable at temperatures between 1300°C and 1600°C. At lower temperatures, methane becomes more thermodynamically stable and its equilibrium concentration increases with decreasing the temperature.

Phenomena of tungsten carburization in this case can be described by the following reactions:

$$2C + H_2 = C_2H_2$$

$$2W + C_2H_2 = 2WC + H_2.$$

The equilibrium concentration of hydrocarbons on the surface of the carbon black particles is significantly greater than that on the surface of the growing W_2C or WC particles, so that a continuous transport of carbon through the gas phase occurs. The carburization reaction rate is determined by the rate of carbon diffusion from the outer layer of the growing tungsten carbide particles toward their core regions, which is affected by the temperature of the process and the size of the particles.

The mean grain size of WC powders depends on the size of the initial tungsten powders, process temperature, and process duration. As a rule, finer WC powders form from finer W powders; however, this dependence is characterized by some special features. If the initial tungsten powder is coarse or ultracoarse, several WC single crystals of different sizes and shapes form as a result of carburization of one large tungsten grain. Such WC single crystals are usually significantly smaller than the initial tungsten grain, which is illustrated in Fig. 6.2. This occurs as a result of the formation of high internal stresses within each grain caused by the noticeable difference

Figure 6.2 Appearance of an ultracoarse WC powder: above—HRSEM image, and below—metallurgical cross-section of the powder particles embedded in a copper matrix (light microscopy, etching in the Murakami reagent). *(With the permission of Element Six.)*

in densities of W, W_2C, and WC when the conversion of W into WC occurs according to the transformation sequence W → W_2C → WC. The number of WC single crystals forming from one initial tungsten grain significantly decreases when reducing the size of the tungsten grain. As one can see in Fig. 6.2. (below) about 10 WC single crystals form from very large W grains exceeding 50 μm in in size, whereas nearly 2–4 WC single crystals form from smaller W grains of about 20 μm in size. In many cases, just one WC single crystal forms from a relatively small initial tungsten grain of roughly 5 μm in size (Fig. 6.2). When carburizing ultrafine or submicron W powders, typically, one WC single crystal forms from one tungsten grain. Nevertheless, in this case, the WC powders can be coarser than the initial W powders due to occurring processes of agglomeration, sintering, and growing together of fine and consequently extremely active tungsten carbide grains, which can be seen in Fig. 6.3A.

WC powders with high uniformity of the grain size distribution obtained by use of the standard technology described above play an important role from the point of view of obtaining cemented carbides with the optimal microstructure and properties. In particular, WC powders with extremely uniform grain size distribution, needed for some carbide grades, are fabricated by processes of deagglomeration of WC powders followed by separation of WC fractions having different WC mean grain sizes. Also, the manufacture of ultrafine and submicron WC powders shown in Fig. 6.3A, which must be extremely uniform not containing large WC grains, usually comprises final stages of their deagglomeration typically performed by jet milling followed by separation of coarse-grain fractions in cyclones or by use of other methods. In many cases, fine-grain WC powders are manufactured in form of doped powders containing grain growth inhibitors, which makes it possible to achieve a very uniform distribution of the grain growth inhibitors when producing submicron and ultrafine carbide grades.

The carburization processes are carried out by use of either molybdenum (Mo) wire heated push-type furnaces with ceramic muffle operating at temperatures of up to 1600°C or push-type furnaces with a graphite tube for temperatures of up to 2200°C. The up-to-date equipment for manufacturing W and WC powders is shown in Fig. 6.4.

In addition to the standard WC fabrication technique described above, unconventional technologies for the WC production, for example, the Menstruum process [7] and the WO_3 direct carburization technology developed by the Dow Chemical Company [8], become more and more popular.

Figure 6.3 HRSEM images of (A) fine-grain and (B) medium-coarse grain tungsten carbide powders. *(With the permission of Element Six.)*

According to the auxiliary melt bath technique, also known as the "Menstruum process," the tungsten ore concentrates or other sources of tungsten (WO_3) are reduced by aluminum as a reducing agent and CaC_2 as a carbon source. As a result, extremely coarse-grain WC powders with a grain size of up to 100 μm are fabricated. Such powders can be milled to various sizes to produce, for example, ultracoarse WC powders needed for the fabrication of ultracoarse grades for mining and construction applications.

The technology of the WO_3 direct carburization, which is nowadays employed by a number of WC manufacturing companies after expiring the

Figure 6.4 Modern furnaces employed for the fabrication of W and WC powders: above—a reduction furnace and below—a carburization furnace. *(With the permission of the Joint Stock Company "Kirovgrad Hard Alloys Plant" (Russia).)*

Dow Chemical Company patent [8], allows one to produce nano or near-nano WC powders with mean grain sizes varying from about 100 to 200 nm (Fig. 6.5). The technology includes two stages:

1. Mixing WO_3 and carbon followed by granulation of the mixture and carburization in atmospheres of N_2 or Ar at temperatures between 1000°C and 1660°C to produce WC (which may contain traces of oxygen).
2. The product obtained on the first stage is treated in hydrogen at temperatures between 1440°C and 2000°C after provisional adding some carbon to completely remove the traces of oxygen and obtain the stoichiometric WC powder.

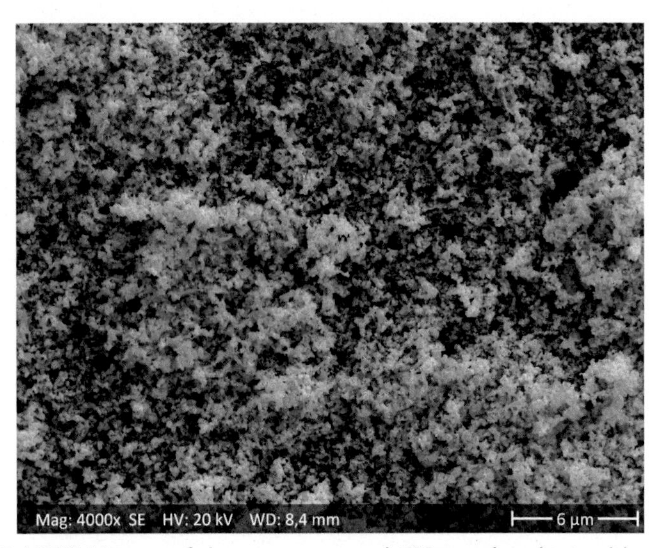

Figure 6.5 HRSEM image of the nanostructured WC powder obtained by the direct carburization of tungsten oxide. *(With the permission of Element Six.)*

These two stages are usually carried out in two separate rotary carburization furnaces.

The extremely fine (nanostructured) and uniform WC powder obtained in such a way, which is shown in Fig. 6.5, still has a limited application due to its extremely high activity during liquid-phase sintering (see Chapter 11). Also, there are problems with the storage of such nanostructured powders, as they are relatively fast oxidized being in contact with air, especially at high humidity.

6.2 Technological processes for fabrication of (W,Ti,Ta(Nb))C powders and powders of grain growth inhibitors

Cubic (W,Ti,Ta)C carbides (which can also contain Nb) having various compositions, which are characterized by different values of the WC/(Ti,Ta)C ratio, are fabricated and employed at different carbide companies. It is well established that if this ratio is close to 70:30 by weight, the maximum solubility of WC in the cubic carbide is achieved, so that its composition remains constant during sintering of (W,Ti,Ta)C or (W,Ti,Ta,Nb)C containing cemented carbides. Nevertheless, some companies produce the mixed cubic carbides which are characterized by this ratio equal to 50:50 [4]. The employment of cubic carbides containing niobium is caused mainly by significantly higher prices of pure TaC in comparison with mixed (Ta,Nb)C carbide, as, generally, niobium carbide is significantly cheaper than tantalum carbide.

Mixed cubic (W,Ti,Ta)C carbides can be produced by mixing the corresponding oxides with carbon black or mixing oxides of Ti and Ta with tungsten metal and carbon black followed by the subsequent carburization reaction at temperatures between 1800°C and 2200°C. They can alternatively be fabricated by diffusion annealing of the carbides at nearly 2000°C according to the following reaction:

$$WC + TiC + TaC \rightarrow (W,Ti,Ta)C.$$

Chromium carbide Cr_3C_2 is widely used in the cemented carbide industry as a grain growth inhibitor; it is characterized by high solubility in Co- and Ni-based binders. Chromium carbides exist in three chemical compounds: Cr_3C_2, Cr_7C_3, and $Cr_{23}C_6$, among which the hardest and most commonly used compound is Cr_3C_2. Chromium carbide is produced by the direct carburization of chromium by carbon black or by the carbon reduction of chromium oxide. The synthesis of Cr_3C_2 starts at a temperature of about 1150°C through the stage of forming $Cr_{23}C_6$ and Cr_7C_3. A product consisting of pure Cr_3C_2 contains some free carbon forming as a result of increasing the temperatures up to 1500°C–1600°C; the process is performed in a hydrogen atmosphere.

Vanadium carbide (VC) is the strongest grain growth inhibitor; it is also widely employed in the carbide industry. VC is produced by heating vanadium oxides with carbon at temperatures above 1000°C in an inert atmosphere.

TaC, NbC, and TiC are known to be refractory carbides, which have some inhibiting effect with respect to the WC grain growth during sintering; sometimes, they are employed as grain growth inhibitors in industrial WC-Co carbide grades (especially TaC).

Tantalum carbides form a range of binary chemical compounds of tantalum and carbon with the empirical formula TaC_x, where x usually varies between 0.4 and 1. Typically, mainly stoichiometric TaC having a yellow-brown color is employed as a grain growth inhibitor. It is produced by heating a mixture of tantalum and graphite powders in a vacuum or inert gas atmosphere (argon). The carburization process is performed at a temperature of about 2000°C. An alternative technique is based on the reduction of tantalum pentoxide by carbon in a vacuum or hydrogen atmosphere at temperatures of 1500°C–1700°C.

Niobium carbide can be produced by heating niobium oxide with carbon black in a vacuum at 1800°C.

Synthesis of TiC from TiO_2 occurs due to its reaction with carbon black in either hydrogen or vacuum according to the following reaction:

$$TiO_2 + 3C = TiC + 2CO$$

The process occurs through stages of the formation of different titanium oxides (TiO_2, Ti_2O_3, and TiO) and results in obtaining a solid solution of oxygen in titanium carbide. When using some excess of carbon (3%) and a temperature of 1750°C, one can obtain stoichiometric TiC from titanium oxide. Due to difficulties related to obtaining oxygen-free TiC from TiO_2, the fabrication of TiC as a result of interaction of titanium or titanium hydride with carbon in a vacuum is presently widely spread to produce high-quality TiC powders.

Mo_2C is also sometimes employed as a grain growth inhibitor in the cemented carbide industry, although its inhibiting effect is relatively insignificant compared to, for example, vanadium and chromium carbides. The fabrication of Mo_2C is usually performed by reaction of Mo oxide with carbon black at temperatures between 1350°C and 1800°C in a hydrogen atmosphere. Mo carbide can also be obtained by reaction of ammonium molybdate with hydrogen and carbon monoxide at a temperature between 550°C and 600°C.

6.3 Technological processes for fabrication of cobalt powders

The main ores of cobalt are cobaltite, erythrite, glaucodot, and skutterudite, but most cobalt is obtained by reducing cobalt by-products of nickel and copper mining and smelting. Intermediate products obtained as a result of separation of Co compounds from Cu and Ni compounds are transformed into cobalt oxide Co_3O_4. This oxide is reduced to compact metal by the aluminothermic reaction or reduction with carbon in a blast furnace.

Relatively coarse-grain Co powders are produced by hydrogen reduction of cobalt oxide. Ultrafine Co powders with a grain size of 1–2 μm, which are now believed to be essential for the fabrication of high-quality cemented carbides, are produced by pyrolysis of a Co salt, such as cobalt oxalate. This compound is made by reacting oxalic acid with cobalt chloride.

Nowadays, Co powders are usually employed in the cemented carbide industry in form of porous granules to reduce dusting.

6.4 Technological processes for fabrication of cemented carbide graded powders

6.4.1 Milling in attritor mills

In general, attritor mills are devices for mechanical disintegration of solid particles by intense agitation of a slurry containing particles of materials being milled and mixed in a milling medium. The word "milling" as a designation of process stage of fabrication of graded WC−Co powders is presumably not the most suitable term describing this process stage. The major objective of milling is not reducing the grain size of WC powders during this processing stage. Nevertheless, although the WC mean grain size is reduced as a result of milling, the WC mean grain size of original WC powders largely defines the grain size of the WC−Co powder mixture and consequently the WC mean grain size of sintered carbide articles. The high-energy attritor milling process ensures an even cobalt distribution within the graded WC−Co powder. The cobalt powder is subjected to forging and "smearing" on the surface of WC grains as a result of shearing and impact forces during milling. As a result of the even cobalt distribution, a uniform microstructure of cemented carbide articles after final liquid phase sintering is obtained. Excessive milling of WC−Co powders tends to lead to the formation of a great number of fine and nano WC grains, having active surfaces, and an increased concentration of crystal lattice defects in both the WC phase and binder phase [9], which can result in the growth of abnormally large WC grain as a result of liquid phase sintering (see Section 6.7). Milling conditions must be precisely adjusted to different cemented carbide grades with various WC gain sizes and metal binder contents. In general, grades with coarser grain sizes require shorter or less intensive milling. The number of solid particles in the slurry has a strong effect on its apparent viscosity. According to Ref. [10], examining the effect of cobalt content on the specific WC surface area for slurries containing 22 vol.% solid particles, an increase in cobalt content results in higher apparent viscosity of the slurries and decreased milling efficiency.

The specific energy input during attritor milling can be calculated by use of the following equation:

$$E^* = E/M = 1.18 \times 10^{-5}(T \times N \times t / M)$$

where E^* is the specific energy consumed by the mill per unit mass of powder charged [kWh/kg], E is the total energy provided to the mill [kWh], M is the total mass of the solid particles being milled [kg], T is the torque supplied by the agitator shaft [in-lb], N is the rotation velocity of the agitator arm [rpm], t is the milling time [hours], and the constant 1.18×10^{-5} reflects the units conversion factor. Controlling the slurry apparent viscosity obtained as a result of milling is an important factor determining the spray drying process of the slurry needed for its drying and producing granulated powders.

An attrition mill shown in Fig. 6.6 consists of a metal support frame with two detachable running rails, on which the double wall milling vessel is located. The support frame comprises an electric explosion-proof engine characterized by power values varying from 24 to 60 KW and a direct drive or a gearbox ensuring the rotation of the wear protected milling arm. The sound protection is recommended to reduce the high noise level generated by the intensive agitation of the slurry. A recirculation system consists of double-walled pipes, an air pressure—powered diaphragm pump, filter cartridge, and several valves to continuously pump the milling slurry from the bottom region of the attrition mill vessel back to its upper part in order to prevent the so-called "dead spots," where no milling occurs; there are also special devices for controlling the recirculation flow. Slurry recirculation volumes usually vary from approximately 2 to 8 m^3/h. The slurry volume flow is strongly dependent on the compressed air pressure, suction height, pump height, and value of the apparent slurry viscosity.

Circulating cold or hot water through the jacket of the attritor vessel and the recirculation system allow effective cooling or heating the vessel. Controlling the slurry temperature prevents extensive evaporation of the milling liquid and protects the milled powders against oxidation.

The milling vessel is filled up to a certain level with cemented carbide milling balls (grinding media) leaving enough volume for powders of WC, metal binder, carbon black, tungsten metal, grain growth inhibitors, organic additives, and the milling liquid. The diameter of the carbide milling balls can vary from 5 to 15 mm [11]. They usually consist of a carbide grade with a WC mean grain size that must be as close as possible to the grain size of the milled WC—Co powder to prevent cross-contamination due to balls' wearing. Attritors designed for powder batches of 500 kg contain nearly 2500 kg carbide milling balls. This corresponds to roughly 250.000 milling balls with a diameter of 11 mm, if the balls consist of the WC—10%Co carbide grade. The corresponding milling balls' surface is equal to about 95 m^2 ensuring intensive mixing and some milling of WC—Co mixtures.

Powders of WC, cobalt, nickel, etc., are usually supplied in drums, containers, or big bags. One can employ manual, semiautomated, or fully

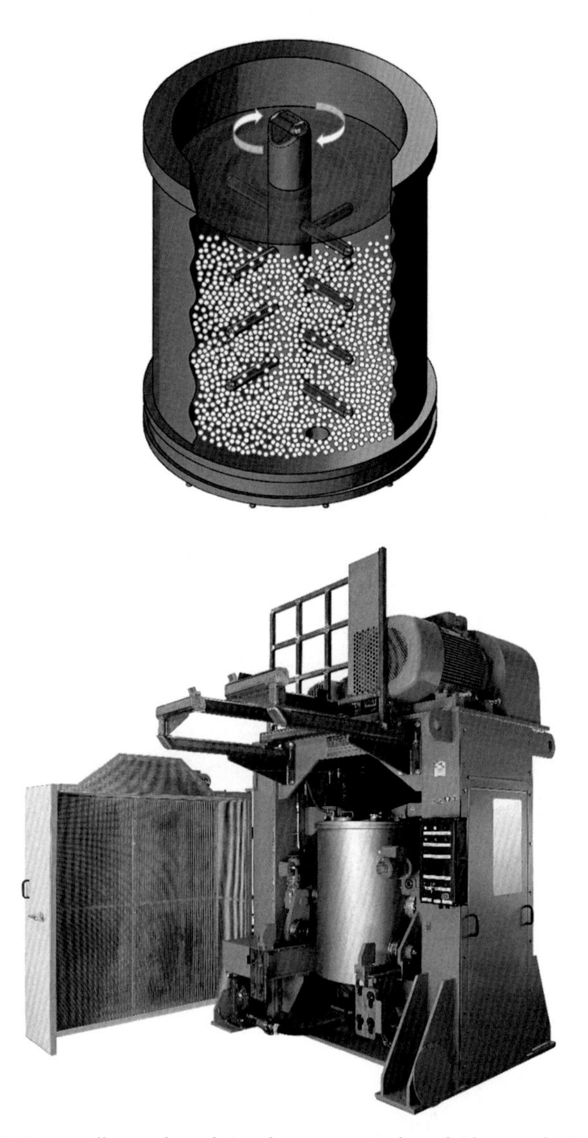

Figure 6.6 Attritor mill employed in the cemented carbide production: above—schematic diagram illustrating the milling process in attritor mills and below—photo of a typical attritor mill. *(With the permission of Tisoma Anlagen und Vorrichtungen GmbH.)*

automated weighing systems for filling a template container with mixtures of the needed composition. The template container is connected with the attritor vessel, allowing a dust emission–free powder transfer, followed by filling up the mill with the milling liquid. Typical milling liquids used in the cemented carbide industry are ethyl alcohol, hexane, acetone, water, etc., employed in ratios to the powder varying from 1:4 to 1:6 to achieve the optimum efficiency of mixing/milling. The ratio of the milling balls to powders varies from 4:1 to 6:1, and milling times are typically 2–24 h depending on the WC grain size and binder metal content. The optimum slurry temperature (50°C–70°C) depends on the organic binder melting point, when milling liquids do not dissolve the organic binder materials. In this case, the homogeneous organic binder distribution within the milling slurry is achieved due to the emulsion formation. The slurry temperature also depends on the boiling point of the milling liquid. Testing the viscosity of milling slurries is usually employed in the manufacture to guarantee reliable and consistent milling conditions. For the spray drying granulation process following the milling process, the viscosity measurement results ensure the optimal throughput, yield, and the best granulate quality.

It is crucial for the reliable and consistent carbide production to precisely steer and control the milling process. Monitoring and controlling the milling time, ball-to-powder ratio, suspension viscosity, and slurry temperature are essential for the quality control. The dust-free usage of attritor mills and reducing or complete elimination of manual contacts with the materials being milled become more and more important.

6.4.2 Milling in ball mills

Ball mills are cylindrical, conical, or tube units, containing grinding media, which are employed to mill or mix different powders or substances by rotating along their longitudinal axis. The milling balls are lifted as a result of ration till they drop due to their self-weight against the centrifugal force obtained as a result of their contact with mill walls; they also roll or fall back to the starting point according to Ref. [12]. As it is mentioned in the previous subchapter, the word “milling” is not an appropriate term describing this process stage, as the major objective of the milling stage is not reducing the grain size of WC powders. The fabrication of WC powders with certain values of the WC mean grain size prior to the milling process primarily defines the WC grain size in the WC–Co mixture and consequently the WC mean grain size in the sintered WC-Co articles. The high-energy ball milling process is intended mainly for the deagglomeration of the WC powders and obtaining uniform cobalt or nickel distribution within the graded powders. The uniform cobalt distribution ensures the consistency of chemical, physical, and metallurgical properties of different batches of WC–Co powders.

Drum, conical, or tube ball mills belong to a group of rotating ball mills. The milling effect is generated by a rolling and/or falling effect of the grinding media (milling balls). Beside the drum rotation speed, the milling ball loading and the friction between the milling balls and the milling drum wall ensure the rolling process. Milling balls with sizes varying from 12 to 20 mm in diameter are typically employed for different applications [9]. In ball mills, the grinding efficiency is increased as the balls' diameter decreases. The milling efficiency increases at a higher value of the ball-to-material ratio [9]. When the grist just fills the free volume between the balls, both the process efficiency and production efficiency become optimal.

The milling balls preferably consist of a WC–Co grade with a WC mean grain size close to that of the milled WC–Co powder. Milling balls' diameter should be precisely adjusted to the diameter of the milling drum, rotation speed, grist grain size, metal binder content, and apparent viscosity of the slurry. The continuous improvement of the ball milling technology over the last decades was a key factor for a reliable production of high-quality submicron WC-based powder mixtures.

Ball mills consist of a metal support frame, carrying a horizontal axial-mounted double wall milling vessel containing a certain amount of cemented carbide milling balls, which is illustrated in Fig. 6.7. Powerful electric explosion-proof engines of 24 KW up to 60 KW are usually equipped with belts ensuring the rotation of the milling drums. A frequent exchange of such drums is needed due to their high wear rates, as, conventionally, stainless steel milling drums without any wear protection are employed in the carbide manufacture. An overall sound enclosure is recommended to reduce the high noise level generated by the intensive agitation of the grinding media and the slurry within the drum.

Circulating cold water through the jacket of the ball mill drum allows effective cooling the mill walls, the temperature of which is usually high due to the generated frictional heat. Controlling the milling slurry temperature in relation to the milling liquid evaporation point is an important measure for preventing an increase of the gas pressure inside the milling drum.

Figure 6.7 Typical ball mill employed in the cemented carbide industry. *(With the permission of Tisoma Anlagen und Vorrichtungen GmbH.)*

Initial powders (WC, cobalt, or nickel powders, etc.) are usually supplied in drums, containers, or big bags. As the first process step, manual, semiautomated, or fully automated weighing the initial powders according to the required composition is performed; the mixtures are loaded into template container. The template container is connected with the mill drum, allowing a dust emission–free powder transfer, followed by filling up the mill with the required milling liquid volume. The milling liquid-to-powder ratios vary from1:4 to 1:6 to achieve an optimum powder mixture dispersion and uniformity. After completing the milling process, a pressured air diaphragm pump can be used for the drainage of the milling slurry into spray dryer feed tanks or vacuum dryers.

The recommended grinding media size for standard ball mills can be calculated in accordance to the following equation:

$$d_{dia}/18 > b_{max} > d_{dia}/24$$

where

d_{dia}—inner diameter of the ball mill drum [m].

b_{max}—maximum ball mill diameter [m].

Often due to the high density of the cemented carbide milling balls, a milling drum/balls' diameter ratios of $d_{dia}/50$ to $d_{dia}/80$ are employed.

Ball mills contain milling balls in amount varying in the range of 20–35 vol.% and a filling rate varying from 15 vol.% up to 25 vol.% with respect to the drum volume.

The centrifugal force C depends on the milling ball mass, milling drum rotation speed, and inner diameter of the milling drum according to the following equation:

$$C = m(2 \times \pi \times n)^2 \times d_T/2.$$

where C - centrifugal force [N], m - mass of the milling balls [kg], n - rotational speed of the drum [1/s], and d_T - inner diameter of the drum [m].

When centrifugal force C is equal to gravity, the milling balls are not detached from the drum wall; the rotation speed corresponding to this phenomenon is designated as $n_{critical}$

$$C = m \times g$$

where C - centrifugal force [N], m - mass of the milling balls [kg], and g - gravity [m/s^2].

The critical rotational speed can be calculated by the following equation [9]:

$$n_{critical} = 1/(2 \times \pi)(2 \times g/d_T)^{-2} = 0.255(g \times /d_T)^{-2}.$$

where $n_{critical}$ - critical rotational speed of the drum [1/s], g - gravity [m/s^2], and d_T - inner diameter of the drum [m].

When the rotational speed is lower than 60% of $n_{critical}$, balls can form the so-called "immovable zones." At such a low rotational speed of the drum, a sliding state occurs, at which the milling balls have almost no stirring effect on the material resulting in the significantly reduced milling efficiency (Fig. 6.8A).

In practice, the commonly used rotational speeds are equal to about 80% of $n_{critical}$, when milling balls move at the optimum mode ensuring them to move together with the drum due to the friction with the drum wall followed by rolling or falling back to the starting point (Fig. 6.8B).

By applying rotational speeds equal to or higher than 80% of $n_{critical}$, centrifugal force becomes dominant causing the phenomenon that milling

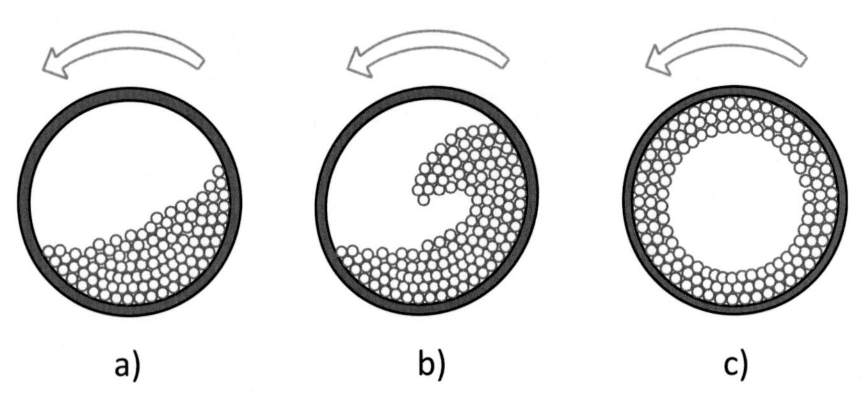

Figure 6.8 Illustration of milling ball cascading in the ball mill drum at different rotational speeds: (A) at a rotational speed lower than 60% of $n_{critical}$, (B) at a rotational speed of 75%–80% of $n_{critical}$, and (C) at a rotational speed of 80% of $n_{critical}$ or higher.

balls are not detached from the drum wall; at this speed, no grinding nor mixing takes place (Fig. 6.8C).

For example, a milling drum with an inner diameter of 1 m is characterized by the value of $n_{critical}$ calculated as $0.255(g\times/d_T)^{-2}$, which is equal to 0.8 1/s.

Ball mills in the cemented carbide industry operate at rotational speeds of 75%–80% of $n_{critical}$ (Fig. 6.8B) to achieve the optimum mixing/milling efficiency. Usually, the ball mill rotation speed for a ball mill having an inner diameter of 1 m ball mill could slightly differ from the recommended rotational speed of 36 rpm when taking into account special features of the milling process in the cemented carbide industry.

The ball-to-powder ratio can vary from 4:1 to 6:1; some ball mills contain milling balls with a surface area bigger than 80 m^2. Ball weights are usually in the range of nearly 25–60 g depending on the material density. The milling times vary from 8 h to 48 h depending on the carbide powder grade. Typical batch weights from 5 kg (laboratory mills) up to 1000 kg (production mills) are used in the cemented carbide industry. Slurry temperatures usually lie in the range of 50°C–80°C depending on the boiling point of the used milling liquid.

Monitoring and controlling the milling time, ball-to-powder ratio, suspension viscosity, and temperature are needed to produce high-quality graded WC-Co mixtures. Regular maintenance ensures an optimized overall equipment efficiency.

6.5 Granulation of graded powders

6.5.1 Build-up granulation

Granulation in the case of WC—Co graded powders is a process of building up granules of the optimum size suitable for their consequent pressing. The granules must have a high flowability and low apparent density. The granulation process occurs when a bed of solid particles moves with simultaneous mixing in the presence of a melted organic binder like paraffin wax. This motion provides particle collisions, when the individual particles coalesce and bind together. Granulation in rolling drums is a technique widely used in the cemented carbide industry; in this case, the granulation takes place as a result of collisions of WC/Co grains and agglomerates in a bed of molten wax subjected to rolling motion. The rolling drum is the simplest continuous granulation unit and presumably was used in the cemented carbide industry a long time ago, so this technique is one of the oldest granulation technologies employed in the carbide industry.

Fundamental results on the solid particles' cohesion were first obtained and reported in connection with the agglomeration process in Ref. [13]. Granules show a significantly improved flowing ability in comparison with powders and can be used for further processing, e.g., for pressing green articles. The understanding of the phenomena of the particles' adhesion is of crucial importance with respect to both the granules strength and issues related to their manufacturing. Regarding the agglomeration (granulation from fines) of cemented carbide powders, the main driving force of the granulation is the formation of organic binder bridges among the WC/Co powder particles. The molten wax acts as a liquid bridge among the powder particles, which are stabilized during cooling down to room temperature. The particles' adhesion is strongly dependent on their surface roughness. Moving the individually or partially agglomerated particles against each other leads to the formation of attracting forces, which become greater than the separating forces always present among solid particles. The drum granulation is a combination of agglomerating, rounding, and separation of granulate grains.

The flexibility is the most important benefit of the build-up granulation technology with respect to varying cemented carbide grades being granulated. The careful cleaning the granulation units takes only about 20 min before the carbide grade is changed. Low prices of commercial drum granulation units allow the employment of several units; each of such units is used for granulation of one type of cemented carbides with respect to the WC mean grain size, which eliminates a risk of cross-contamination of granules of different carbide grades. The dust emission is a major concern when using the build-up granulation, as the granulation drums are usually open. The employment of special dust exhausting systems is mandatory to minimize the risk of uncontrolled WC/Co dusts emissions. An exhaust air ventilation volume of 2000 m^3/h up to 4000 m^3/h per drum is required to achieve an optimum dust removal from the exposed areas. Positioning the suction points must be optimized for any drum granulation unit. It should be noted that the successful operation of the drum granulation units requires skilled and experienced operators to obtain consistent properties of WC−Co granulated powders.

A typical drum granulation system consists of a heated granulation drum which slightly inclined with respect to the horizontal axis (Fig. 6.9) usually combined with an automated feeding unit and a vibration cooling section.

Before the drum granulation process, the milling liquid must be removed by evaporation in closed boilers or in rotating double conus vessels. The removal of the solvent is performed at a low residual pressure of gases inside the boilers and vessels. The WC−Co powder is sieved after drying to achieve a homogeneous particle distribution and filled into the adopted feeding system or directly into the granulation drum. In case of the presence of a feeding system, the powder is continuously conveyed into the granulation drum. The drum temperature varies from 70°C to 85°C, drum inclining angle is usually equal to 20°−30°, drum diameter varies from 300 to 600 mm, drum rotation speed is between 16 and 40 rpm, and the dwell time depends on the needed granules' size, being affected by the drum inclining angle, rotation speed, and feed rate of the WC−Co powder. The granules being produced cannot be carried to the top drum edge due to their high density [11]; after completion of the granulation process, they fall on a vibrating cooling track. The granules are cooled down to room temperature followed by their sieving to the grain size classes required for the pressing process. The typical carbide granules obtained by the build-up granulation technology are shown in Fig. 6.10.

Figure 6.9 Typical drum granulation unit. *(With the permission of Tisoma Anlagen und Vorrichtungen GmbH.)*

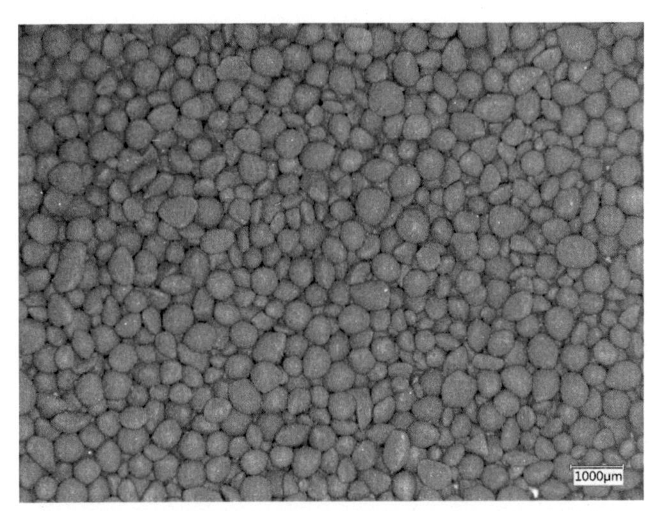

Figure 6.10 Appearance of WC–Co granules obtained by the build-up granulation technology (optical microscopy, ×25). *(With the permission of Element Six.)*

Table 6.1 Major properties of the granulates produced by the build-up granulation technology.

Property	Measuring procedure	Typical granules' properties	Unit
Granule grain size distribution	Laser grain size meter	50–300	[μm]
Fines	Classification, sieving	<50	[μm]
Granule shape (aspect ratio)	Image analysis	0.8–0.95	
Specific surface	Sorption (BET)	–	
Mechanical granule properties	Single granule hardness measurement device	0.3–1.2	[MPa]
Outer structure	Field emission SEM	–	
Inner structure	Iron beam FESEM	–	
Porosity	Mercury porosimeter	0.3–1	[μm]
Residual moisture	IR—Thermo-balance	0.1–03	[%]
Flow behavior			
Bulk density	Scott volumeter	3.0–4.5	[g/cm^3]
Tapped density	Volume measuring cylinder	3.5–5.0	[g/cm^3]
Flow time	Hall-flow-meter	18–30	[s/50g]
Shear behavior	Rheometer	7–17	ffc [–]

The granules produced by the build-up granulation technology are usually slightly larger, harder, and characterized by a higher hall density (Table 6.1) compared to the granules obtained by other techniques (fluid bed granulation and spray drying). Due to the influence of intensive rolling under loads in the granulation drum, the granules are supplementarily compacted in comparison with the other granulation techniques. The larger granules' size and their higher hardness do not necessarily lead to higher loads needed for pressing carbide green compacts. A lower number of granules' contact points allow reducing friction during compaction, which could result in lower pressing forces required for obtaining the needed density values of the green carbide articles.

6.5.2 Fluidized bed granulation

Another granulation technique employed at some cemented carbide companies is fluidized bed granulation. Two phases are involved in the fluidized bed granulation process: a solid and gaseous or liquid fluid, which is hot air in case of cemented carbide powders. If the solid powder bed flows through the granulation unit with the fluid flow having a sufficient speed, the fluidized bed becomes stable, so that individual solid particles are present in a floating state. This state is therefore designated as the "fluidizing state." The resulting fluidized bed behaves (thermodynamically and flow-wise) like a pseudoliquid. As a result of the large contact surfaces between the solid particles and the fluid, heat and material transport processes occur at a significant rate [13]. When the gas flow speed is low, the gas moves through the cavities of the solid bed in such a way that the packing structure almost does not change. If the gas flow speed becomes higher, one gets a state, at which all the solid particles are suspended without permanent contacts with each other. In this state, there is a balance between the compressive force on the solid layer and the gravitational force on the solid in the gaseous fluid. For the consistent granulation of WC—Co powders, a stable fluidized bed state must be achieved.

Fluidized bed granulation units are flexible and reliable. They are completely closed, which ensures a dust-free granulation process; this is a significant advantage of the fluidized bed granulation technique in comparison with the build-up granulation technology described above. Cleaning the whole unit takes nearly 1—2 h before another carbide grade must be granulated, which makes this granulation technique significantly more time consuming compared to the build-up granulation technology. There is also a need for permanent proper maintenance of the granulation units highly skilled and experienced operators. As the fluidized bed granulation units are relatively noisy, a separate room or a full noise enclosure are recommended. The prices of fluid bed granulation units in comparison with the drum granulation units are significantly higher. The above-mentioned disadvantages of the fluidized bed units were presumably the major reasons why they are not employed in the carbide industry on a large scale. Nevertheless, the price of a spray dryer is more than 10 times higher.

A fluidized bed granulator shown in Fig. 6.11 consists of a granulation chamber with a supporting frame on wheels, conical stainless steel vessel, bag filter, and hot air ventilation system. The granulation chamber consists of a vessel, two inflow floor metal discs, screen cloth, woven filter medium layer, and vibrator. The conical metal vessel is equipped with small compressed air hammers to prevent powder sticking on the vessel walls. The bag filter consists of a metal frame covered with the filter tile. Ventilated hot air

Figure 6.11 Typical fluidized bed granulation unit. *(With the permission of Element Six.)*

is transferred via two nozzles into the granulation chamber. The upward hot air flow at the required speed creates the fluidized bed. A downward hot air flow is intended for compaction of the powder/granule bed on the inflow floor metal disc resulting in a higher bulk density. By a sharp upward hot air pressure pulse, the compacted powder/granule bed is loosened followed by a short constant hot air upward flow. Continuously alternating hot air process steps ensure the needed granulation distance within the granulation unit. The needed granulation distance can be achieved within approximately 20 min. Advanced control systems allow producing relatively soft granules with a consistent grains size distribution.

Depending on the unit size, a batch of 20–50 kg of the milled, dried, and sieved WC–Co powder is manually filled into the granulation chamber. The use of two special perforated inflow floor metal discs installed in the lower area of the granulation unit serves as a floor surface for the powder bed. The modular design ensures the hot air upward and downward flow allowing the WC–Co powder granulation. The granulation chamber is finally moved into the attended position under the conical stainless steel vessel. When the granulation chamber is tightly connected to the stainless vessel by lifting it and pressing the seals together, the granulation process is started and controlled by an integrated control system.

As a result of altering upwards and downwards gas flows, each granulate grain travels nearly 20–30 m during the granulation cycle occurring for 20–30 min; this distance is comparable with the distance that the grains travel in a spray tower. After the end of the granulation procedure, the system starts cooling the granules down to room temperature within 5–10 min. After pressure equalization, the granulate is removed from the unit. To ensure an almost continuous granulation process, it is desirable to prepare a second granulation drum in advance. An intense cleaning of the granulation chamber, inflow floor, and back filter is necessary before changing to another cemented carbide grade.

Properly maintained and sufficiently cleaned fluidized bed granulation units can be well controlled and the properties of the cemented carbide granulates are usually quite consistent. The granulate properties can be adjusted by changing and adjusting the following parameters:

- The upward gas flow pressure (which increases the granule size) varies from 5.000 to 10.000 m^3/h
- The upward gas flow temperature varies from 70°C to 90°C
- The downward gas flow pressure (which increases the granule size) varies from 5.000 to 10.000 m^3/h
- The downward gas flow temperature varies from 70°C to 90°C
- The process time is usually 20–30 min per batch
- The cooling time is nearly 5–10 min.

The granules' properties are adjustable in terms of tape density, hardness, and pressability, and their properties are comparable with those obtained by spray granulation.

6.5.3 Spray granulation

The spray granulation is a widely used method to produce cemented carbide granules by spraying a suspension with a hot gas stream as a drying medium in a closed system. As a result of spraying the suspension, spherical drops form that are simultaneously dried in the inert hot gas stream [13]. Due to adhesion forces forming in the fluid bridges among the solid particles, they are held together during granulation. When the granules fall freely in the oncoming hot inert gas stream, at a temperature of approximately 200°C, the liquid in the cavities of the granules is evaporated leading to their shrinkage and solidifying. The temperature and drop time of the spray granulation process must be chosen in such a way that the granules are not superficially encrusted, which ensures the free gas and vapor transport within each granule and allows the organic liquid vapor removal from the core and near-surface regions of the granule. An organic binder, for example, paraffin wax, present in the milling suspension additionally hinders the gas transport within the granule due to its accumulation at the granulate surface, as the organic binder has a higher boiling point compared to that of the organic liquid [14]. The binder content in the granules affects their compressibility during the subsequent pressing process and must be correspondingly adjusted. Spray towers with nozzle atomization systems are usually employed in the cemented carbide industry. The employment of highly flammable and volatile organic liquids such as alcohol, acetone, hexane, etc., requires the usage of closed systems filled with an inert gas. A schematic drawing of a typical spray drying unit is shown in Fig. 6.12.

The spray granulation technique has become the most popular and widely spread technology for granulating cemented carbide powders due to the simultaneous drying and granulation of slurries formed as a result of milling. The major benefit of this technology is the closed loop from such slurries into high-quality granulates without any dust emission. The increasingly stringent laws with respect to the health protection require

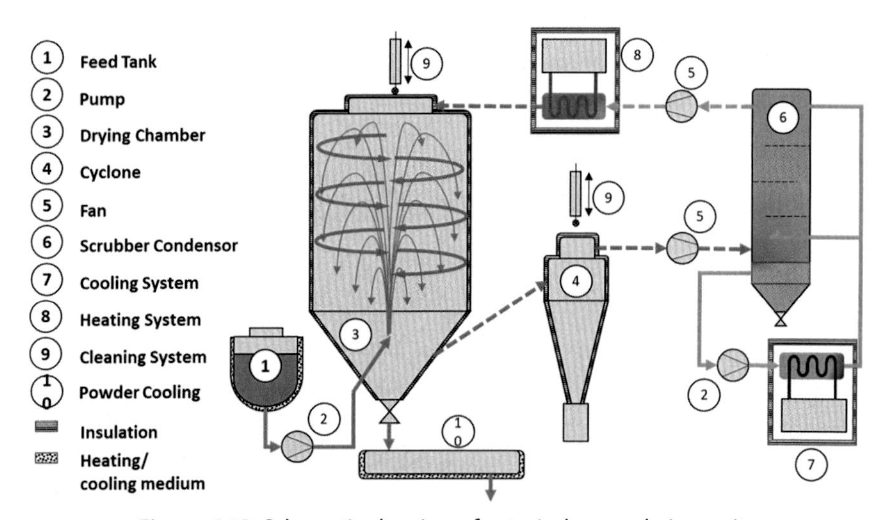

Figure 6.12 Schematic drawing of a typical spray drying unit.

the employment of a fully closed powder production cycle, which can be realized by use of spray granulation. Another big advantage of the spray granulation technique is its high throughput of up to 300 kg granulate per hour combined with a high yield, which increases the productivity and production output of the whole carbide fabrication process. The above-mentioned leads to the significant cost reduction by use of the spray granulation technology and results in the improved working environment compared to the build-up granulation technique characterized by the significant WC–Co dust emission. A higher yield of the granulates obtained by spray granulation in comparison with other granulation methods is also a big advantage of this granulation technique.

The improved quality of cemented carbide granulates obtained by spray granulation compared to the granulates obtained by alternative granulation techniques described above is another big advantage of the spray granulation technique. This granulation technique allows producing finer and softer granules having a high flow rate and consistent hall density. The improved granulate properties ensure pressing carbide green articles by use of lower forces resulting in the reduced wear and prolonged service life of pressing dies. The higher flow rate and consistent tape density secure a constant filling rate of the pressing dies leading to lower reject rates and significant improvements in the tolerance of the green articles.

On the other hand, units for spray granulation are quite expensive, complex, and must be carefully maintained and cleaned prior to changing graded powders of each carbide grade being granulated. Many carbide companies producing different types of cemented carbides, for example, ultrafine grades, coarse-grain grades, grades comprising cubic carbides for metal-cutting etc., must employ a number of dedicated spray towers for each type of cemented carbides to completely avoid their cross-contamination. In contrast, the use of inexpensive and simple units for the build-up granulation allows one to employ one single unit for WC–Co powders of each type eliminating the need to carefully clean them, which excludes a risk of cross-contamination of different carbide grades.

A spray tower consists of the main components including a supporting structure, feed tanks with a high-pressure diaphragm pump, drying chamber with an atomizer system, hot gas system, gas recycling system, scrubber condenser, granule discharge system with a cooling line, automatic cleaning unit, and special PLC/PC control system (Figs. 6.12 and 6.13). The equipment ensures production of fine, uniform, and spherical granules by direct use of milling slurries.

Figure 6.13 Typical spray granulation unit. *(With the permission of Element Six.)*

Different process steps can be outlined of the schematic diagram shown in Fig. 6.12. As a first process step, the fully processed cemented carbide slurry is pumped by the air powered diaphragm pump from the mill (attritor mil or ball mill) into the double walled stainless steel feed tank (1). The feed tank is equipped with an agitator comprising an infinitely variable drive and load cells for the level measurement to ensure a suitable agitator speed, which is dependent on the slurry quantity in the tank. The well-stirred slurry is pumped by a high-pressure oil powder diaphragm pump (2) through high-pressure nozzles shown in Fig. 6.14 into a drying chamber (3), where it is atomized by a stream of a hot inert gas.

Nitrogen, widely employed as the inert drying gas, is indirectly heated with a heat exchanger using hot oil as a heating agent (300 L in a closed system). The gas temperature is controlled by a temperature sensor located in the outlet pipe of the heater and a valve that controls the oil flow to a gas heat exchanger. A special gas disperser creates a spiral rotational movement

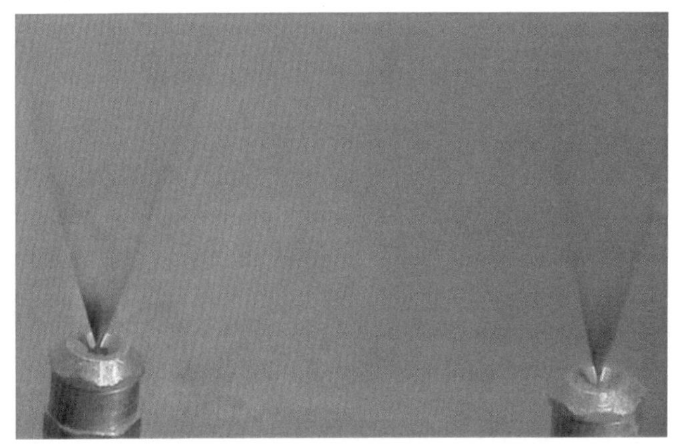

Figure 6.14 High pressure nozzles employed for spraying the milling slurry into the drying chamber. *(With the permission of Element Six.)*

of the hot gas flow directly after the hot gas entry into the drying chamber. This rotational movement is an important tool with respect to obtaining the optimal granulate shape and size. The product discharge is carried out on the chamber floor by a double check valve. The granulate is then fed into the granulate cooler (10) and relatively fast cooled down to room temperature. The granulate discharge is supported by pneumatic hammers attached to the casing of the drying chamber. The appearance of the typical cemented carbide granulate fabricated by spray granulation is shown in Fig. 6.15.

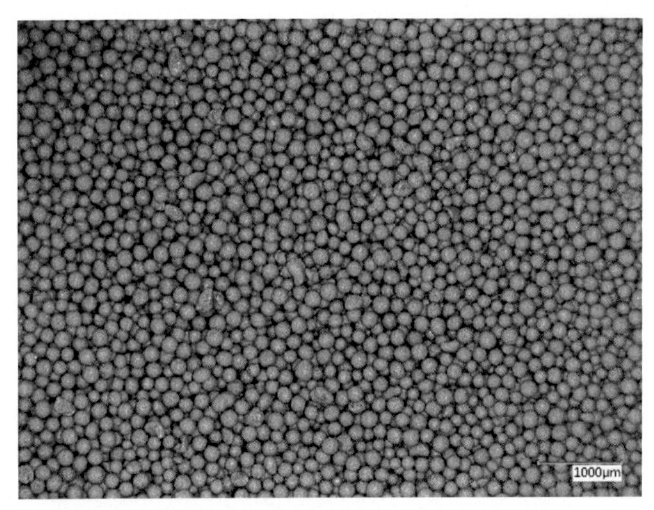

Figure 6.15 Appearance of the typical cemented carbide granulate produced by spray granulation having a particle mean size of roughly 150–200 μm (optical microscopy image). *(With the permission of Element Six.)*

The hot gas recirculation system directs the nitrogen from the heat exchanger through the air disperser, drying chamber, and outlet channel into the cyclone (4). Fine granulate particles that leave the chamber through the outlet channel are separated from the process gas in such a cyclone. A fan blows the circulating gas into the scrubber condenser (5), where the nitrogen gas is cleaned and cooled. After leaving the scrubber condenser, the cold nitrogen gas flows through a heat exchanger before it is fed back into the circuit.

In the scrubber condenser, the nitrogen gas is cleaned, cooled, and a vapor of the milling liquid evaporated in the drying chamber is condensed. The nitrogen gas flows through a series of nozzles into the scrubber condenser. The condensed milling liquid is fed back to the milling liquid tank.

A dedicated cleaning system for the drying chamber, cyclone, and exhaust duct is located between the drying chamber and centrifugal separator, and the feed tanks are installed for effective and fast cleaning of the whole system. The drying chamber and cyclone are cleaned by use of a rotating cleaning turbine. The cleaning fluid is pumped to the cleaning turbine by a high-pressure pump. When the cleaning process is finished, the fluid with the suspended particles is pumped back to the settling basin. The venting and feed tanks are cleaned using a handheld rotating nozzle located at the end of a flexible hose.

The following process parameters must be optimized to achieve the best granulate properties and consistency: the content of powder components in the slurry, type of organic binder, organic binder content, milling liquid composition, slurry temperature and viscosity, etc. The pressure in the diaphragm pump and distribution chamber as well as the nozzles' size play also an important role with respect to producing high-quality carbide granules. The typical distribution chamber size varies from 0.2 to 0.8 mm and the nozzle diameter lies usually between 1.2 and 1.8 mm. Gas inlet temperatures between 170°C and 210°C are typically used [9]. The granule discharge temperature can vary between 50°C and 70°C. The employment of a small spray cone angle is needed to prevent the formation of powder deposits on the drying chamber wall.

The granules' properties can be changed in a wide range by varying the slurry characteristics and spray granulation parameters; the quality and consistency of pressed carbide green articles strongly depend on the granules' properties.

6.6 Technological processes for fabrication of cemented carbide green bodies

6.6.1 Pressing in dies

Uniaxial dry pressing is generally understood as the compaction of powders or granules between two stamps in a rigid die [14]. Uniaxial pressing in dies is a widely used economical compaction technology for fabrication of carbide articles. During powder compaction, the applied pressure induces particle rearrangement and consolidation. The green density and compact strength increase with increasing the compaction pressure [15].

The up-to-date pressing technology has been developed from simple mechanical presses to modern advanced hydraulic presses and servo-electric presses. Advantages and disadvantages of different press systems are illustrated in Fig. 6.16. The precision of dimensions of the pressed parts and number of strokes per minute increased dramatically in the last decade.

The servo-electric press systems were subjected to intensive development in terms of precision and speed in order to utilize the full capability of servo-electric presses. Servo-electric presses are characterized by relatively low noise emission and require less production space compared to all other pressing systems. Due to the increased health and safety requirements with respect to the use of Co-containing powders, fully enclosed handling systems equipped with dust removal devices become more and more important in the cemented carbide industry nowadays.

Parameter	Mechanical press	Hydraulic press	Servo-electric press
Pressing force	↓	↑	⇨
Speed (Strokes/minute)	↓	⇨	↑
Accuracy	↓	⇨	↑
Maintenance	↑	↑	↑
Foot print	↑	↓	↑
Energy consumption	⇨	↑	↑
Safety standards	⇨	⇨	↑

Figure 6.16 Schematic diagram illustrating advantages and disadvantages of different pressing systems.

Typical presses consist of a rigid steel frame, granule feeding system, and adapter system that accommodates the pressing tools. The pressing tools (dies and punches) usually consist of steel and carbide wear parts to ensure the optimum durability and dimensional accuracy of the pressed parts.

Mechanical, hydraulic, or servo-electric presses transmit the force generated by powerful electric motors to the pressing tools. Modern presses are equipped with programmable controller systems and data logs for statistical analysis of pressing date (Fig. 6.17). Nowadays, modern presses are not stand-alone machines, so that they can be integrated into the manufacturing chain.

On the first operation step, cemented carbide granules are manually or automatically loaded into a hopper. From the hopper via a tube connected to a filling shoe, the granules are transferred to the cavity of the pressing tool. The filling volume is controlled by the position of the lower punch in the die. Afterward, the granules are compacted into a green compact in the die as a result of applying forces by the both punches (lower and upper) or by only the upper punch. The lower punch moves upward and ejects the green compact, which is afterward put on a sintering tray by a separate robot system (or manually when pressing a limited number of carbide articles). The tray with the pressed carbide articles is forwarded to the sintering department for subsequent sintering.

Figure 6.17 Typical servo-electric press. *(With the permission of Element Six.)*

If only the movable upper punch is used for compaction with a fixed die, which is illustrated in Fig. 6.18, a significant green density gradient is induced in the axial direction due to the friction between the powder and the die wall [9]. If the both movable upper and lower punches are employed for compaction with a fixed die, the green density gradient in the axial direction is noticeably reduced, which leads to obtaining a neutral zone in the middle of the green compact. For pressing multilevel compacts, several punches must operate independently, so that the different regions of the green compact in the die are equally compacted. In a floating die press, a similar effect can be achieved as a result of the movement (floating) of the die corresponding to the upward movement of the lower punch. In more sophisticated floating die press systems, the punches and the die can move independently in order to reduce density gradients in the green compact. This allows the production of complex parts with tapers, holes, and multiple steps. Cavity filling defects, bulk density variations, hollow granules, or insufficient granule deformation rates taking place as a result of using hard granules can cause significant variations of the compact green density.

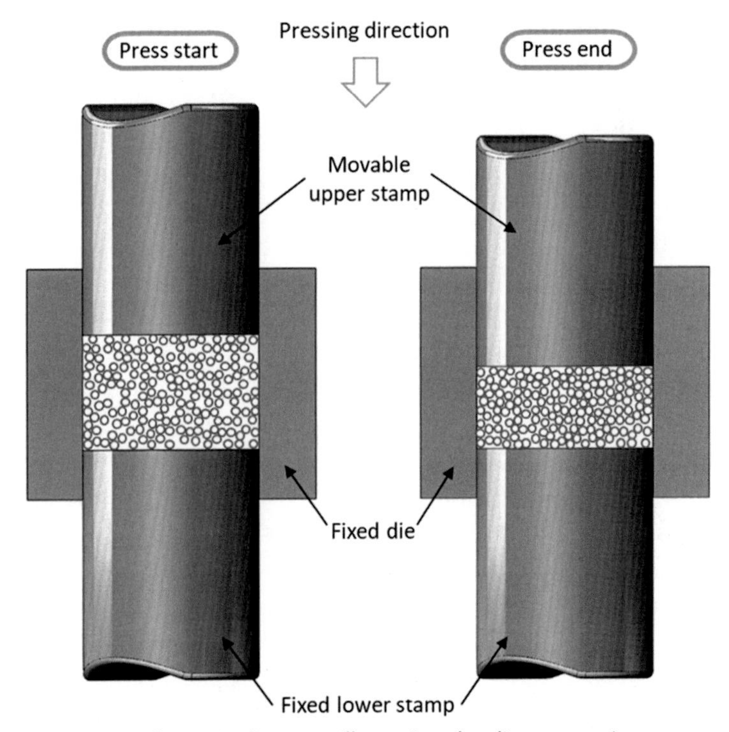

Figure 6.18 Schematic diagrams illustrating the die compaction process.

The employment of granules having a constant granulate tape density is the key factor ensuring the uniform density distribution, which allows obtaining constant shrinkage rates and consistent properties of carbide articles after sintering.

Organic binders (paraffin, PEG, etc.,) lubricate interfaces between the individual particles and the pressing die wall (if they are uniformly distributed in the granules) thus reducing friction forces during pressing. Low green density gradients and optimum green compact densities can be achieved by the employment of lubricants, smooth surface dies, low compression ratios, double-action, or floating die pressing. To reduce differential shrinkage shape distortions during sintering, the green density gradient must be minimized. The use of lubricants additionally reduces the required ejection pressure of the green compacts, which avoids cracking the compacts during ejection.

Modern presses usually operate at pressing forces varying from 100 to 5000 kN. Depending on the shape of the pressed parts, cemented carbide granules commonly require the employment of pressures in the range of approximately 50–150 N/mm^2 for obtaining sufficient values of green density and strength [9]. At such pressures, green compact densities lying in the range of 7–8 g/cm^3 can be achieved. Nowadays, the different pressing systems work at a high speed with strokes of 5–50 per minute depending on the size and shape of the green compacts like indexable cutting inserts, inserts for mining tools, burrs, etc. By using multiple pressing molds, the output of a uniaxial press can be multiplied, thereby achieving a greatly increased efficiency.

6.6.2 Cold isostatic pressing (wet and dry bag pressing)

The cold isostatic pressing technique is based on the concept of isostatic powder compression, according to which a hydrostatic pressure is applied simultaneously to all external surfaces of a die containing a powder. As a result, the die wall friction is significantly reduced or eliminated, which results in a high packing uniformity [16]. Cold isostatic pressing in the cemented carbide industry is mainly performed in two ways: (1) Wet-bag pressing and (2) Dry-bag pressing, which are described below. The large-scale employment of cold isostatic pressing in the cemented carbide industry is related to the need of producing large carbide articles with a complex geometry, which cannot be fabricated by the conventional uniaxial pressing technology.

6.6.2.1 Wet isostatic pressing (wet bag)

By use of wet isostatic pressing, the powder is compacted in an elastic mold sealed with a lid; the mold is usually a flexible rubber bag, which is immersed in the pressure fluid. Due to filling effects of the flexible rubber bag, the green compacts are typically characterized by inexact external shapes and dimensions. The wet bag pressing technology allows green parts with weights of up to 800 kg to be pressed. This technology is typically employed for producing large complex carbide articles. The possibility of automation of the wet bag pressing process is limited.

Generally, wet bag pressing systems are more flexible in terms of shapes and sizes of carbide green compacts in comparison with other pressing techniques. Costs of wet bag presses are lower and they occupy less space than dry bag presses.

Wet bag presses consist of a thick-walled cylindrical pressure vessel made of stainless steel, pressure valves, powerful hydraulic pumps or large pressure pistons, pressure equalizing vessels, and control systems, which are shown in Fig. 6.19. Usually, the pressure vessel is closed and sealed by a massive threaded lid. Pressures varying from 200 MPa up to 1000 MPa are applied to the rubber bags. The pressure vessels with an inner diameter varying

Figure 6.19 Typical wet isostatic press. *(With the permission of Element Six.)*

from 200 mm up to 1500 mm, working length of up to 1500 mm, and wall thickness of up to 400 mm are employed for producing carbide green compacts.

The pressure is built up by hydraulic pumps or pressure cylinders. Hydraulic pump systems pumping the pressure fluid into the pressing chamber are considered to ensure a better controllable technological process, which allows obtaining carbide green compacts with improved quality and consistency. The preparation procedure of WC—Co powders for isostatic pressing is typically the same as or very similar to that of WC—Co powders for uniaxial pressing. WC—Co powders with low wax contents of up to 0.5 wt.% are generally used to prevent the significant organic binder evaporation during the subsequent debinding process. Inadequate wax contents in carbide green compacts can lead to high wax vapor pressures within the green compacts during the debinding process resulting in a high risk of cracking.

On the first step of the pressing process, the WC—Co powder is manually filled into the mold (rubber bag) located outside the press. A surrounding support steel structure keeps the mold in shape, which ensures obtaining the uniform pressure distribution on the entire surface of the mold. In addition, processes of vibrating and evacuating the mold containing the WC—Co powder lead to uniform powder filling, which results in a high bulk density and optimal green compact shape; the green compact shape and dimensions must be as close as possible to those of the final carbide article [14]. Using rigid mandrels allows pressing hollow carbide compacts (such as tubes), which ensures reducing the amount of WC—Co powders needed for fabrication of hollow carbide articles (such as valves, components of pumps, etc.). The inner diameter of such hollow cylindrical compacts (tubes) is characterized by higher precision in comparison with the outer diameter. After completion of powder filling, the mold must be carefully closed and sealed to prevent any penetration of the pressure transfer fluid. The fluid is subjected to pressure and the pressure is transmitted through the flexible wall of the mold (rubber bag) to the powder, which results in its compaction. Improper sealing of the molds can lead to local liquid inclusions in the green compacts. Green compacts with such liquid inclusions are no longer suitable to produce high-quality cemented carbide parts and can be further used only as swarf.

On the second stage of the process, the pressing mold filled with the WC—Co powder is located into the pressing vessel and subjected to the needed pressure. When the compaction process is finished, the fluid pressure in the vessel is released at a controlled release rate to minimize tensile stresses, which form in the green compacts and can lead to their cracking during the pressure release process.

On the third stage of the pressing process, after releasing the pressure, the mold is removed from the pressing vessel and, afterward, is carefully dried. The green carbide compact is retrieved from the mold, followed by its presintering or premachining. Depending on the green compact geometry, it is subjected to different types of 3D machining, such as turning or sawing, to produce cylinders, pipes, or quarters. The shrinkage occurring as a result of sintering is decisive with respect to obtaining the final dimensions of carbide articles and must therefore be calculated in advance before processing the green compacts. Shrinking behavior tests of the respective green compacts ensure obtaining the optimal final dimensions of the carbide articles after sintering. The proper heat treatment process (presintering) of green compacts is crucial with respect to achieving the high robustness and fracture resistance of the green compacts needed for subsequent handling and machining [3]. Processing brittle carbide green compacts requires intensive training of operators and workers dealing with them.

6.6.2.2 Dry isostatic pressing (dry bag)

Dry bag isostatic pressing is an efficient method for producing relatively small carbide articles with axisymmetric geometry. The pressure is built up by means of a high-pressure pump; it is transmitted to the elastic pressing mold filled with the WC—Co powder radially via a membrane installed in the bore of the pressing chamber. The major advantage of the dry bag presses is the possibility of their automation, allowing the cost-effective mass or semimass production of carbide articles with complex geometry, for example, rods, tubes, bushes, balls, plungers, drills, screws, nozzles, etc. With the aid of dry isostatic pressing, it is also possible to press almost net-shape hollow green compacts with the precise outer and internal dimensions resulting in the significantly reduced amount of machining such green compacts, which leads to the consequently reduced swarf generation. The WC—Co powder quality and pressing mold design play an important role with respect to the highly effective fabrication of carbide green compacts.

Dry bag presses usually consist of a pressing frame, pressure vessel, pressing mold change system, powerful oil pressure pump, cleaning device, and modern computer control system, which are shown in Fig. 6.20.

Figure 6.20 Typical dry bag pressing unit. *(With the permission of Element Six.)*

Different options regarding the design of the pressing molds (integrated in the pressing vessel or exchangeable) allow improving the flexibility of the pressing process. The integrated pressing mold is preferable when the carbide articles' dimensions remain constant for a certain period. A flexible pressing mold system is a better option when a constant change of the molds is required. The molds are loaded with WC—Co powders manually or automatically. The pressure vessel together with the pressing mold filled with the WC—Co powder is inserted into the press frame and securely fixed. After the completion of the compaction process, the green compacts are removed from the mold and can be further processed in the same way as after wet bag pressing. One press run lasts nearly 5—10 min and ensures obtaining one green compact. Usually, rods or tubes with a diameter of up to 200 mm and length of up to 600 mm can be pressed by dry bag systems. It is quite common to press tubes with a wall thickness of 2 mm or more by the use of dry bag pressing. The inner and outer diameters of such tubes can be pressed with a tolerance varying in the range of 0.1–1 mm depending on their dimensions. The values of the green compact strength of both the dry bag pressed and wet bag pressed green compacts are comparable. The press mold material required for dry bag pressing must differ from that employed for wet bag pressing. In the case of the dry bag pressing molds, a thixotropy effect is essential for the proper pressure transformation. The viscosity of the press mold material must decrease under pressure allowing the uniform transfer of pressure to the powder in order to achieve the required compaction. When the pressure is released, the viscosity of the mold material increases back to the initial value, so that, afterwards, the following press cycle can be conducted.

6.6.3 Extrusion

Extrusion is a widespread production method for fabricating cemented carbide rods and similar products with or without central or helical cooling channels as blanks for carbide drills or end mills. Several carbide companies have been employing this method over the last decades. The carbide drills or milling tools are used in the automotive and aerospace industries, machine building, electronics, etc., on a large scale. Nowadays, approximately 25%–30% of the world cemented carbide production by weight are just the rod blanks.

For the extrusion of the rod blanks, a production line with numerous special units is required. Additions of a special plasticizer are required to lower the yield stress, as the WC–Co powder does not exhibit plastic flow even at high pressures. Commonly used plasticizers are polyvinyl alcohol, polyethylene glycol (PEG), starch solution, paraffin, synthetic resin, etc. Ideally, each particle of the WC–Co powder must be separated from another particle by a film of such plasticizers. On application of pressure, the material begins to flow through the die orifice when the critical yield stress is reached. The critical yield stress is a function of the plasticizer material properties, its concentration, temperature of extrusion, and degree of mixing [9]. Increased plasticizer concentrations lead to reducing the green density of the extruded blanks, which results in the need of the employment of special debinding and sintering processes. Beside the standard powder preparation equipment, kneaders, and cutting machines for the feedstock production, extrusion machines for the fabrication of rods, handling systems, drying chambers, debinding furnaces, special sintering furnaces, and quality control systems are required. Depending on the needed product characteristics, such as the diameter or type of cooling

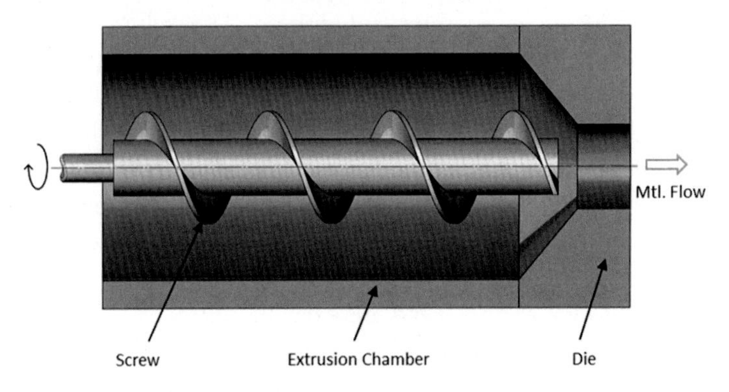

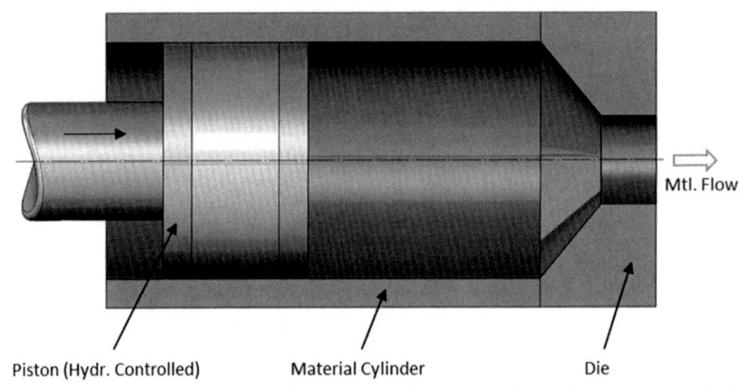

Figure 6.21 Schematic diagrams illustrating the principles of operation of the screw and piston extruders.

channels, rotating screws or piston extruders, which are shown in Fig. 6.21, are used. Both systems are characterized by some advantages and disadvantages. The piston extruders operate in a batch mode due to their design and require a sensitive filling level—dependent pressure control. The flow velocity difference between the cylinder wall and the material stream center could lead to the formation of textures in the extruded rods [14]. The employment of high-pressure piston extruders operating at forces of up to 500 t is a good option to produce carbide rods with large diameters and rods with a single, diametral channel. The extruders with a rotating screw are the preferred units for fabricating carbide rods with smaller diameters and rods with helical cooling channels.

On the first production step, the fully processed and dried cemented carbide powder is mixed with the organic binder and, afterward, filled into a special kneader needed for performing an intensive kneading process. The kneaded mass can be stored in small vacuum-sealed plastic bags until its filling into the extrusion press. The extruded rods are transferred onto special grooved graphite trays. The trays with the rods are moved into the drying chamber. A special wet optical test with respect to the presence of cracks is conducted afterward to prevent premature tool failures when fabricating and employing the carbide drills and end mills. After drying, the rods are cut to the needed length and charged into the sintering furnace for debinding, presintering, and final sintering followed by centerless grinding to the final dimensions and tolerances.

Cellulose, paraffin, and PEG are ususally employed as plasticizers. As a guideline, for small rods of less than 8 mm in diameter 5–8 wt.% paraffin wax are used and for bigger rods of more than 8 mm in diameter 1–8 wt.% cellulose are employed. The intensive kneading process, which lasts 1–2 h, is usually conducted at a low temperature to prevent the feedstock oxidation. The feedstock density increases from nearly 4 g/cm^3 up to approximately 8 g/cm^3 as a result of kneading. The optimum gap of 1.5–3 mm between the shear arm and the vessel ensures obtaining suitable kneading results. This completes the preparation of the feedstock, which then can be used as a starting material for the extrusion of green rods. In order to achieve the optimal extrusion results, the process parameters such as the feedstock temperature, rotation speed of the feeding and extrusion screw, pressure, angle of extrusion, etc., are needed to be precisely monitored and controlled. The extruded rods are pushed onto special diameter-matched grooved graphite trays. Compressed air fed through small holes in the trays combined with their optimized surface roughness and quality ensure the friction to be reduced when pushing the rods onto the trays.

When organic solvents or other plasticizers are used, the rods must be carefully dried in drying chambers at temperatures of up to 250°C for a period varying from 12 h up to 40 h depending on the rods' diameter to prevent cracking followed by their cutting to the final length. The dewaxing step is time consuming because a high amount of organic binder must be carefully removed from the green compacts.

Because of the required high organic binder content, the green compacts' density is approximately 50% lower than that of green compacts produced by uniaxial pressing. As a result, the well adopted presintering and sintering processes last a long time to achieve the full debinding and obtain a certain rate of diffusion between WC grains and Co inclusions and interlayers. The follow-up grinding is mandatory to achieve the tolerances required for each concrete application. A uniform microstructure combined with adequate physical, mechanical, and metallurgical properties are essential for the high performance of the drilling tools produced from the sintered and ground carbide rods.

6.6.4 Shaping green carbide compacts

Nowadays, various specially adapted machining methods such as turning, grinding, drilling, etc., are used to produce articles of complex geometries from green compacts [17]. The green compacts' integrity, strength, and toughness are significantly increased as a result of presintering, which allows their subsequent machining. The organic binder must be completely removed from green compacts before their machining, as it builds-up at the tool cutting edges resulting in a significant performance reduction and increase of the needed machining time. Processes of machining the green compacts are highly flexible and adaptable to achieve the desired geometries of carbide articles. High brittleness of the green compacts must always be considered when performing processes of machining. Damage of the green compacts can be caused by the tool vibration, tool breakages, deflection, etc. Tools' wear, insufficient green compact hardness, and inclusions of hard particles significantly affect the quality of the green compacts after their machining. The significant shrinkage of the green compact during sintering also strongly affects the shape and dimensions accuracy of the finally sintered carbide articles. The precise shrinkage factor of each carbide grade and batch can be calculated by the exact measurement of test samples in the green state and after their sintering, which allows adjusting the following machining steps. It should be noted that the shrinkage factors in all the three spatial directions might be different.

Dust emission generation is one of the major issues in the processes of green machining of carbide green compacts and must be controlled and minimized. Advanced dust exhaust systems consisting of an industrial extractor fan and dust filtration unit with a dedicated sinter membrane filter should be used for all the machines employed for machining the green compacts. Remaining green compact sections, removed debris, and dust collected in the dust filter should be processed by mechanical or thermal processes to achieve a high rate of powder reuse.

The following provides an overview of the most common methods for machining the cemented carbide green compacts such as turning, milling, drilling, grinding, and sawing.

Turning is a machining process in which a cutting tool, typically a nonrotary tool bit, defines a helix toolpath by moving nearly linearly while the workpiece rotates. Turning lathes enable rapid finishing the internal and external surfaces of rotationally symmetric carbide parts. Turning allows machining carbide green parts in the pressed or presintered state. Modern programable turning machines are fully encapsulated and equipped with dedicated dust collection systems to prevent any dust emission. Dust protection wipers for guideways and compressed air shut-offs for measuring units and spindle bearings are necessary when turning abrasive carbide articles subjected to presintering in order to minimize wear.

Rotationally symmetrical workpieces are mechanically or pneumatically clamped in the lathe chuck and subsequently turned at a depth of 5–10 mm per pass and a feed rate of 0.1–0.2 mm/rev. Adjusting the speed in a range of 100–600 rpm, peculiar tool angle and unimpeded debris removal result in a high dimension precision of the turned carbide articles and prolonged tool life. High rotational speeds of the workpiece during machining should be avoided, as high centrifugal forces can cause breakages of the presintered carbide green parts characterized by relatively low toughness and strength.

Part inspection includes (beside a close dimensional check) the examination of surface defects such as pores, pull outs, cracks, chipping, etc. Parts machined to final dimensions are stored on dedicated trays after their inspection followed by their sintering.

Milling is the process of machining using rotary cutters to remove material by advancing a cutter into a workpiece. This can be done with varying the direction of one or several axes, cutter head speed, and pressure. Milling covers a wide variety of different operations and machines, on scales from small individual parts to large, heavy-duty milling operations. It is one of the most commonly used processes for machining cemented carbide parts to precise tolerances. Modern programable milling machines are fully encapsulated and equipped with dedicated dust collection systems to prevent any dust emissions. 5-axis technologies illustrated in Fig. 6.22 combined with special high-speed spindles enable highly productive shaping of complex green parts. Spindle speeds of up to 18.000 rpm combined with a feed rate in the range of 200–400 mm/min are typically employed, which allow quick machining at relatively low cutting forces minimizing the risk of chipping. Machining large green compacts weighing up to 800 kg with complex geometries is employed nowadays in the cemented carbide industry on a large scale. The milling machine can be programed with the

Figure 6.22 Typical cabinet of a 5-axis milling machine for machining cemented carbide green compacts. *(With the permission of Element Six.)*

necessary process parameters via a direct download from the CAD (computer aided design) system. A simulation cycle can be performed before the first workpiece is manufactured to avoid tool collusion and machine damage. The operator's task is mainly limited to fixing the green compact into the holder chuck, closing the machine and starting the program. After machining the green compacts for different times varying from 30 s up to 24 h, a fully machined part is removed and forwarded to the sintering department.

Drilling is a machining process used to produce holes. Since the cutting geometry of the drill is known, it counts in the classification of the manufacturing processes for machining with geometrically determined cutting edge, which also includes turning and milling. The drilling equipment employed in the cemented carbide industry typically consists of ordinary drilling machines or very specific advanced drilling systems, which can produce micro or macro holes. A so-called "dead man's switch," which must be continuously pressed by the operator's foot at a certain pressure, is still one of the most important safety device employed in drilling machines. When drilling the green compacts, high drilling speeds and low feeds are preferable to reduce sharp edges on the input side and minimize the presence of burrs on the exit side. The spiral markings on the inner surface of the borehole resulting from the drilling process should be minimized in order to maintain the stability of the carbide article being drilled. Drilling the carbide green compacts is preferably conducted at a rotation velocity of 500–700 rpm and feed rate of about 30 mm/min. The debris removal is one of the main challenges of the drilling process, whereby drilling from below is ideal for the unimpeded debris removal by gravitation. Special drills enable obtaining borehole diameters as small as approximately 0.5 mm. Drills with a diameter of the order of roughly 2 mm enable obtaining bores of up to 200 mm in length. Drilling is often employed together with other machining methods to obtain carbide parts with complex geometries in a single machining cycle.

Grinding is an abrasive machining process that uses a grinding wheel as the cutting tool. The employed grinding devices for green machining can consist of ordinary grinding bucks or very special systems, which can produce precise plan-parallel faces. Grinding is also used to obtain final heights of all types of suitable green compacts. Grinding machines are usually equipped with an integrated dust removal system in combination with a controlled workpiece table. During grinding, green carbide articles are passed below the grinding wheel under a certain contact pressure, generating the corresponding material removal. Galvanically bonded diamond grinding wheels are preferably employed for grinding the cemented carbide green compacts. Lower surface speeds ensure the needed time for removing debris from the surface, preventing overheating of the green compact and grinding wheel. The hardness achieved by the presintering process has a significant influence on the grinding process parameters. Grinding hard green compacts causes a risk of overheating as a result of spark formation. For grinding carbide green compacts, the wheel rotation speed typically varies in the range of 4000–6000 rpm combined with a feed rate of about 0.4–0.5 mm. The advance rate of a workpiece under the grinding wheel depends on the removal rate, green compact properties, and correspondingly generated counterforces.

The sawing equipment employed in the cemented carbide industry typically consists of a rigid, low vibrating band saw equipped with a diamond bladed saw band and electrically movable worktable. The green compacts are placed on the movable worktable and sawed to obtain the needed geometries. Saw band cutting speeds varying between 22 and 150 m/min are usually employed; they can be adjusted in accordance with the green compact geometry and properties.

Figure 6.23 Typical cemented carbide articles fabricated by use of machining green carbide compacts. *(With the permission of Element Six.)*

Complex cemented carbide articles, which are shown in Fig. 6.23, are manufactured by use of the machining technologies described above; they are employed in different industries such as mechanical engineering, automotive, aerospace, oil and gas, civil engineering, etc.

6.7 Technological processes for sintering cemented carbide articles

6.7.1 Dewaxing and presintering

Dewaxing and/or presintering can be performed in a separate cycle in a special furnace or as part of the complete sintering process including final liquid-phase sintering. Dewaxing can be considered as a process of removing the organic binder from green carbide parts. Thermal dewaxing is a common method used to remove the organic binder from such green carbide parts. An alternative debinding method is based on the fluid or chemical extraction and employed in cases of high organic binder contents or complex binder systems. Thermal dewaxing is usually performed in a vacuum and or sweep gas, e.g., nitrogen, argon, or hydrogen, to remove vapors of the organic binder from the furnace. By using hydrogen, a process of reduction of thin films of oxides and adsorbed oxygen takes place in the bulk of the carbide green bodies on the initial stage of presintering/dewaxing when the porosity is open and the gas permeability is high. The carbon content in the initial WC–Co powders and consequently carbide green bodies must be precisely adjusted to the applied dewaxing or presintering process, which can be either carburizing or decarburizing.

Dewaxing is usually performed at relatively low temperatures because it must result in the full removal of the organic binder without its thermal decomposition, whereas the presintering process is conducted at higher temperatures to achieve a certain level of mechanical properties of the carbide green bodies needed to perform their machining.

Special furnaces are used for dewaxing or presintering of the green carbide parts, if there is a need to perform dewaxing and presintering in a separate process. Such furnaces usually comprise a double-wall heat- and scale-resistant steel vessel with a water-cooling system, sealings, isolation, inner graphite box, heaters, and several thermocouples to control temperature in different parts of the furnace. In addition, the furnace is equipped with a vacuum pump and a gas sweeping system comprising an organic binder condenser, which are steered by an electronic process control system. The visualization of the process flow and detailed furnace operation parameters secures consistent properties of the carbide green articles after the dewaxing or presintering processes. Dewaxing can be performed using sweep gases mentioned above, in a vacuum or by use of the so-called partial pressure flow. The employment of hydrogen as a sweeping gas requires additional safety equipment to avoid a risk of explosions (Fig. 6.24).

On the first process step, the inner graphite box, often equipped with graphite roller tracks, is charged with the green articles located on robust graphite trays. The graphite roller tracks allow precise positioning the

Figure 6.24 Typical presintering/dewaxing furnace equipped with electronic systems for steering the residual pressure, sweep gas flow and temperature distribution within the furnace. *(With the permission of Element Six.)*

graphite trays with the green compacts within the graphite box. After closing the inner graphite box and the outer furnace door, the entire furnace vessel is evacuated followed by a vacuum check. The precise thermal processing of the green articles according to the required temperature profile and process parameters is started via a control panel. At the end of the furnace run, a fluting with an inert gas, usually nitrogen, takes place to speed up the cooling process and to achieve a pressure balance between the vessel and atmospheric pressure. The furnace load is discharged manually by pulling the graphite trays from the furnace via the graphite roller tracks directly onto dedicated lifting carts with steel roller tracks. Nowadays, the time-consuming process of cooling small carbide green parts down to room temperature can be significant accelerated by tailored cooling systems. Optimum inert gas inlet combined with flexible systems for opening the graphite box door are vital components allowing one to minimize the cooling time.

Typically, the removal of different organic binders is finished at a temperature within the range of 450°C—500°C. Heating rates depend mainly on the WC grain size and Co content, which in turn affect the residual porosity and size of open channels needed for removal of organic binder vapors. Presintering cycles are conducted at higher temperatures varying from nearly 650°C to 900°C, which depend on the cobalt content in the green articles and the needed level of their mechanical properties. Hardness and toughness of the presintered green articles strongly influence upon the follow-up process of their machining.

The duration of the dewaxing/presintering cycles typically varies from 12 to 60 h depending on the size of the green articles, their geometry, Co content, WC mean grain size, pressure and temperature regimes, sweeping gases, and the furnace design. The dewaxing/presintering processes are typically performed in a vacuum with a residual pressure of 20—100 mbar, which allows the temperature to be significantly reduced. Usually, sweep gas flows lie in the range of 12—36 L/min. The commonly used organic binders are either short chain paraffins with melting points of 45°C—65°C and evaporation temperatures of nearly 250°C—350°C under normal pressure, or PEG with chain lengths of 3000—5000 carbon atoms with melting points of 45°C—65°C and evaporation temperatures of 270°C—320°C depending on the chain length. During the early stage of heating, such long molecules of PEG decompose forming gaseous species such as CO, H_2O, CO_2, and CH_4, which must be removed from the furnace in a hydrogen flow to protect the furnace, pipes, and vacuum pumps against corrosion.

The evaporation of paraffin wax in a vacuum starts at temperatures below 250°C first of all from the near-surface region of the green parts. When the green parts are heated for dewaxing, the heating rates must be chosen in such a way that significant temperature gradients within the green parts are avoided to minimize a risk of formation of the thermal expansion mismatch, which can lead to cracking the parts, and to eliminate the formation of elevated pressures of wax vapors in the core region of the green parts. The wax vapor pressure within the green parts must be carefully controlled to minimize a risk of cracking or "exploding" the green parts on the stage of the temperature increase above the wax melting point. During the dewaxing process, the wax vapor is continuously removed from the furnace by use of the sweeping gas via the vacuum pump and is collected in a condenser. After finishing the dewaxing or presintering cycle, the organic binder collected in the condenser can be molten by hot water and the molten wax can be accumulated in a separate vessel. Controlling the color of the condensed wax allows evaluation of the dewaxing process, so that it is possible to estimate whether the dewaxing process was performed properly if the color of the condensed wax remains nearly the same as that of the virgin wax. It is crucial to remove the organic binder completely from the green articles at temperatures below 450°C, as the paraffin wax can decompose above this temperature resulting in the formation of free carbon thus significantly affecting the total carbon content in the green articles. Thus, the color of the condensed wax is considered to be a reliable indicator showing that the organic wax did not decompose during the dewaxing process (Fig. 6.25). Controlling the entire mass of the

Figure 6.25 Condensed wax after a proper dewaxing cycle. *(With the permission of Element Six.)*

removed and condensed wax in relation to the furnace load can also be helpful to check whether the dewaxing process was performed properly. The properly performed dewaxing process is important with respect to obtaining the optimal carbon content in the fully sintered carbide articles.

The influence of the presintering process on the density and mechanical properties of green carbide articles is directly related to the processes of solid-phase sintering of WC—Co materials at temperatures of above 50% of the eutectic temperature (melting point) [18].

Presintering conducted at temperatures of above 900°C can result in noticeable shrinkage of the green articles as a result of solid-state sintering occurring due to the material transport through the surface, volume, and grain boundary diffusion in the carbide green articles. The material transport through the plastic or viscos flow, as well as that occurring in the gas phase, can also be considered as important mechanisms of solid-state sintering. Usually, almost all the phenomena of the material transport mentioned above occur simultaneously [11].

The phenomena occurring during solids-state state sintering of two contacting powder particles illustrated in Fig. 6.26 include the following processes of the material transfer from the particles' volume and surface regions toward the neck between the particles. The phenomena (1—3) occur without significant shrinkage of the green articles during presintering; the phenomena (4—6) lead to shrinkage of green bodies during presintering [11].

1. Evaporation—condensation;
2. Surface diffusion;
3. Volume diffusion;
4. Diffusion at grain boundaries;
5. Diffusion along grain boundaries;
6. High-temperature creep (not relevant for presintering at temperatures below 800°C)

The solid-state sintering occurring during dewaxing and presintering results in prehardening the carbide green articles. The prehardening is one of the main enablers for conducting the final machining steps of the presintered green articles. In addition to the prehardening effect, the toughness of the green articles is also noticeably increased (by roughly by one order of magnitude) as a result of presintering. The higher toughness of the presintered green articles is vital for their final handling and machining.

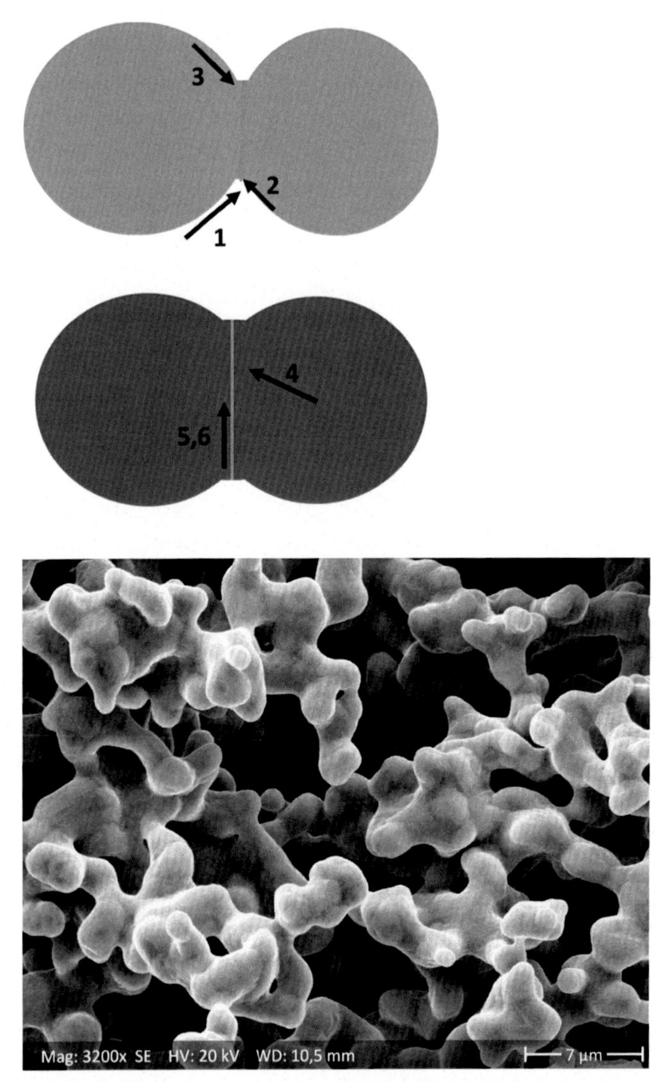

Figure 6.26 Above: mechanisms of material transfer toward the neck between two powder particles occurring during solid-state sintering: 1—evaporation–condensation; 2—surface diffusion; 3—volume diffusion; 4—diffusion at grain boundaries; 5—diffusion along grain boundaries; 6—high-temperature creep. The phenomena (1–3) lead to the growth of necks between the powder particles without shrinkage of green bodies; the phenomena (4–6) lead to shrinkage of green bodies [11]. Below: SEM image of Co particles after sintering at 800°C for 30 min. *(With the permission of Element Six.)*

6.7.2 Introduction on theoretical bases of liquid-phase sintering

Theoretical bases of sintering phenomena are summarized in the well-known books of R.M.German and W.Schatt [18,19], so this chapter contains only a very brief overview of basic principles of sintering theory.

Solid-phase sintering of WC–Co materials occurs when heating a carbide green body up to the eutectic temperature of the W–C–Co system, which depends on the total carbon content (see Chapter 3). WC–Co materials are characterized by the significant densification of green bodies on the stage of solid-phase sintering. The densification of green bodies obtained from finer WC powders started at much lower temperatures than those produced from coarse-grain WC powders, which leads to high densities of such fine-grain green bodies (up to 90% of the theoretical density) achieved at sintering temperatures below 1298°C. The phenomena occurring during solid-state sintering of two contacting powder particles are outlined in Section 6.7.1, as such phenomena occur during presintering of carbide green articles.

At temperatures above nearly 1298°C–1299°C, a liquid phase forms in the pre-sintered cemented carbide green body and the very rapid densification takes place as a result of liquid-phase sintering.

Major principles of liquid-phase sintering were formulated by Cannon and Lenel, and Kingery and Narasimhan [20,21]. Densification during liquid-phase sintering occurs in the three following stages:

1. Liquid-viscous flow and rearrangement of grains of the solid phase in the direction of their denser packing;
2. Dissolution of small grains in the liquid phase and growth of larger grains at the expense of the small grains (recrystallization through the liquid phase) or so-called "Ostwald ripening."
3. Additional densification due to coalescence of the solid-phase grains among themselves.

Kingery and Narasimhan indicated that a certain minimum amount of a liquid phase is required for occurring the liquid-viscous flow process [21]. This amount is equal to ~35 vol.% for spherical particles of the solid phase.

The consequence of phenomena occurring during liquid-phase sintering can be described as the following:

- the appearance of a liquid phase as a result of melting a low-melting point component;
- its propagation through open porous forming capillary channels;
- filling the pores due to wetting their surfaces by the liquid phase;

- dissolution of small particles of the solid phase in the liquid phase;
- obtaining an equilibrium rate of dissolution of small solid particles followed by the growth of large particles;
- possible coalescence of grains of the solid phase.

Driving forces of the processes occurring during liquid-phase sintering are related to aligning values of the chemical potential of all the components in the body being sintered as well as reducing the total surface of the solid—gas and liquid—gas interfaces followed by their full disappearance.

Justifying the three stages of liquid-phase sintering mentioned above, Kingery and Narasimhan suggested models that can potentially describe the sintering densification processes for the first two stages [21]. However, the role of the phenomena of wetting and capillary effects during liquid-phase sintering was not sufficiently taken into consideration resulting in difficulties with respect to applying these models to real sintering processes, as capillary phenomena significantly affect the rate of liquid-phase sintering.

Capillary forces cause phenomena of movement and rearrangement of the solid particles being in contact with liquid, which leads to their dense package resulting in the densification and shrinkage of the sintered body. The capillary pressure of liquid in channels between the solid particles can significantly vary depending on the amount of the liquid phase in the sample and wettability of the solid particles by the liquid. When the amount of liquid phase increases up to a certain level, the capillary forces "pulling together" solid particles noticeably grow. However, as the pores become filled with the liquid, the capillary pressure and consequently capillary forces start decreasing, so the influence of the liquid phase amount on the densification process becomes less distinct.

The presence of capillary forces attracting solid particles and liquid phase interlayers at the junctions of the particles leads to regrouping and rearranging the solid particles. These processes occur very quickly and make the main contribution to the shrinkage, the rate of which depends on the volume fraction of the liquid phase in case if the wettability of the solid phase by liquid is complete or nearly complete. However, it is difficult to explain the fast compaction in systems with a low content of a low-melting point component only by the rearrangement of solid particles. It is suggested that under the action of significant surface tensions at solid/liquid interfaces, high capillary forces are transmitted to areas of contacts among the solid particles. As a result, the solubility rate of these areas in the liquid phase increases leading to the enhanced dissolution of the solid phase in such contact areas. As a result, processes of growing and rearranging the shape of the solid particles are significantly accelerated. The process of dissolution of fine particles in the liquid phase followed by diffusion of the dissolved atoms in the liquid phase and their precipitation on the surface of large particles can take place at an early stage of sintering processes resulting in the grain growth due by the dissolution—precipitation mechanism (Ostwald ripening), which is particularly important for sintering of cemented carbides.

6.7.3 Processes occurring during sintering of WC—Co cemented carbides

Liquid-phase sintering of cemented carbides is characterized by the three following major features:

1. The solid phase (WC) partially dissolves in the Co liquid phase, which results in achieving its equilibrium eutectic composition at the sintering temperature;
2. Complete wettability of the solid phase (WC) by liquid takes place in cemented carbides with medium and low carbon contents, so that in this case, the contact angle of wetting tungsten carbide by the liquid Co-based phase is equal to 0. For cemented carbides with a very high carbon content corresponding to the free carbon formation, the contact angle is equal to nearly 15 degrees indicating incomplete wetting (see Chapter 9).
3. The WC—Co system is characterized by a low dihedral angle leading to the fact that the motion of the boundary through the small grain increases the surface energy, as the area of the grain boundary increases in this case, so that coalescence does not play an important role.

According to results of Sandford and Trent [22], shrinkage of WC—6 wt.%Co and WC—10 wt.%Co green bodies begins at a temperature of about 1150°C and ends at nearly 1320°C, i.e., at the temperature nearly corresponding to the liquid phase formation. Shrinkage occurs to a large extent already at temperatures of solid-state sintering. The densification in the solid state before the appearance of the liquid phase is accompanied by the diffusion of tungsten and carbon from WC grains into cobalt leading to the formation of solid solutions, so that up to nearly 10 wt.% WC can dissolve in cobalt in the solid state.

At temperatures exceeding the melting point of the W—C—Co eutectics, a density close to theoretical is achieved in a very short time (1—4 min). When taking into account the rapid shrinkage just after the appearance of the liquid phase, further densification occurs due to fluid flow, rearrangement, and movement of carbide grains under the influence of capillary forces and surface tension [23], especially when the amount of liquid phase is at least 20–30 vol.%. At lower liquid contents, the densification is enhanced by the process of recrystallization of carbide grains through the liquid phase by the mechanism of dissolution–precipitation, which leads to adjusting WC grains' shape to that of the neighboring WC grains. This is believed to create conditions needed for a more effective action of the capillary and tension forces. According to results of Nelson and Milner [24], the processes of dissolution–precipitation starts playing an important role in the densification process enhancing the penetration of liquid between solid particles just from the very beginning of liquid-phase sintering. In should be noted that in systems similar to the WC–Co system characterized by a significant rate of the solubility of solid grains in liquid and complete or nearly complete wettability, all the three stages of liquid-phase sintering formulated by Kingery and Narasimhan [21] occur usually almost simultaneously.

According to Gurland [25], the final formation of the microstructure takes place on the next stage. It includes the formation of contacts between growing WC grains, dissolution of small grains leading to further recrystallization, and WC grain growth. The driving force of these processes is related to reducing the surface energy at WC/WC and WC/Co grain boundaries. It was believed earlier that a carbide skeleton forms on this stage and WC grains grow together forming direct WC/WC grain boundaries, which results in the formation of a WC skeleton. Nevertheless, according to the results obtained in the last decade by TEM and HRTEM, there are Co segregations or interlayers at most WC/WC grain boundaries (except for the $\Sigma 2$ grain boundaries, see Chapter 9). Therefore, in general, the carbide skeleton, as it was understood about 4 decades ago, does not form on this sintering stage. The contact areas of the growing WC grains are believed to be subjected to high surface tensions leading to the higher solubility of WC in the liquid phase occurring in these contact areas, which in turn results in changing and mutual adjusting the grains' shape ("grain shape accommodation") to achieve their highest package rate, which can be seen in Fig. 6.27.

The following major mechanisms of WC coarsening in the process of liquid-phase sintering can be distinguished as follows:

1. The WC grain growth can occur due to the crystallization of tungsten and carbon dissolved in the liquid Co-based binder on the WC surface as a result of cooling and solidification of the liquid binder phase. It is

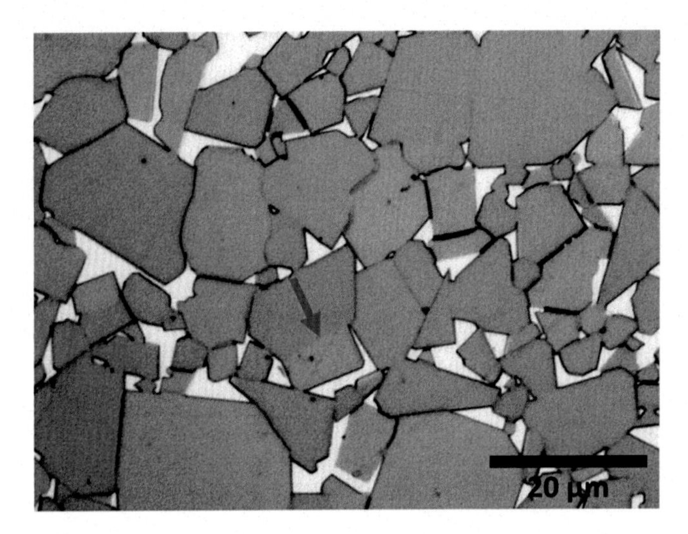

Figure 6.27 Microstructure of the ultracoarse WC–6%Co grade sintered at 1490°C for 75 min (light microscopy after etching in the Murakami reagent). The WC/WC grain boundary, that is likely to be the Σ2 grain boundary, is indicated by an arrow. *(With the permission of Element Six.)*

well known that significant amounts of tungsten and carbon can dissolve in the liquid binder at sintering temperatures, so that the liquid binder contains up to 40% dissolved WC. The saturation of the liquid binder phase with tungsten and carbon occurs due to the dissolution of fine grains of tungsten carbide during sintering. Upon cooling, significant amounts of W and C atoms dissolved in the liquid binder precipitate on undissolved tungsten carbide grains. However, as a result of this process, one cannot expect a significant increase in the size of the WC grains in conventional carbide grades containing usually nearly 6–15 wt.%Co. For example, the liquid binder of the WC–10 wt.%Co cemented carbide dissolves only about 6% of the total amount of WC present in the alloy, so the precipitation of such an insignificant amount of W and C on the undissolved WC grains upon cooling cannot lead to their noticeable growth.

2. Some growth of tungsten carbide grains is possible due to the mechanism of coalescence of grains with a close crystallographic orientation. Intergranular boundaries of grains of this type are designated in literature as the "Σ2 grain boundaries," one of which is shown in Fig. 6.27, but their share in the total number of WC/WC grains boundaries is relatively insignificant. In general, coalescence plays an important role in

systems, in which there are low misorientation angles among grains and a high dihedral angle apart from large grain size differences. However, for the WC–Co system characterized by a low dihedral angle, the motion of the boundary through the small grain increases the surface energy, as the area of the grain boundary increases, so there is an energy barrier for impending coalescence [9].

3. The growth of tungsten carbide grains occurs mainly due to recrystallization through the liquid phase. This is the main mechanism for the growth of the WC phase. With an increase in the content of cobalt in coarse-grain grade up to 10%, a noticeable increase in the size of the WC grains is observed, especially if the sintering conditions are not optimal. At a very large amount of liquid binder, the WC growth rate decreases due to an increase in the thickness of the Co interlayers and, consequently, an increase in the distance that tungsten and carbon atoms diffusing in the liquid phase must overcome. However, this phenomenon usually does not take place in submicron and fine-grain cemented carbides characterized by very thin Co interlayers. According to the W–C–Co phase diagram (see Chapter 3), when a mixture of tungsten carbide and cobalt is heated, the process of dissolution of tungsten and carbon in cobalt begins already in the solid state, which lowers the melting point of the cobalt-based solid solution in comparison with pure cobalt, characterized by the melting point of 1493°C. The appearance of a liquid phase and further increasing the temperature promotes the acceleration of diffusion processes and an increase in the amount of the liquid phase of the eutectic composition.

For submicron cemented carbides with the initial broad grain size distribution of WC grains, especially for those containing much WC nanofraction, the process of WC coarsening proceeds continuously. This process occurs as a result of dissolution of small grains having a higher free energy and consequently higher solubility, which significantly depends on the radius of the grains (surface curvature), followed by diffusion of the dissolved W and C atoms in the liquid phase toward the surface of large grains having a smaller free energy and lower solubility resulting in the growth of such grains. Nevertheless, if the amount of nano fraction is very insignificant and large WC grains in coarse-grain cemented carbides have similar sizes, the primary intensive grain growth interrupts after initial 5–10 min of sintering and, afterward, almost no WC coarsening occurs even for a very long time. In this case, after very fast dissolution and recrystallization of the nano WC fraction, a driving force for the WC grain growth becomes very insignificant due to a small difference between grain sizes of the coarse WC grains. As a result, the WC grain growth of such grains due to Ostwald ripening almost does not occur even for very long sintering times.

According to the results of Müller, Konyashin et al. [26], the WC grain size and grain size distribution of the original WC powders strongly affect the grain growth in WC–Co cemented carbides sintered for a long time (up to 24 h). Fig. 6.28 shows STEM images of a submicron WC powder with a broad grain size distribution comprising much nanograin fraction of below 100 nm in size (up to 20 vol.%) and those of an extremely uniform medium grain WC powder with a narrow grain size distribution almost not containing the nanograin fraction. As it can be seen in Fig. 6.29, the process of WC coarsening in the WC–Co material produced from the submicron powder proceeds steadily and fast, and its microstructure becomes very coarse and contains many abnormally large WC grains after sintering for 24 h. In contrast, when sintering cemented carbide from the medium-grain powder with the narrow grain size distribution, the WC grain growth occurs mainly only in the initial 5 min of sintering and afterward almost ends off. As one can see in Fig. 6.29, there is only an insignificant difference between microstructures of this cemented carbide sintered for 5 min and 24 h indicating that the process of WC coarsening due to Ostwald ripening is almost completely suppressed in this case.

According to Ref. [26], the limiting stages and consequently mechanisms of the WC grain growth for cemented carbides produced (1) from the submicron WC powder with a broad grain size distribution comprising much nanograin fraction and (2) from the medium-coarse WC powder with a narrow grain size distribution shown in Fig. 6.28 are completely different.

In the first case, the WC grain growth is fast and thought to be limited by the rate of interfacial reactions of dissolution of fine and nano WC grains in liquid cobalt. The diffusion of W and C atoms dissolved in liquid is believed to be faster than the rate of the interfacial reactions due to the presence of great concentration gradients between the large or abnormally large WC grains and fine/nano WC grains resulting in very high diffusion rates at such concentration gradients according to the Fick's second law. Thus, the mechanism of grain growth during liquid-phase sintering in this case is related to interfacial reactions at WC/liquid boundaries, which was established based on the value of apparent activation energy of the WC coarsening process.

Figure 6.28 STEM images of the submicron WC powder with a broad grain size distribution comprising much nanograin fraction of below 100 nm in size (above), and the extremely uniform medium-coarse WC powder with a very narrow grain size distribution almost not comprising the nano-grain fraction (below). *(Redrawn from Ref. [26].)*

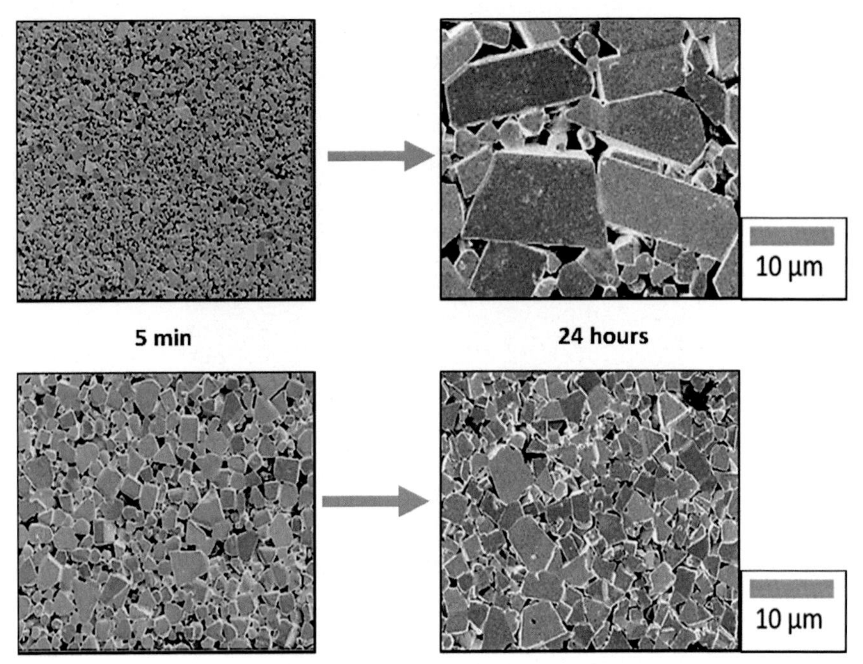

Figure 6.29 HRSEM images of WC–10%Co cemented carbides produced from different WC powders: above—the submicron WC powder with a broad grain size distribution comprising much nanograin fraction of below 100 nm in size after sintering at 1400°C for 5 min and 24 h, and below—the extremely uniform medium-grain WC powder almost not comprising the nanograin fraction after sintering at 1400°C for 5 min and 24 h. *(Redrawn from Ref. [26].)*

In the second case, the sintering process is characterized by another mechanism of the WC grain growth. The grain size distribution in the original WC powder is very narrow and almost all the WC grains have nearly the same size. If even there are very insignificant amounts of fine or nano grains in the initial WC powder, they dissolve in liquid very fast and reprecipitate on the surface of larger grains within a very short initial period of sintering. Consequently, all the WC grains have very similar sizes during sintering leading to insignificant concentration gradients among such WC grains, which results in very low values of diffusive flux and diffusion rate of W and C atoms dissolved in liquid according to the first and second Fick's laws. As a result, the process of WC coarsening due to Ostwald ripening is almost completely suppressed in this case, which is illustrated in Fig. 6.29 below. Thus, the diffusion of W and C atoms (mainly tungsten atoms having a lower diffusivity) in the liquid phase is the limiting stage of the WC grain growth process as a whole, which was established based on the value of apparent activation energy of the WC coarsening process.

The influence of capillary forces and surface tension forces ensures obtaining porosity close to zero during vacuum sintering typically employed in the carbide industry. However, the remaining insignificant porosity is present as a limited number of large pores that cannot be removed as a result of vacuum sintering. Such pores can be pressed out only by use of sintering under pressure of inert gases (mainly Ar), so according to the present state of the art in the cemented carbide industry, carbide articles are sintered in vacuum followed by hot isostatic pressure (HIP) sintering in fully automated Sinter-HIP furnaces, one of which is shown in Fig. 6.30.

Carbide green articles are located on graphite trays coated with powders or flakes of different compounds; for example, alumina. The trays are assembled into stakes which are manually located into the furnace. Sintering conditions (temperature and time) in the production furnaces depend on the WC mean grain size and Co content and usually vary from about 1360°C for high-Co grades to 1490°C for ultracoarse grades and from half an hour to an hour and a half. Typical sintering cycle includes dewaxing described above, heating up to the sintering temperature at a rate of 2—4° per minute in a vacuum, sintering in a vacuum for 45 min followed by HIPing in Ar at pressures of 40—50 Torr for 30 min. These general sintering conditions can be varied in the production of special carbide grades, for example, functionally graded cemented carbides. Also, modern Sinter-HIP furnaces allow performing presintering processes in different gas

Figure 6.30 Industrial Sinter-HIP furnaces employed in the carbide production. *(With the permission of Element Six.)*

atmospheres, for example, in methane, if the carbide green articles are needed to be carburized. Cooling from sintering temperatures is performed in Ar at rates of 3–5° per minute down to 1000°C and at higher rates (up to 10° per minute) at lower temperatures. When the temperature becomes as low as nearly 40°C, the furnace is opened and the stakes with the sintered carbide articles are manually unloaded.

Summarizing the main processes occurring during the sintering process of WC–Co cemented carbides, the following major features can be outlined:

- noticeable shrinkage of green bodies begins during solid-state sintering; it starts in the temperature range of 1050–1150°C and leads to the formation of a solid solution of tungsten and carbon in cobalt;
- upon reaching the eutectic temperature (melting of the cobalt-based phase), capillary forces start acting, which causes rearranging the WC grains and obtaining their dense packing resulting in the rapid densification to almost full density within the first 3–5 min of liquid-phase sintering;
- with a further increase of the temperature or sintering duration, the shape of WC grains changes as a result of their mutual adjustment and rearrangement;
- processes of WC coarsening and forming the WC/Co grain boundaries and WC/WC grain boundaries (including the "genuine" Σ2 WC/WC grain boundaries not containing Co and WC/WC grains boundaries containing different amounts of Co) occur.
- upon cooling, the tungsten and carbon atoms dissolved in the liquid phase precipitate on the existing WC grains, which leads to some increase in their grain size, filling the gaps between the grains and creating additional WC/WC grain boundaries, thus resulting in the formation of the final cemented carbide microstructure.

6.7.4 Special features of sintering WC–TiC–TaC–(NbC)–Co cemented carbides

The basic processes occurring during sintering of WC–TiC–TaC–(NbC)–Co cemented carbides remain nearly the same as those for WC–Co materials. When green bodies are heated up to the melting point, a Co-based liquid phase containing dissolved W, C, Ti, Ta, and Nb forms. Its melting point can be considered as eutectics in the pseudoquaternary system W–C–Ti–Ta (Nb)–Co, which forms at nearly 1360°C for cemented carbides with a medium carbon content; the eutectic temperature also depends on the total carbon content [27], as in the case of WC–Co cemented carbides. Nevertheless, in practice, the carbide grades containing grains of cubic carbides are sintered at slightly higher temperatures compared to the corresponding WC–Co grades due to incomplete wettability of cubic carbides by liquid binders. The formation of cobalt-based liquid phase leads to the shrinkage of the original green body and results in almost full densification in a short initial period of sintering, in spite of the fact that the wettability of cubic carbide grains by liquid is incomplete. As in the WC–Co materials, the liquid phase of the eutectic composition of the Co–W–C–Ti–Ta/Nb system completely decomposes upon cooling and the dissolved atoms precipitate on the existing grains of WC and cubic carbides leading to their insignificant growth. The grains of cubic carbides have an irregular rounded shape related to their incomplete wetting by the liquid Co-based binder, which differs them from the faceted grains of the WC phase.

Some special features of the sintering process of WC–TiC–TaC-(NbC)–Co cemented carbides are associated with the lower solubility of cubic carbides in liquid cobalt compared to tungsten carbide. Nevertheless, in general, the densification mechanisms of the cemented carbides containing cubic carbides are nearly the same as for the WC–Co materials.

Coarsening the grains of cubic carbides occurs by mechanisms of both coalescence and recrystallization by dissolution–reprecipitation in the liquid phase. The coarsening process becomes faster with increasing the sintering temperature, which is related to enhancing both the coalescence process and the dissolution–reprecipitation process, as the solubility of cubic carbides in liquid binders increases with increasing the sintering temperature.

The experimentally established noticeable hindrance of the growth of the cubic carbides' grains in the presence of fine grains of the WC phase can be explained by the fact that such WC grains suppress the coalescence of cubic carbides' grains due to their "mechanical" separation.

According ref. [27], the solubility of cubic carbides with different TaC contents in liquid binders with medium carbon contents decreases from 8 wt.% down to 3 wt.% when increasing the percentage of TaC in the cubic carbide from 5 wt.% up to 22 wt.%. The solubility of cubic carbides at a high carbon content (in the presence of free carbon in the microstructure) becomes much lower (about 1 wt.%) and almost does not depend on the TaC content in the cubic carbides.

It should be noted that the mean grain size of WC grains in cemented carbides containing cubic carbides is lower compared to the straight WC—Co grades with the same Co content, which is related to the inhibiting effect of Ta, Ti, and Nb dissolved in the liquid binder with respect to the WC grain growth.

6.7.5 Special features of sintering cemented carbides with alternative binders

Sintering of cemented carbides with Fe—Ni binders can be conducted at noticeably lower temperatures than that of WC—Co materials (see Fig. 3.12). According to the W—C—Fe—Ni phase diagram (see Chapter 3) at the Fe:Ni ratio of 85:15, the difference between eutectic temperatures at low and high carbon contents is quite significant (correspondingly 1190°C and 1330°C) and noticeably greater than that of the WC—Co materials. The "carbon window" in the WC—Fe—Ni alloys is shifted toward high carbon contents when increasing the iron content [28]. Nevertheless, sintering of cemented carbides with Fe—Ni binders on an industrial scale is not challenging, as the width of the "carbon window" in such cemented carbides is comparable with that of WC—Co alloys. It must be noted that the properties of WC—Fe—Ni cemented carbides are strongly dependent on the amount of WC dissolved in the binder, which in turn depends on the total carbon content. According to Ref. [27], the solubility of WC in Fe—Ni binders does not exceed 8 wt.%.

Sintering parameters of cemented carbides with Fe—Ni—Co binders depend on the Fe:Ni:Co ratio with respect to both the carbon content and sintering temperature, which are usually optimized empirically. In the case of fabrication of coarse and ultracoarse grades, there can be significant difficulties with respect to obtaining the needed coarse or ultracoarse microstructure, as the process of dissolution—recrystallization of fine WC fraction is strongly suppressed due to partial substitution of Co by Ni and Fe, which is illustrated in Fig. 6.31. As it can be seen in Fig. 6.31, the microstructures of the carbide grades with the FeNiCo binders are noticeably finer as a result of using the Fe—Ni—Co binder instead of pure Co. For example, the WC mean grain size of the ultracoarse WC—Co grade, the microstructure of which is shown in Fig. 6.31, is roughly 5.2 μm, whereas that of the WC—FeNiCo grade sintered at the same conditions is equal to nearly 2.9 μm. Also, the control of carbon and WC mean grain size by measurements of magnetic properties (magnetic saturation and coercive force) of WC—FeNiCo alloys is complicated due to completely different dependences of the cemented carbides' properties on their magnetic characteristics, especially at various ratios among Fe, Ni, and Co.

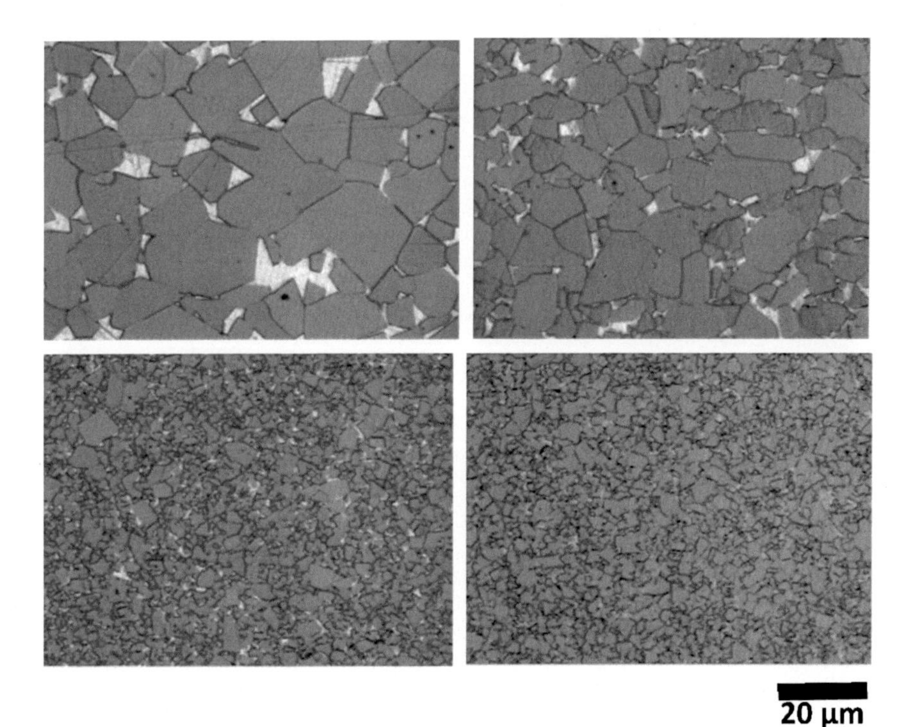

Figure 6.31 Microstructures of ultracoarse grades (above) and medium-grain grades (below) with 6 wt.% binder: right—with the Fe—Ni—Co binder at a ratio of Fe:Ni:Co = 40:40:20 (wt.%) and left —with the Co binder. *(With the permission of Element Six.)*

References

[1] E. Lassner, W.-D. Schubert, E. Lüderitz, H.-U. Wolf, Tungsten, Tungsten Alloys, and Tungsten Compounds. Ullmann's Encyclopaedia of Industrial Chemistry, Wiley-VCH, Weinheim, 2000, https://doi.org/10.1002/14356007.a27_229.

[2] E. Lassner, W.-D. Schubert, Tungsten: Properties, Chemistry, Technology of the Element, Alloys, and Chemical Compounds, Kluwer Academic/Plenum Publishers, New York, 1999.

[3] A.N. Zelikman, L.S. Nikitina, Tungsten, Metallurgiya, Moscow, 1977.

[4] V.S. Panov, I. Konyashin, E.A. Levashov, A. Zaitsev, Hardmetals, MISIS, Moscow, 2019.

[5] I. Konyashin, A. Anikeev, V. Senchihin, V. Glushkov, Development, production and application of novel grades of coated hardmetals in Russia, Int. J. Refract. Metals Hard Mater. 14 (1996) 41—48.

[6] I. Konyashin, T. Eschner, F. Aldinger, V. Senchihin, Cemented carbides with uniform microstructure, Z. Metallkd. 90 (1999) 403—406.

[7] C.J. Terry, Makrocrystalline Tungsten Monocarbide Powder and Process for Production, US patent 4,834,963, 1989.

[8] S. Dunmead, W. Moore, A. Weimer, G. Eisman, J. Henley, Method for Making Submicromeier Carbides, Submicromieter Solid Solution Carbides, and the Material Resulting Therefrom, US Patent 5,380,688, 1995.

[9] S. Gopal, Upadhyaya, Cemented Tungsten Carbides, Production, Properties and Testing, Noyes Publications, New Jersey, 1998.
[10] J.A. Payne, D.W. Smith, Influence of Cobalt on the attritor milling efficiency of WC-Co powder blends, Met. Powder Rep. (December 1990) 847–851.
[11] H. Kolaska, Powder Metallurgy of Hardmetals, German Powder Metallurgical Association (FPM), Hagen, 1992 (in German).
[12] Vauk/Müller, Grundoperationen Chemischer Verfahrenstechnik, 6. Auflage, VEB Deutscher Verlag für Grundstoffindustrie, 1981, pp. 252–255.
[13] H. Schubert, Grundlagen des Agglomerierens, Chem.-Ing.Tech, vol. 51, Verlag Chemie, 1979, pp. 266–277.
[14] J.G. Heinrich, Introduction to the Principles of Ceramic Forming, Verlag CFI, Baden Baden, 2014.
[15] F. Lemoisson, L. Froyen, Understanding and Improving Powder Metallurgical Processes, Fundamentals of Metallurgy, Woodhead Publishing, 2005.
[16] M. Trunec, K. Maca, Chapter 7: Advanced ceramic process, in: Advanced Ceramics for Dentistry, 2014, pp. 123–150.
[17] T.E. Easler, Green Machining, Encyclopaedia of Materials: Science and Technology, second ed., 2001.
[18] R.M. German, Sintering: From Empirical Observations to Scientific Principles, Elsevier, Waltham, MA, 2014.
[19] W. Schatt, Bases of Sintering, VDI Verlag Düsseldorf, 1992 (in German).
[20] H.S. C Cannon, F.V. Lenel, Some Observations on the Mechanism of Liquid Phase Sintering Plansee Proc, 1952, pp. 106–122. Reutte.
[21] W.D. Kingery, M.D. Narasimhan, Densification during sintering in the presence of a liquid phase. II. Experimental, J. Appl. Phys. 30 (1959) 301–306.
[22] E.G. Sandford, E.M. Trent, Symposium of Powder Metallurgy, The Iron and Steel Institute, Special Report, 38, 1947, pp. 84–93.
[23] J. Gurland, Structure and sintering mechanism of cemented carbides, Trans. Am. Inst. Min. Metall. Petrol. Eng. 215 (1959) 601–608.
[24] R.J. Nelson, D.R. Milner, Densification processes in the tungsten carbide-cobalt system, Powder Metall. 15 (30) (1972) 346–363.
[25] J. Gurland, A study of the effect of carbon content on the structure and properties of sintered tungsten carbide-cobalt alloys, J. Met. 6 (1954) (1954) 285–301, 6(AIME Trans. 200).
[26] D. Müller, I. Konyashin, S. Farag, B. Ries, A.A. Zaitsev, P.A. Loginov, WC coarsening in cemented carbides during sintering. Part I: The influence of WC grain size and grain size distribution, Int. J. Refract. Met Hard Mater. (2021) 105714. https://doi.org/10.1016/j.ijrmhm.2021.105714.
[27] V.I. Tretyakov, Bases of Materials Science and Production Technology of Sintered Cemented Carbides, Metallurgiya, Moscow, 1976.
[28] I.N. Chaporova, K.S. Chernyavsky, Structure of Sintered Cemented Carbides, Metallurgiya, Moscow, 1975.

CHAPTER 7

Major methods for controlling cemented carbides in the manufacture

7.1 Examination of cemented carbide microstructures

Investigations of the microstructure of cemented carbides are carried out on metallographic cross-sections, the manufacturing technique of which is described in detail in the monograph [1]. The metallographic examination of the porosity and microstructure of cemented carbides is performed in accordance with the ISO 4499: 2008 standard corresponding to the standard ASTM 6276-52. According to the ISO standard, metallographic studies can be carried out using an optical or scanning electron microscope (SEM) with conventional resolution or high resolution.

Despite the fact that modern technologies of vacuum sintering in a vacuum followed by sintering under pressure in Ar (SinterHIP) make it possible to eliminate the porosity of cemented carbides, the porosity still needs to be controlled, as its presence can be due to, for example, large inclusions of impurities in carbide green articles. The residual porosity is determined as a degree of porosity at a magnification of $\times 100$ or $\times 200$ on large areas of metallurgical cross-sections. The degree of porosity is estimated relative to the area of the pores, expressed as a percentage. The following scales are used to estimate porosity: A - for pores with a size of up to 20 μm and B - for larger pores of up to 40 μm in size. In the conventional end control of carbide products, the determination of density by hydrostatic weighing is also used, which makes it possible to evaluate the porosity of carbide articles (if it is quite significant).

The presence and content of free carbon is determined on unetched metallurgical cross-sections with a magnification of $\times 200$. Usually, the entire surface of the cross-section is examined, since the formation of free carbon can occur locally (for example, in the near-surface zone of a carbide article). Fig. 7.1A shows typical inclusions of free carbon in medium-coarse WC$-$Co cemented carbide.

Cemented Carbides
ISBN 978-0-12-822820-3
https://doi.org/10.1016/B978-0-12-822820-3.00012-4

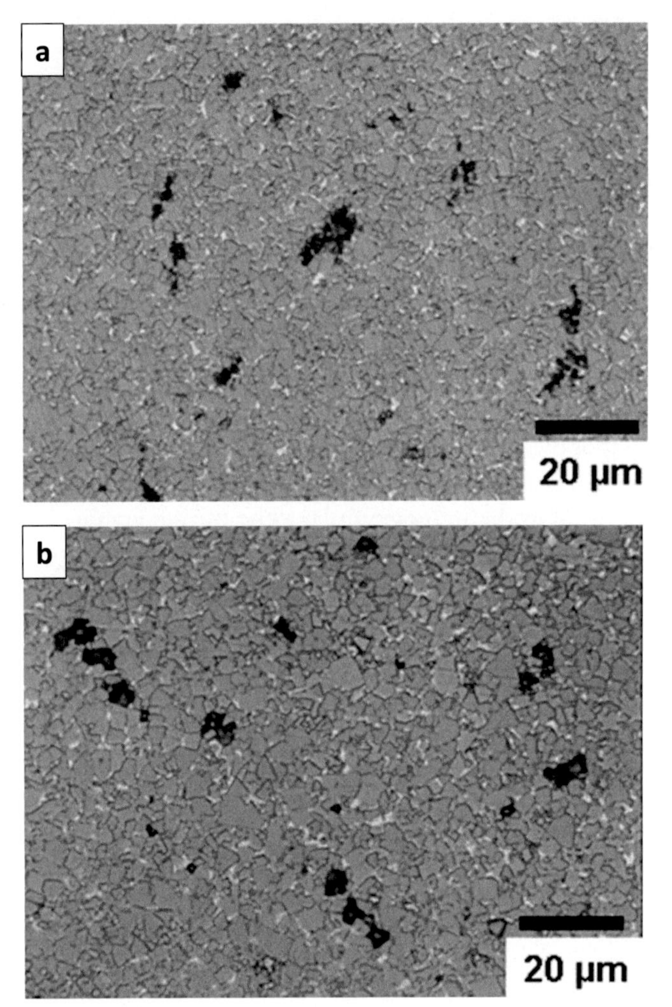

Figure 7.1 Microstructures of medium-coarse cemented carbide comprising typical inclusions of (A) - free carbon, and (B) - η phases (light microscopy, a - before etching in the Murakami reagent, b - after full etching in the Murakami reagent). *(Redrawn from Ref. [2].)*

Afterwards, the cross-section is examined with respect to the presence of η-phase, first, after short etching (for 3—5 s) in a freshly prepared mixture of 10 vol.% calcium hexacyanoferrate and 10 vol.% sodium hydroxide in distilled water (the Murakami reagent). After the short etching, the cross-section is washed in water and placed on a filter paper for drying. It is important to very carefully dry the cross-section surface, eliminating any friction on the filter paper, since the colored inclusions of the η-phase are present in a very thin near-surface layer of the cross-section and can be easily removed by rubbing against the filter paper. The cross-section is inspected on an optical microscope, first at a magnification of ×100 or ×200, and then at a larger magnification (×1000 or higher), since small inclusions of η-phase are often formed along the WC/Co grain boundaries and are therefore not visible at small magnifications. If the microstructure comprises inclusions of η-phase, they become yellowish as a result of the short etching.

After the short etching, longer etching (usually 3–5 min depending on the WC mean grain size and binder content) is carried out to examine the uniformity of distribution of the WC phase and binder phase in WC–Co alloys as well as the phases of cubic carbides (Ti,Ta,W)C and (Ti,Ta,Nb,W)C in the carbide grades for steel-cutting. The η-phase inclusions are completely etched away and become dark after such long etching (Fig. 7.1B). According to the ISO standard, the magnification is chosen based on the WC mean grain size of the carbide grade, so that at least 10–20 carbide grains are visible in the area being examined. It is necessary to examine the entire surface of the cross-section, as η-phase can be formed locally. The shape of the η-phase inclusions (pool-like, dendritic, along grain boundaries, etc.) must be distinguished, as it usually depends on the carbon content in the carbide sample, which can be seen in Fig. 7.2.

It is possible to employ optical microscopy or SEM by the use of computer software to determine the WC mean grain size. Nevertheless, the most accurate method for determining the mean grain size of carbide phases is the electron backscattering diffraction (EBSD) method. This method allows the separation of grains (single crystals) of carbide phases with high accuracy and eliminates inaccuracies in determining the grain size of the carbide phases due to the inability to precisely detect WC–WC grain boundaries by the use of an optical microscopy and SEM. Fig. 7.3 shows EBSD images of the microstructure of two different carbide grades having various WC grain sizes. Usually, the employment of even high-resolution SEMs at high magnifications does not allow the exact detection of separate WC single crystals. In many cases, two adjacent WC grains are visible as one single crystal, so that the value of WC mean grain sizes of carbide samples obtained by use of SEM is noticeably larger than the value precisely established by the EBSD method. The difference between the values of WC mean grain sizes obtained by these two methods is quite significant (nearly 30%–40% depending on the type of cemented carbides being examined).

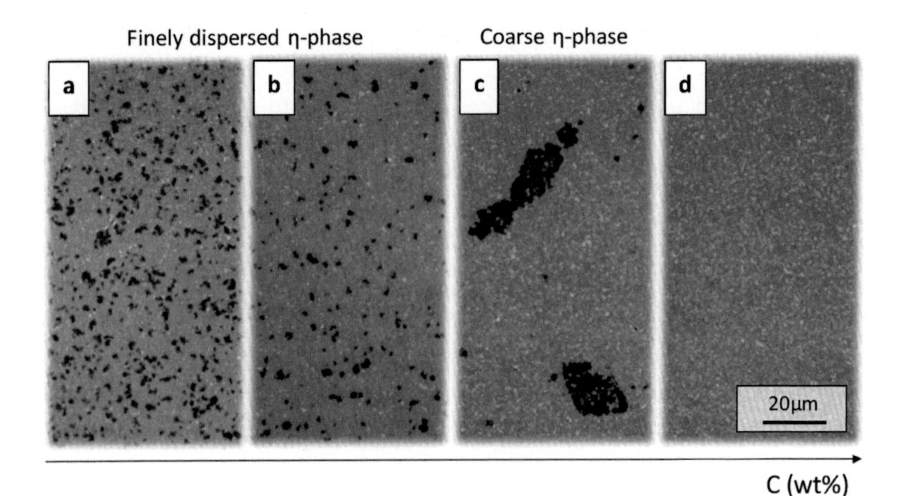

Figure 7.2 Microstructures of WC–8% Co cemented carbides having different carbon contents. Finely dispersed η-phase is present in the samples at a very low (A) and medium-low (B) carbon content. Large dendritic inclusions of η-phase frequently form in the microstructure at a reduced carbon content close to the border line corresponding to the η-phase formation (C). The η-phase free microstructure is shown in (D) for comparison. *(Redrawn from Ref. [3].)*

To precisely examine the phase of cubic carbides (Ti,W)C, (Ti,Ta,W)C, and (Ti,Ta,Nb,W)C in the carbide grades for steel-cutting, etching in a mixture of hydrofluoric acid and nitric acid in a ratio of 1:2 for 15–20 min is used [1].

In the recent times, transmission electron microscopy (TEM) and high-resolution transmission electron microscopy (HRTEM) became very important tools for examining fine structure of cemented carbides on the nano- and atomic level. They allow the investigation of crystal lattice defects present in both the carbide and binder phase as well as at the WC/Co and WC/WC grain boundaries. HRTEM ensures obtaining images of single crystals and grain boundaries with atomic resolution enabling interfacial effects including the segregation of grain growth inhibitors at the grain boundaries and thin structurally ordered interface layers of few atomic monolayers usually designated in literature as complexions. An HRTEM image of a carbide sample containing chromium carbide is shown in Fig. 7.4 [5]. More sophisticated TEM techniques, for example, in situ examinations of deformation processes in the carbide and binder phase of cemented carbides directly in a TEM are described in Chapter 15.

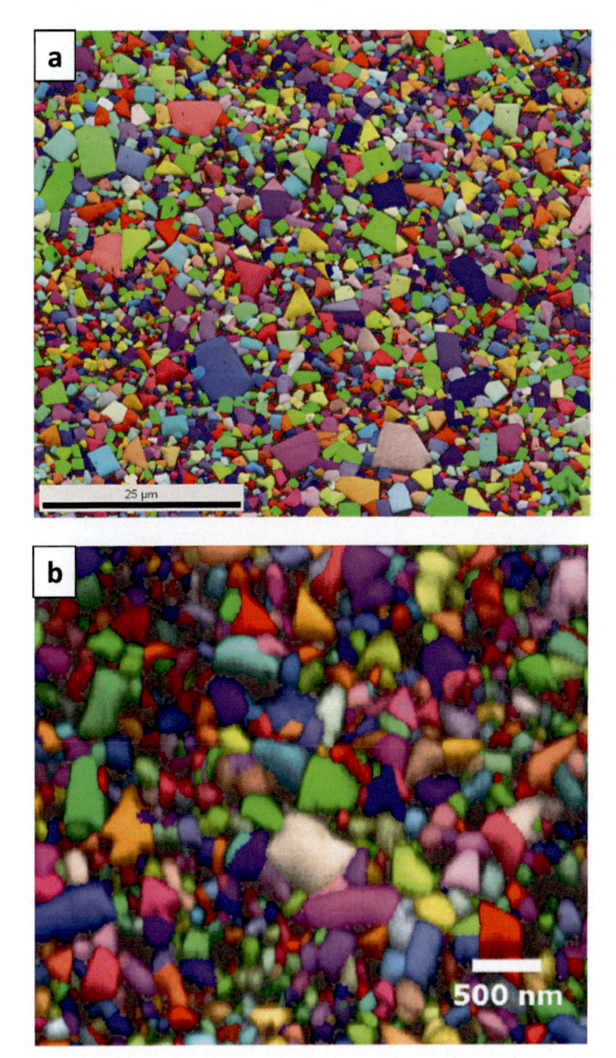

Figure 7.3 EBSD images of the microstructures of (A) medium-grain carbide grade and (B) near-nano carbide grade. *(Redrawn from Ref. [4].)*

7.2 Examination of magnetic properties

7.2.1 Measurement of the coercive force

Traditionally, the coercive force is the main magnetic property characterizing the structure of cemented carbides. Its value depends directly on the thickness of cobalt binder interlayers in cemented carbides and is therefore

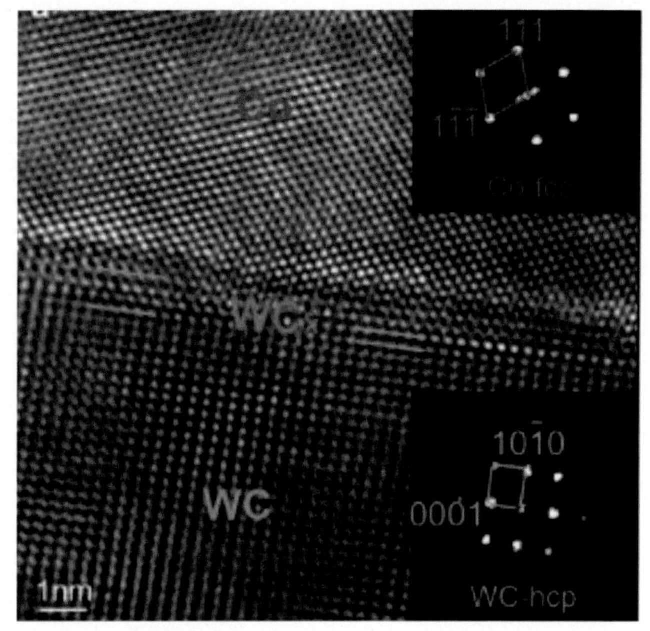

Figure 7.4 HRTEM image of the WC–Co cemented carbide with an addition of chromium carbide quenched from 1390°C indicating a complexion at the WC/Co interface. *(Redrawn from Ref. [5].)*

used to estimate the WC mean grain size at a certain Co content. The coercive force is widely used as a nondestructive quality control method to evaluate the quality of serial products in the carbide industry.

The mean grain size of tungsten carbide in WC–Co alloys can be calculated using the equation obtained in Ref. [4]:

$$d = \frac{236.7(\text{Co wt.} - \%)^{-1.2}}{H_c - 3.3}$$

where d is the average grain size in μm, Hc is the coercive force in kA/m, Co wt.% is the percentage of Co by weight.

It should be noted that the above equation can be used to determine the WC mean grain size only in cemented carbides having a uniform microstructure characterized by the absence of abnormally large carbide grains. If the microstructure is characterized by a high degree of bimodality, i.e., it comprises both a fine-grained WC fraction and very large WC grains, this equation has a limited application. Also, as it was mentioned above, the absolute value of the WC mean grain size strongly depends on the method employed for its measurement (HRSEM, EBSD, etc.).

7.2.2 Measurement of the magnetic moment or magnetic saturation

Another important parameter widely used for the production control of cemented carbides in the carbide industry is the magnetic moment or magnetic saturation. The specific magnetic moment of pure cobalt σ_{Co} (magnetic moment per unit mass) is 16.1 mT m^3/kg, and the specific magnetic saturation of cobalt $4\pi\sigma_{Co}$ is, respectively, 201.9 mT m^3/kg. When Co is mixed with nonmagnetic grains, which are in the case of WC−Co materials tungsten carbide grains, or alloyed with a nonmagnetic metal, the magnetic moment/magnetic saturation of cobalt decreases when increasing the amount of the nonmagnetic grains or the nonmagnetic metal. A powder of WC−Co mixture containing 10 wt.% pure Co is therefore characterized by the magnetic moment equal to one tenth of that of pure Co, i.e., 1.61 mT m^3/kg. After sintering a carbide article, the Co-based binder contains various amounts of dissolved nonmagnetic tungsten, which in turn are determined by the carbon content in the article in such a way that higher concentrations of dissolved tungsten correspond to lower total carbon contents. When tungsten is dissolved in the binder phase due to a decreased carbon content, the magnetic saturation of the binder decreases and becomes equal to the value calculated according to the equation given in Ref. [6]:

$$\sigma_{binder} = \sigma_{Co} - 0.275\ M_W$$

where σ_{binder} is the specific magnetic moment of the Co-based binder, σ_{Co} is the specific magnetic moment of pure cobalt, and M_W is the concentration of tungsten dissolved in the binder in weight percent.

Thus, the magnetic moment (magnetic saturation) of the binder and, accordingly, of the sintered carbide article decreases when increasing the concentration of tungsten dissolved in the binder and, correspondingly, decreasing the carbon content. When the carbon content decreases and reaches the value corresponding to the formation of the η-phase in the microstructure, a further increase in the concentration of tungsten dissolved in the binder does not occur anymore and a further decrease in the magnetic moment seems not to take place. In fact, with a further decrease in the carbon content after reaching the value corresponding to the formation of the η-phase, the decrease in the magnetic moment of the carbide article continues. This occurs due to the fact that part of Co is bound in form of the nonmagnetic η-phase, the amount of which increases as a result of a decrease in the carbon content. In the extreme case of a carbide sample with a very low carbon content, when all Co is bound to the nonmagnetic η-phase, the value of the magnetic moment drops down to zero.

With the increase in the carbon content in cemented carbides toward the value corresponding to the free carbon formation, the magnetic moment increases as a result of a decrease in the content of tungsten dissolved in the binder. After reaching the carbon content value corresponding to the formation of free carbon, the content of tungsten dissolved in the binder becomes very low and remains unchanged; as a result, the magnetic moment also almost does not change.

When taking into account the mentioned above, measurements of the magnetic moment (magnetic saturation) of cemented carbides can be employed to evaluate the total carbon content in carbide samples after sintering, which is very important with respect to obtaining the optimal carbon contents for different types of cemented carbides (see Chapter 9).

For cemented carbides with partial or complete replacement of cobalt by nickel, the evaluation of the total carbon content by measuring the magnetic saturation is impossible or limited due to its complicated dependence on the carbon content [6].

7.3 Examination of mechanical properties in the carbide manufacture

Among numerous tests of examining different mechanical properties of cemented carbides summarized in the monograph [7], only the most important test procedures for examining the major mechanical properties of cemented carbides that are widely used in the carbide industry are described below. These tests can be supplemented by various types of measurements of mechanical properties of cemented carbides (for example, the determination of compressive strength according to ISO 4506). They are usually commonly employed in research laboratories for the development of new carbide grades for special applications.

7.3.1 Transverse rupture strength

Determination of the transverse rupture strength (TRS) is conducted in accordance with the ISO standard ISO 3327: 2009. Historically, according to the original version of the ISO standard, the TRS was tested on samples of two types—type A and type B. Samples of type A had dimensions of 5 mm × 5 mm × 35 mm, whereas samples of type B had dimensions of

6.5 mm × 5.25 mm × 20 mm. Later, the ISO standard was supplemented with cylindrical samples of type C with a length of 25 mm and a diameter of 3.3 +/− 0.5 mm. At the majority of carbide producers, the TRS of each carbide batch is determined.

7.3.2 Vickers hardness

Vickers hardness is determined in accordance with the ISO standard ISO 3878. Usually, loads of 20 or 30 kg are used. In special cases (for example, in the production of functionally graded cemented carbides), the employment of smaller loads (10 kg or less) is possible.

7.3.3 Rockwell hardness

The Rockwell hardness test is performed in accordance with the ISO standard ISO 3738-1.

7.3.4 Palmqvist fracture toughness

The determination of fracture toughness is based on the results of measuring the total length of the Palmqvist cracks formed near the corners of the Vickers indentations according to the ISO standard ISO 28079: 2009. For most carbide grades, loads of 30 kg are used. For coarse and ultracoarse grades characterized by high values of fracture toughness, it is necessary to use higher loads (of 100 kg and more) to obtain the measurable Palmqvist cracks. According to the ISO standard, before the indentation, the metallurgical cross-sections must be annealed at 800°C for 1 h in a vacuum or an inert atmosphere to remove the residual compressive stresses resulting from the grinding process.

7.4 Examination of wear resistance of cemented carbides

Determination of abrasive wear resistance of cemented carbides is carried out in accordance with ISO standard ISO 28079: 2009. The standard includes two types of testing the wear resistance of carbide specimens corresponding to (1) the ASTM G65 standard and (2) the ASTM B611 standard. The test of the first type includes abrading the surface of carbide specimens by round particles of quartz sand, supplied to the surface of a rotating rubber wheel; the test procedure is schematically shown in Fig. 7.5. The test of the second type comprises abrasion of the surface of carbide

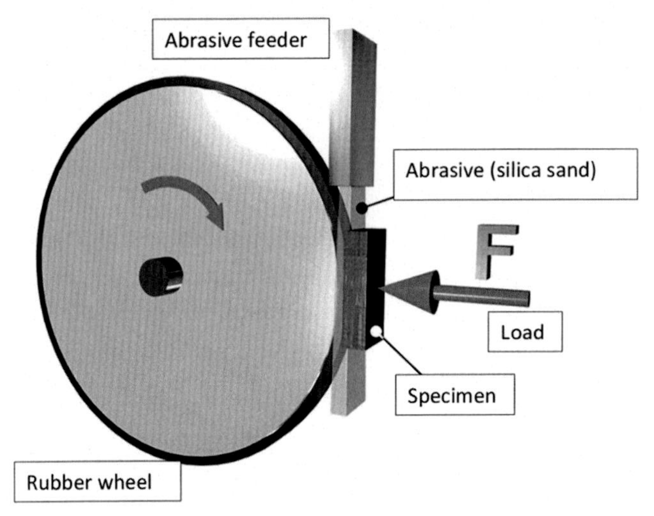

Figure 7.5 Scheme of the test on wear resistance in accordance with the ASTM G65 standard.

specimens using a steel wheel that is partially immersed in a suspension of alumina grit in water. The steel wheel is designed in such a way to provide the supply of alumina particles to the surface of the carbide specimen during rotation; the test procedure is schematically shown in Fig. 7.6.

It should be noted that the wear test of the first type is very gentle, as it includes scratching the carbide surface by rounded silica particles at low

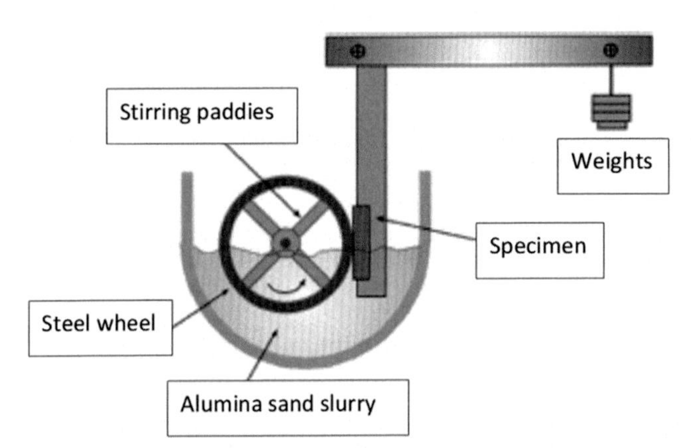

Figure 7.6 Scheme of the test on wear resistance in accordance with the ASTM B611 standard.

loads; it can therefore simulate the performance of cemented carbides in wear applications characterized by the absence of impact loads. In contrast, in the wear test of the second type, the tested specimens are subjected to relatively high loads and severe fatigued due to high-energy impacts of alumina particles, so that this test can simulate the performance of cemented carbides under high impact loads, for example, in mining and construction applications.

7.5 Examination of high-temperature properties of cemented carbides

Different tests of high-temperature properties of cemented carbides are described in Ref. [7] in detail. Almost all the examinations of mechanical properties including the TRS, compressive strength, etc., can be conducted at elevated temperatures, so that they are employed at research laboratories for developing new carbide grades operating at high temperatures, for example as anvils and dies in the diamond and cubic boron nitride production, for cutting Ni-based superalloys performed at high cutting temperatures, etc. In this respect, the most important high-temperature tests are those on examining the hot hardness at temperatures of up to 800°C, high-temperature compressive strength at temperatures of up to 1000°C, and creep behavior at temperatures of up to 1200°C; typical experimental conditions employed for such tests of WC—Co cemented carbides and tests' results can be found in Refs. [8,9].

References

[1] I. Chaporova, K. Chernyavsky, Structure of Sintered Cemented Carbides, Metallurgiya, Moscow, 1978.

[2] I. Konyashin, B. Ries, F. Lachmann, A.T. Fry, Gradient hardmetals: theory and practice, Int. J. Refract. Metals Hard Mater. 36 (2013) 10—21.

[3] J. García, V. Collado, C.A. Blomqvist, B. Kaplan, Cemented carbide microstructures: a review, Int. J. Refrac. Metals Hard Mater. 80 (2019) 40—68.

[4] I. Konyashin, B. Ries, D. Hlawatschek, et al., Wear-resistance and hardness: are they directly related for nanostructured hard materials? Int. J. Refract. Metals Hard Mater. 49 (2015) 203—211.

[5] X. Liu, X. Song, H. Wang, X. Liu, F. Tang, H. Lu, Complexions in WC-Co cemented carbides, Acta Mater. 149 (2018) 161—178.

[6] B. Roebuck, Magnetic moment (saturation) measurements on hardmetals, Int. J. Refrac. Metals Hard Mater. 14 (1996) 419—424.

[7] G.S. Kreimer, Strength of Hard Alloys, Consultants Bureau, New York, USA, 1968.

[8] B. Roebuck, S. Moseley, Tensile and compressive asymmetry in creep and monotonic deformation of WC/Co hardmetals at high temperature, Int. J. Refrac. Metals Hard Mater. 48 (2015) 126–133.
[9] I. Konyashin, S. Farag, B. Ries, B. Roebuck, WC-Co-Re cemented carbides: structure, properties and potential applications, Int. J. Refract. Metals Hard Mater. 78 (2019) 247–253.

Further Reading

[1] A. Love, S. Luyckx, N. Sacks, Quantitative relationships between magnetic properties, microstructure and composition of WC–Co alloys, J. Alloys Compd. 489 (2) (2010) 465–468.

CHAPTER 8

Final processing of carbides articles and deposition of wear-resistant coatings

8.1 Grinding

According to the standard DIN 8589 [1] (the third main group "Cutting" under the subtitle "Processing with geometrically indeterminate cutting edge"), grinding can be designated as a cutting process employing tools comprising multicutting edges. The grinding process occurs as a result of cutting workpieces by nonregularly formed and randomly oriented grains of hard or superhard materials, which are bound by a binder matrix. The grinding tools rotate resulting in microcutting processes due to the interaction of the abrasive grains with the workpiece thus removing the workpiece near-surface layer at high rates. The shape of the abrasive grains continuously changes, and after certain wear, they are detached from the grinding tool, which affect the dimensional accuracy and surface quality of the workpiece. Only about one third of the abrasive grains located in the surface zone of the grinding tool (grinding wheel) are used in the process of material removal of the workpiece, while about one third of the abrasive grains rub over the surface without removing the workpiece near-surface layer, and the remaining abrasive grains cause the plastic deformation of the workpiece surface layer without its cutting [2] (Fig. 8.1). It was established that predominantly negative cutting edges of the abrasive grains cause the microcutting processes [3].

Grinding carbide articles by diamond grinding wheels ensures obtaining high-dimensional accuracy and smooth surfaces. Such diamond grinding wheels consist of fine diamond particles bonded by metallic, resin, or ceramic binder matrixes. Usually, only near-net shape carbides parts are subjected to grinding due to high production costs of the grinding process, which allows reducing grinding times and consequently production costs.

Cemented Carbides
ISBN 978-0-12-822820-3
https://doi.org/10.1016/B978-0-12-822820-3.00006-9

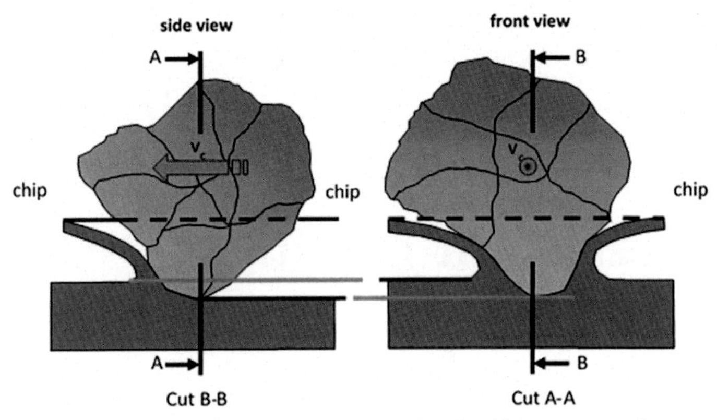

Figure 8.1 Process of chip formation as a result of rubbing, squeezing, and microcutting the workpiece by abrasive particles [2].

Grinding of cemented carbide articles by nonregularly formed, randomly oriented diamond grains illustrated in Figs. 8.1 and 8.2 leads to simultaneous rubbing, squeezing, and microcutting actions in the surface layer of the workpiece. This effect results in obtaining numerous chips having different shapes and sizes affecting the roughness of the workpiece surface and formation of abrasion marks.

The presence of uneven protruding, irregularly shaped, and aligned abrasive grains with uneven cutting edges leads to the constantly changing material removal behavior during grinding illustrated in Fig. 8.2 resulting in a continuous change in the chips' size and geometry. Generally, it is necessary to adapt the grinding wheel material to the cemented carbide

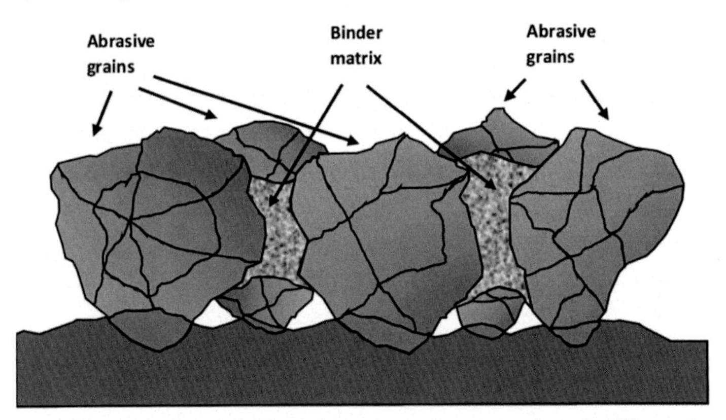

Figure 8.2 Schematic diagram illustrating irregular shaped and differently aligned abrasive grains embedded in the binder matrix; steadily changing cutting edges of such grains results in the constantly varying performance of the grinding wheel [2].

properties, which ensures the optimal material removal rate affected by processes of breakages and wear of the abrasive grains under the influence of severe fatigue and high loads occurring during grinding. Fresh sharp cutting edges of the abrasive grains forming due to their microchipping allow reducing cutting forces during grinding, which ensures consistent process conditions. The abrasive grains are firmly mechanically bound in the binder matrix until their significant wear rate is achieved. As a result of detachment of the worn abrasive grains from the upper layer of the graining wheel, the lower layer starts working leading to the "self-sharpening effect" which is an important factor affecting the effectiveness of the grinding process. This effect must be taken into account when selecting the optimum wheel material.

Conventional grinding methods employed in the cemented carbide industry can be divided into the following main groups:

- Inner diameter grinding
- Outer diameter grinding
- Flat grinding
- Tool grinding
- Double disc grinding
- Centerless grinding
- Honing

Various special, process-specific grinding systems, such as CNC (computerized numerical control) grinding machines for internal and external grinding, machines for flat grinding, tool grinding machines, double disc grinding machines, centerless grinding machines, and honing machines, are available on the market. One of such machines is shown in Fig. 8.3 (above); Fig. 8.3 (below) also illustrates various methods of grinding cemented carbide articles and tools mentioned above. The CNC internal and external grinding machines are equipped with special grinding wheels, control systems, and mainly employed to grind round or elliptical workpieces. Flat grinding machines are equipped by a moving workpiece mounting table located under the grinding wheel; the table moves longitudinally and/or transversally and is equipped with magnetic and/or mechanical clamping devices. The grinding wheels used in flat grinding are generally relatively large in order to obtain high material removal rates. Double disc grinding machines are equipped with two large horizontal grinding wheels with special profiles and orbital systems that ensure the workpiece to be rotated between the grinding wheels. The high productivity of the double discs grinding machines enables highly effective and economical processing carbide parts with large surface areas. The tool

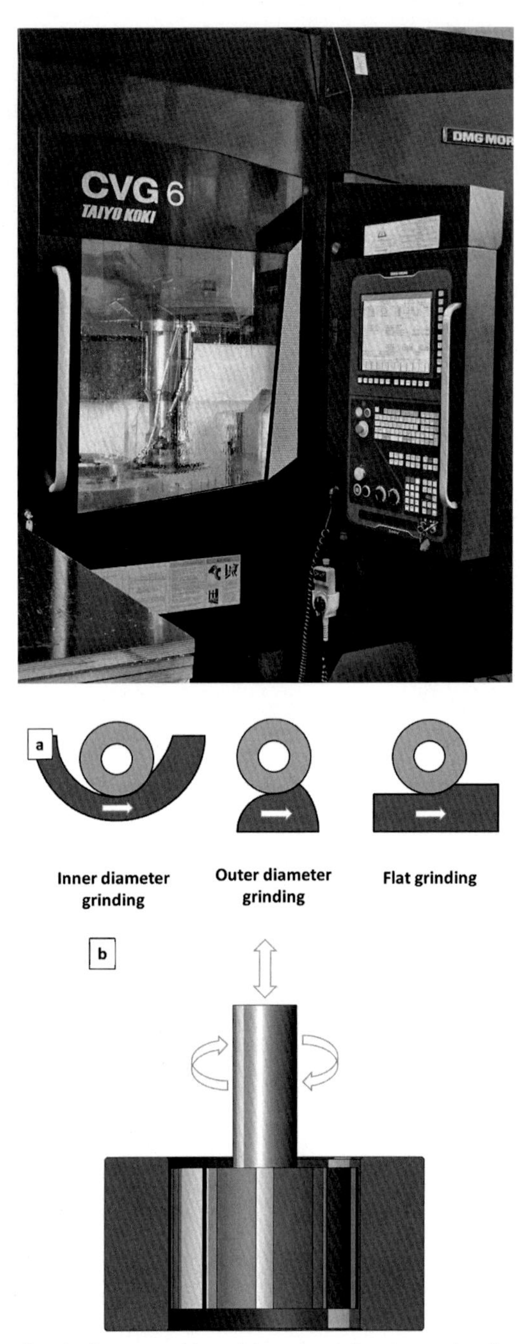

Figure 8.3 Typical grinding machine (above) and schematic drawings illustrating various technologies of grinding (below): (A) grinding by a single grinding wheel, (B) honing, (C) centerless grinding, (D) tool grinding, and (E) double-disk grinding. *(With the permission of Element Six.)*

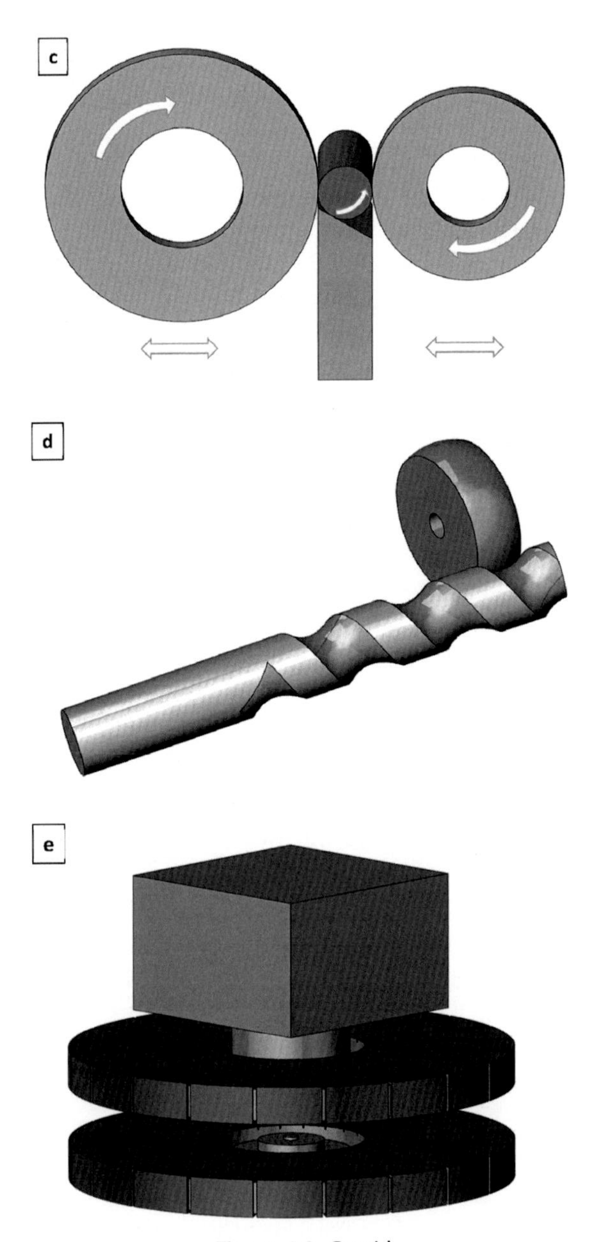

Figure 8.3 Cont'd

grinding machines comprise complex workpiece clamping systems, grinding tool exchangers, high pressure cooling systems, and modern electronic control systems. The combination of the clamping system with tool exchanger allows grinding of carbide parts having complex geometries in a single operation. The honing machines comprise rigid holder systems and rotating spreadable honing devices, which are specially designed to improve the precision and surface quality of carbide parts being ground. The honing machines are employed to effectively grind holes and borings that must have a high precision of diameter and surface roughness. The centerless grinding machines comprise a rigid frame, grinding wheel, and control wheel; they can be equipped with a manual or automated parts' feed device. Grinding between two rotating wheels (the grinding and control wheel) is usually employed for processing cylindrical and rotationally symmetrical carbide parts.

It should be noted that the operation of machines for grinding high-quality carbide part with high tolerances and surface quality requires well-trained and experienced operators. The design of the grinding machine combined with the process parameters optimized with respect to the concrete carbide part, the type of the grinding wheel used, and process control combined with the appropriate quality assurance measures ensure the basis for producing high-quality carbide articles with complicated geometry as a result of grinding.

The workpieces are mechanically or pneumatically clamped into the grinding machine chucks and subsequently ground. Grinding processes can be grouped by the relative movements of the grinding wheel and workpiece. Either the workpiece moves forward in the direction of the rotating grinding wheel building up pressure on the workpiece surface as a result of its interaction with the grinding wheel or the rotating grinding wheel moves toward the workpiece and pushes against it with at a certain pressure and feed rate. In some cases, mostly with the use of the double-sided planar grinding machines, the removal rate is determined by the grinding pressure.

Fast rotation of the grinding wheels allowing high surfaces speeds, which are employed in the majority of applications, ensures high rates of the material removal from the surface layer of carbide parts. The content, shape, abrasive grains' size, abrasive grains' concentration, and binder material determine the performance of the grinding wheels. The presence of coarse abrasive particles enables higher material removal rates resulting in a rougher workpiece surface combined with an increased heat input, thus resulting in shorter life of the grinding wheel. An increased concentration of abrasive grains embedded in the binder matrix requires higher grinding forces resulting in higher contact temperatures and prolonged life of the grinding wheel, which leads to obtaining smoother workpiece surfaces.

Efficient cooling when grinding carbide parts is crucial for obtaining high-quality surfaces as a result of grinding. Depending on the application the cooling systems employing oil, special emulsions or water are used. Cooling is typically achieved by spraying the grinding wheel surface with the needed cooling fluid under pressure. The pressure should be sufficiently high and the cooling fluid must be fed to the most important area of the carbide part being ground to reach the optimal cooling effect. Insufficient cooling at the grinding wheel/workpiece contact point can result in damaging the carbide part.

Modern grinding machines can be equipped with integrated measuring systems allowing the precise measurement of the carbide parts in the clamped state, which enables the possibility of direct and uncomplicated regrinding if necessary. After grinding and subsequent measurement, the carbide parts are removed from the grinding machine, cleaned, and prepared for final inspection.

Grinding parameters and types of the grinding wheel must be chosen and optimized with respect to each grinding process to achieve the best characteristics of the grinding process (productivity, efficiency, surface quality of carbide parts, etc.). Nowadays, typical rotation speeds of grinding wheels employed in the cemented carbide industry vary from nearly 20 m/s up to 120 m/s. Technically, it becomes now possible to achieve very high values of the rotation speed—on the level greater than the speed of sound [4]. Such high-speed grinding processes could allow obtaining extremely higher productivity rates combined with prolonged tool life of the grinding wheels. Nevertheless, such highly efficient grinding processes require perfectly balanced grinding wheels, higher cooling fluid pressures, and safe housing in the event if the grinding wheel breaks.

Nowadays, process simulations (digital twins) are widely employed for the grinding process optimization. The process simulation leads to an optimal grinding strategy, including all relevant process parameters with the expected process duration, which allows improved process reliability. An example of the process simulation and optimization is illustrated in Fig. 8.4.

Grinding wheels for grinding carbide parts often comprise diamond grits varying from 15 to 256 μm in size at concentrations varying from 15 vol.% to about 32 vol.%; such diamond grits are embedded in the binder of synthetic resin, ceramics, or metals. The forces induced by the grinding wheel during the grinding process result in obtaining compressive stresses in the near-surface zone of carbide parts being ground and have a significant impact on their durability, toughness, and transverse rupture strength; such

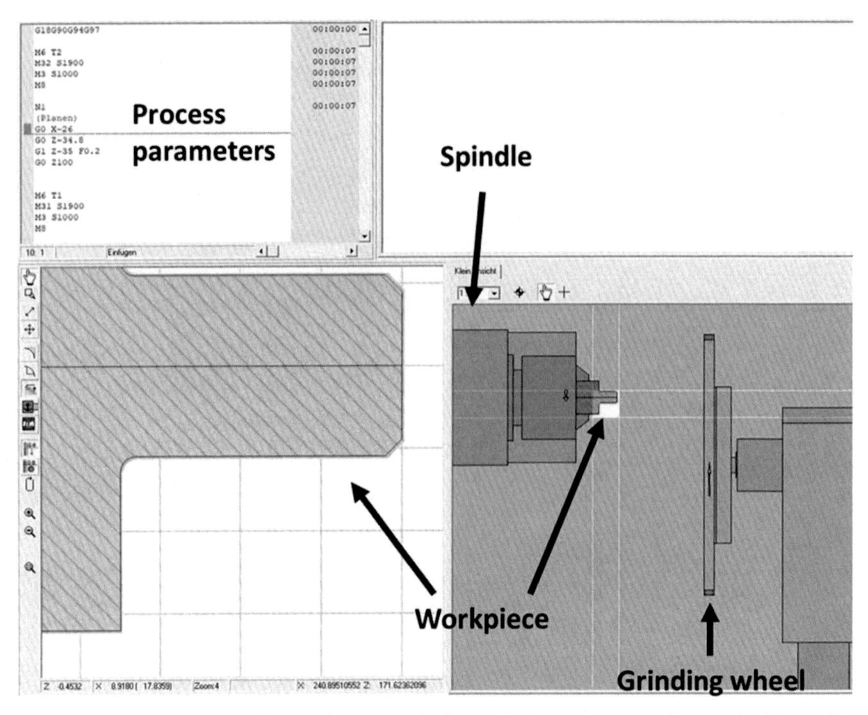

Figure 8.4 Illustration of simulations of the grinding process for optimizing the grinding process parameters. *(With the permission of Element Six.)*

compressive stresses can be measured by XRD. Typical values of the residual compressive stresses achieved as a result of grinding are in the range of approximately 1800 MPa according to Ref. [5].

Modern grinding technologies allow producing carbide parts with complicated geometries shown in Fig. 8.5 and ensure obtaining the very high precision of detentions (of the order of several micrometers) and exceptionally low surface roughness (of the order of 0.1 micron).

Complex carbide wear parts, which are shown in Fig. 8.5, produced with the aid of the grinding systems described above are employed in numerous branches of industry such as mechanical engineering, automotive, aerospace, oil and gas, civil engineering, etc.

8.2 Brazing and assembling

Several joining technologies, allowing steel and carbide articles to be joined, are employed nowadays in the carbide and tool industries on a large

Figure 8.5 Typical ground cemented carbide parts with complex geometries. *(With the permission of Element Six.)*

scale. Each joining technology is characterized by certain advantages and disadvantages depending on the needed properties of the final product.

The following list presents summaries of the commonly used brazing and assembling methods employed in the cemented carbide industry:

- Brazing
  - furnace
  - induction
  - flame
- Shrink fitting
- Gluing

8.2.1 Brazing

Generally, brazing is a process of joining two articles of metals, in which a layer of a low melting point nonferrous alloy (braze alloy) is introduced between the articles and melted; as a result of its cooling and solidification, the articles are joint to each other. In composite tools and wear parts, a cemented carbide article must be brazed to a steel body. The melting

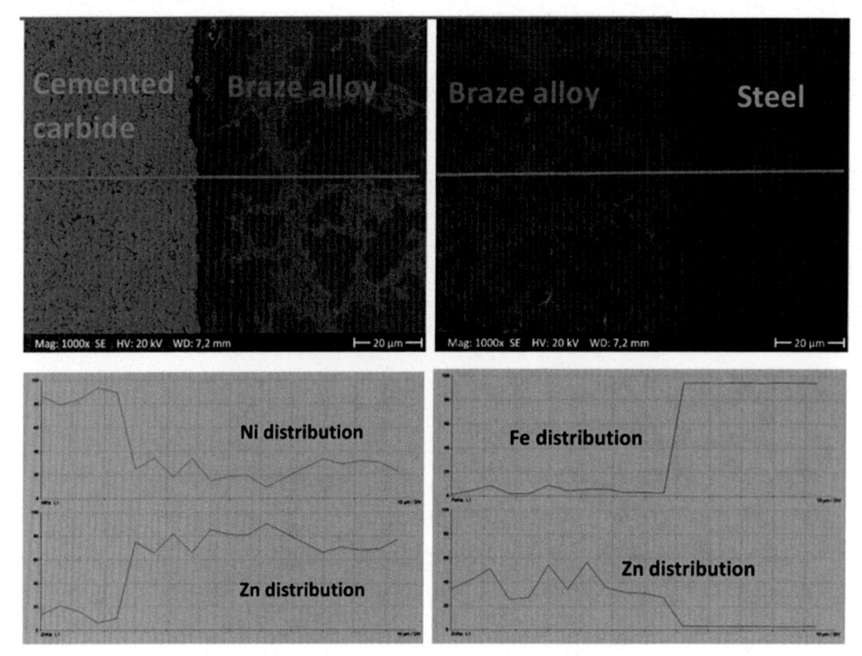

Figure 8.6 SEM images of a WC–Ni cemented carbide article and steel component as well as a layer of braze alloy between them (above), and the corresponding curves indicating the distributions of Zn, Ni, and Fe at the interfaces: cemented carbide/braze alloy and steel/braze alloy (below). The curves were obtained by EDX through the scan line shown on the SEM images by red color (gray in print). *(With the permission of Element Six.)*

temperature of the applied braze alloy lies in the range between about 680°C and 1100°C in the case of cemented carbides' soldering, which is noticeably lower than the melting point of the W–C–Co eutectics. The braze alloy is distributed between the surfaces being brazed by capillary forces. In order to obtain a strong joint between the dissimilar cemented carbide and steel articles, there is a need for diffusional processes. Fig. 8.6 shows SEM images of the WC–Ni cemented carbide article and steel article, and a layer of braze alloy between them as well as the corresponding curves indicating the distributions of Zn, Ni, and Fe at the cemented carbide/braze alloy and cemented carbide/steel interfaces. One can see significant rates of diffusion of the Zn-based braze alloy into the cemented carbide (left) and steel (right).

Brazing is widely employed as a standard process intended for flexible producing high-integrity cemented carbide/steel composite tools and wear parts. Brazing in the cemented carbide industry is usually referred to as "hard brazing" as the typically employed braze alloys are characterized by high melting points and elevated strength values. The optimal join gap design between the carbide and steel articles being brazed guarantees the high shear strength of the brazed joint ensuring high performance properties and reliability of the brazed tools and wear parts.

When heating the steel/braze/carbide composite having different material-related thermal expansion coefficients of its components at temperatures lying above the braze alloy melting point, the components expand resulting in their different lengths. However, after the solidification of the braze alloy during the cooling process, a firm bond between the carbide and steel articles is achieved. Since the steel and carbide articles contract at different rates, some distortion of the carbide/steel composite occurs, which causes tensile stresses in the carbide article. Therefore, failures of the cemented carbide article, such as cracks or breakouts, can occur. In order to reduce the residual stresses in the carbide articles, suitable buffer layers having a thermal expansion coefficient in-between of steel and cemented carbide can be applied [6–9].

An intensive surface cleaning which allows removal of oil, dirt, or oxide films, and the optimum clearance between the mating articles are essential to achieve a strong brazing bond. Instead of using a protective gas atmosphere, flux can be employed to protect the mating surfaces against oxidation and improve their wettability by the liquid braze alloy.

Typical shear forces of braze joints, which are achievable when conducting the brazing process at optimal conditions, lie between about 150 and 300 N/mm^2 depending on the braze alloy, braze joint thickness, cemented carbide grade the steel grade as well as the surface conditions of the articles being brazed. In general, the wettability of cement carbides by different braze alloys becomes better when the binder content increases. Optimized cooling conditions allow obtaining high-performance properties of the braze joints leading to improved performance of the brazed tools and wear parts during operation.

Cemented carbide composite tools are successfully employed under high shear stresses in applications characterized by high impact loads and severe fatigue, such as tools for mining, stone crushing, mineral processing, soil processing, road milling, etc., which are shown in Fig. 8.7. Industrial brazing systems can be grouped into (1) furnace brazing, (2) flame brazing, and (3) induction brazing. Brazing furnaces are typically comparable with vacuum sintering furnaces.

FEA (finite element analysis) modeling (see Chapter 14) is frequently applied to optimize the braze joint design of different brazing processes.

Figure 8.7 Typical induction-brazed mining and road-planing tools. *(With the permission of Element Six.)*

The quality of brazing is strongly affected by the production conditions that must be optimized for each type of tools or wear parts being brazed. Special destructive tests, like shear or impact tests, are useful methods for controlling the quality of the braze joints. Ultrasonic control of the braze joints is also possible; however, it requires the careful preparation, probe selection, and optimized test conditions.

Over the past decades, a variety of braze alloys were developed in order to cover different applications of brazed tools and wear parts characterized by specific requirements (the melting point, braze thickness, shear resistance, corrosion behavior, etc.). Table 8.1 provides a general overview of typical braze alloys for producing cemented carbide tools and wear parts operating at different temperatures and having various compositions and properties.

8.2.2 Furnace brazing

Furnace brazing is widely used for processing carbide composite tools mainly due to its well controllable conditions. The optimal brazing conditions allow furnace brazing to be employed for the mass production of brazed tools and wear parts on a large scale. Slow temperature up- and down ramp cycles in a vacuum or inert gas atmosphere ensure brazing large parts and obtaining reduced macrostresses and surface oxidation.

In most cases, special brazing furnaces are applied for joining cemented carbide/steel parts. Such furnaces usually consist of a double wall

Table 8.1 General overview of typical braze alloys for producing various steel/carbide composites.

Braze alloy	Working temperature [°C]	Typical compositions	Properties
Silver braze	680–710	Ag (40%–70%); Cu; Zn; Mn; Ni	Good flow behavior, low melting point, good wettability, possible copper interlayer to reduce stress
Eco silver braze	700–730	Ag (20%–30%); Cu; Mn; Ni; In	Cost efficient, good wettability, low melting point, possible copper interlayer to reduce stress
Copper based braze	980–1100	Cu; Mn; Co	Low costs, hardening after brazing possible
Brass braze	890–950	Cu; Zn; Ni; Mn; Co	Cost efficient, lower melting point in comparison with copper braze
Palladium braze	890–910	Pd; Ni; Si	Very high shear strength

heat-resistant steel vessel made of high-alloyed steel with a water-cooling jacket, sealings, isolation, inner recipient, and heater systems controlled by a number of thermocouples. In addition, vacuum pumps and/or inert gas flooding systems steered with modern electronical process control systems are employed. The visualization of the process flows and furnace process parameters secures repeatable and consistent brazing processes. When employing hydrogen as a gas atmosphere the use of safety burn-off systems with a pilot flame is mandatory. Typical brazing cycles are conducted at temperatures between 700°C and 1100°C depending on the braze alloy and its melting point. Brazing cycle times usually vary from 12 h up to 48 h depending on the size, wall thickness, material properties, pressures, temperature regimes, protective gases, and the products to be brazed as well as the furnace design. Vacuum in the range of 50–100 mbar is typically employed to avoid oxidation. Since the components to be brazed cannot be moved in the furnace, a good fit or fastening the cemented carbide articles is crucial for the successful brazing success. In addition, careful positioning the components is important to prevent a sliding out from the needed position, when the braze alloy melts and forms a viscose liquid film. The main advantage of the furnace brazing process in terms of the quality of the braze joints is that no flux is required.

The parts finally arranged for brazing are put on trays, which are loaded into the furnace. After closing the recipient and outer furnace door, the entire furnace vessel is evacuated followed by a vacuum check. The precise brazing process of the compacts according to the required temperature profile and detailed process parameters is afterward started via a control panel. After completing the braze process, the furnace is cooled down and discharged manually.

8.2.3 Induction brazing

Induction brazing systems consist of a high-frequency generator, brazing coil with a cooling devise, brazing table, and electronic control system; a typical induction brazing unit employed for the fabrication of cemented carbide tools is shown in Fig. 8.8.

A source of high frequency electric power is used to drive a large alternating current through the coil. The passage of current through such a coil generates a very intense and rapidly changing magnetic field in the

Figure 8.8 Typical induction brazing unit employed for the fabrication of cemented carbide tools. *(With the permission of Element Six.)*

space within the coil. The workpiece subjected to heating is placed within the magnetic field; the heating process is very efficient, as the heat is generated inside the workpiece [6,7].

Braze coils are usually made of round or square copper tubes connected to an inductor applying the magnetic field. The alternating magnetic field induces a current flow in the conductive workpiece generating heat. The effectiveness of the induced current depends on the electrical conductivity of the workpiece material, so that materials with a higher electrical conductivity are heated up slower than those with a low electrical conductivity. Depending on the frequency range, the eddy currents generated on the surface induce the heat deeply in the workpiece material [3]. Magnetic fields with lower frequencies penetrate the workpiece deeper than those with higher frequencies and lead to the better heat distribution within the workpiece.

To prepare the tools for brazing, the base body is tipped with a braze foil and covered with flux and, afterward, the cemented carbide article is adjusted; alternatively, a brazing paste can be employed. Such assemblies are placed within or close to the induction coil.

Steel parts and braze alloys have a higher electrical conductivity compared to the cemented carbide parts, which results in their different heating rates. It is important to consider this effect when adjusting and designing the coil to avoid local overheating of the components. The brazing process is finished when the braze is completely melted and the cemented carbide article is joined to the steel part in the needed position. Brazing times vary in a wider range depending on the applied frequencies, used energy level (kW), and mass of the brazed articles. Brazing times lie usually in between few seconds and some 10 min. The brazing process temperature can be controlled by a pyrometer or thermographic camera and is dependent on the braze alloy melting point. Braze alloys with a melting point from approximately 680°C up to 1100°C are commonly employed for brazing carbide articles.

The common power range of brazing machines employed in the cemented carbide industry lies between 10 and 100 kW and is selected depending on the size of the parts to be brazed. The typical applied brazing frequency lies between 10 and 350 KHz. Flux or a protective gas atmosphere is required to prevent the surface oxidation and increase the wettability of the workpieces by the braze alloys. Depending on the employed brazing alloy and its melting point, the flux must be appropriately selected. The most commonly used fluxes for brazing cemented carbide/steel composites are boric acid (H_3BO_3) and borax ($Na_2B_4O_7$). During the brazing process, boron oxide (B_2O_3) forms, which is transformed into a low-melting slag as a result of its reaction with metal oxides preventing further oxidation.

Controlled cooling rates after brazing play an important role for reducing residual stresses resulting in the improved braze joint strength. The process of final slow cooling of brazed articles in a graphite powder or preheated furnace reduces thermal stresses in the braze joint of large articles. For relatively small article, cooling in atmosphere down to room temperature is typically used. Cooling (quenching) in liquids, such as water or oil, is harsh and allows direct hardening of the steel parts after brazing, which is possible when brazing articles of cemented carbide grades with high fracture toughness and high thermal conductivity. The reduction of the hardness of the steel parts due to the high temperatures of brazing is one of the main disadvantages of the brazing technology. It is essential to control the steel tempering and softening rates as precisely as possible. Residual stresses caused by the thermal expansion coefficient mismatch between the cemented carbide and steel parts are critical and cause limitations of sizes of the articles being brazed. Employing low melting point braze alloys reduces the thermal stress level and allows brazing large carbide articles.

8.2.4 Flame brazing

The main parts of a flame brazing unit are a torch and fixation system for the articles being brazed. The process control is more complicated and inaccurate in comparison with the furnace or induction brazing technologies, as the temperature of steel and carbide articles being brazed measured by pyrometers is strongly influenced by the presence of flame.

The general preparation steps for flame brazing correspond to those required for inductive brazing. The steel base body is tipped with a brazing foil and subsequently coated with flux; alternatively, a brazing paste can be employed.

The flame is directed toward the items to be brazed with the objective to achieve the highest flame temperature spot at the braze joint. The brazing process is finished when the braze is fully molten and carbide articles are positioned on the needed area of the steel base body. The flux protects the carbide and steel articles from oxidation and improves their wettability with the braze alloy. The combustion of acetylene with oxygen creates a flame with a maximum working temperature of about 3300°C, which is applied to heat the components being brazed and melt the braze alloy.

8.2.5 Assembling

Shrink fitting and gluing, in addition to brazing, are widely used assembly processes for fabrication of cemented carbide/metal composites. The joining methods mentioned above have their specific advantages and disadvantages. Depending on the thermal stability required for the operation conditions, the necessary shear strength, impact energy, and corresponding operation conditions, one must choose the optimal assembling method.

8.2.6 Shrink fitting

Shrink fitting is a technique allowing one to assemble steel and carbide components typically as a result of shrinkage. The shirinkage of one component is achieved due to its preliminary heating or cooling followed by obtaining room temperature, so that phenomena of thermal expansion followed by cooling are employed for joining steel and carbide articles [10].

Shrink fitting is widely used for percussive drill bits and can also be employed for complex cemented carbide/steel composites. Typical examples are rotation symmetric liners, valve seats of guides, etc. Obtaining strong joints by shrink fitting requires precise grinding and accurate machining of both the steel and carbide part. Modern CNC or optical measuring instruments enable precise measurements and shape control to be performed to ensure a perfect fit and stress distribution.

When calculating the shrinkage allowance, the heating temperature should exceed the anticipated working temperature by 100°C–200°C and different thermal expansion coefficients of the steel and carbide material must be taken into account to ensure a tough and durable joint. In order to prevent an overload of the steel part surrounding the carbide insert, it is important not to exceed the steel yield strength. For some applications, it is recommended not to exceed 70% of the maximum steel yield strength limit.

The procedure begins with loading the precisely machined and cleaned steel component into a furnace with subsequently heating up to the needed temperature. After the steel components are heated to the temperature leading to their thermal expansion, they are removed from the furnace followed by careful inserting carbide parts or buttons. The contraction of the steel components during cooling down to room temperature results in secure fitting of the components. Compressive stresses induced during cooling result in an increased fracture toughness of the carbide parts, which leads to a further improvement of their reliability and performance during operation.

In certain cases, shrinking steel parts in cemented carbide bores is needed, which must be performed with caution to avoid overstressing the cemented carbide parts due to the resulting tensile stresses.

8.2.7 Adhesive joining

Gluing also known as adhesive joining is a bonding technique employed for different materials by means of organic or inorganic adhesives; if necessary, it is conducted under pressure at room temperature or slightly elevated temperatures [9]. The binding mechanisms are of crucial importance for the bonding process. In the case of synthetic resin—based adhesives, adhesion between the adhesive and parts being joined is the main factor determining the bonding strength. The adhesion as a result of mechanical anchoring of the adhesive to the parts' surfaces is considered to be less important [11].

Reaction adhesives enable the realization of joining with the highest strength. The solidification of such adhesives is caused by curing plastic resin epoxy, polyester, polyurethane, or silicone-based compounds. In order to initiate the process due to the polymerization of the reaction adhesives (polyaddition and polycondensation), the necessary components are mixed shortly before the bonding process. The resulting chemical reaction of curing the adhesives occurs either exothermically leading to the heat generation or endothermically requiring the supply of external heat. The adhesion strength of the composites produced by use of this technique is mainly determined by the temperature and time of curing [12].

The following factors are decisive to generate the best adhesive joints:

- Optimal and uniform wetting of the surfaces being joined by the adhesive
- Low internal stresses in the parts being joint after completing the joining process
- Avoiding the presence of gas or air inclusions in the joint
- Perfectly cleaned joining surfaces

Adhesive joints are preferred if the expected shear forces and thermal stresses during operation are moderate. Shear strength values of the adhesive joints are only about 10%—20% of those of the conventional brazing joints. The strength of the adhesive bond is further reduced as a result of high temperatures. A major advantage of the adhesive joints is their wide range of applications, which is suitable for obtaining uniform, large-area, low-residual stress joints. When taking into account low values of shear resistance of small articles, adhesive connections in holes, slots, or squares are preferable and ensure achieving the best results in different applications. Adhesive joints are chemically relatively inert and do not affect the integrity and properties of the steel base body, which is one of the major advantages of adhesive joining in contrast to hard brazing.

Surface quality requirements for flatness and roughness are not very demanding when employing two-compound synthetic resin-based adhesives. The thickness of the adhesive joints typically lies in the range of 0.2–0.6 mm.

In general, the preparation steps for adhesive jointing are relatively simple. After an intensive cleaning process, the base body is completely coated with the premixed adhesive followed by exact positioning the base body and carbide article and the solidification of the organic compounds at room temperature or elevated temperatures in a drying chamber. If required, a cleaning or sandblasting process can be applied as the last process step.

It is recommended to install a dedicated adhesive area to avoid impurities that may reduce the joins' adhesion. The room, where the adhesive joint process is conducted, must be equipped with ventilation systems to guaranty the full evaporation of organic solvents' vapor.

8.3 Coatings obtained by chemical vapor deposition (CVD)

As it is mentioned in Chapter 1, the implementation of TiC thin coatings obtained on indexable cutting inserts by chemical vapor deposition (CVD) by Sandvik in 1969 led to revolutionary improvements in performance and prolongation of lifetime of cutting tools. The next large step was made by Benno Lux's group in Switzerland and was related to developing and patenting a technology for deposition of α-Al_2O_3 thin films; in 1971, Sandvik took over the IP rights on this technology [13] and in 1974 started the production of cutting inserts with the TiC–Al_2O_3 coatings. Unexpectedly, thin films of nearly 5–10 μm in thickness allowed the very significant tool life prolongation in metal cutting (by up to 5 times). Later, the mechanism of the coatings' influence on tool lifetime was established; it is related to special features of rank wear and flank wear of cutting inserts in machining steels and cast iron, which are described in detail in Chapter 12.

The present state of the art in the field of research, development, and fabrication of wear-resistance coatings is summarized in the review publications [13,14], so only bases of and major approaches to the deposition of thin coatings on carbide cutting inserts are reported in this chapter.

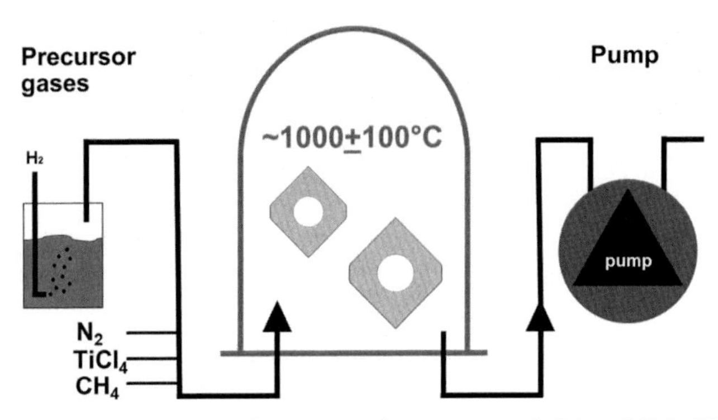

Figure 8.9 Schematic diagram illustrating the process of CVD of TiC, TiN, or TiCN coatings.

The CVD processes include the evaporation of volatile compounds of metals (in the case of Ti-containing coatings—$TiCl_4$) either at room temperature or elevated temperatures (Fig. 8.9). The compounds' vapor is mixed with a carrier gas (mainly hydrogen of high purity) and fed together with other gases, which are sources of carbon, nitrogen, or oxygen, into a deposition chamber heated to a temperature of around 1000°C. The heated surface of indexable cutting inserts or other substrates located on holders (usually metallic grits or graphite trays with holes) interacts with the gas mixture leading to the deposition of coatings (thin films) consisting of metal carbides, nitrides, carbonitrides, or oxides. Titanium refractory compounds and alumina form according to the following reactions:

$$TiCl_4 + H_2 + CH_4 \leftrightarrow TiC + HCl$$

$$TiCl_4 + H_2 + N_2 \leftrightarrow TiCl_2 + HCl$$

$$TiCl_4 + H_2 + CH_4 + N_2 \leftrightarrow TiCN + HCl$$

$$AlCl_3 + H_2 + CO_2 \leftrightarrow Al_2O_3 + HCl + CO$$

Titanium tetrachloride is a relatively stable compound, but it slowly hydrolyses in contact with water vapor, so it is used as a precursor for deposition of TiC, TiCN, and TiC. In contrast, aluminum chloride is highly hygroscopic, having a very pronounced affinity for water, so it must be freshly chemically synthesized to be used as a precursor for deposition of alumina coatings. The $AlCl_3$ synthesis is conducted by reaction of aluminum with HCl directly during the deposition process in a separate reactor connected with the CVD unit.

It must be noted that the reaction of hydrogen reduction of $TiCl_4$ does not lead to deposition of titanium at temperatures of around 1000°C [15], so the equilibrium of the reaction

$$TiCl_4 + H_2 \leftrightarrow Ti + HCl$$

is shifted to the left; titanium can be easily chlorinated by HCl at such temperatures. The role of methane and other hydrocarbon or nitrogen in CVD of TiC, TiCN, or TiN is therefore important, as their presence enables binding Ti atoms forming due to interaction of titanium tetrachloride with hydrogen as a result of the formation of titanium carbide, carbonitride, or nitride. Once these refractory compounds form, they are not subjected to chlorination by HCl, thus, shifting the equilibrium of the chemical reactions revealed above to the right, which results in the formation of titanium refractory compounds. Also, Fe group metals penetrating the coating from carbide substrates as a result of reaction of chlorination redeposition significantly affect the morphology and microstructure of carbide coatings [16]. This is related to the fact that Fe group metals are catalysts of reactions of the hydrocarbons' decomposition thus affecting the deposition process as a result of the formation of highly reactive carbon-containing species. The gas mixture going through the vertical reactor chamber is depleted with respect to the reactive components due to deposition of coatings on numerous cutting inserts and holders located inside the reactor. This depletion is usually compensated by creating a temperature gradient, so the temperature near the point of entering the gas mixture is lower than that near the point of its removal from the chamber.

Instead of methane and nitrogen, it is possible to use organic compounds containing both elements. Usually, acetonitrile is used as such a compound, which allows the TiCN deposition temperature to be noticeably decreased (down to nearly 800°C) leading to the possibility of conducting medium-temperature CVD (MT-CVD). Typically, the duration of one deposition cycle is nearly 1.5–3 h and flow rates of the reactive gases strongly depend on the chamber volume and the number of indexable cutting inserts being coated. The total pressure, flow rate, and concentration of the reaction gases are important factors influencing the deposition rate. An increase in the gas flow rate leads to an increase in the growth rate of the coatings; however, the deposition efficiency decreases when increasing the gas flow rate. The dependencies of the deposition rate on the concentration of the gas components, for example, methane for CVD of TiC or TiCN, are characterized by the presence of maxima, so there is the optimal concentration of each reactive gas with respect to obtaining the maximum deposition rate at a certain combination of pressure and temperature.

The gaseous reaction by-products are removed from the deposition chamber in the flow of the carrier gas, usually with the aid of a separate pump.

Another important aspect of the CVD processes is the possibility of elimination of the substrate decarburization leading to the formation of the η-phase underlayer in the carbide substrate. As it was shown in Ref. [17], just the presence of such an underlayer and a region with a reduced carbon content below this underlayer can lead to a decrease of transverse rupture strength of CVD-coated cemented carbides. The formation of a decarburized surface layer can occur already on the stage of heating cutting inserts to the deposition temperature in hydrogen, so the purity of hydrogen must be exceptionally high. Nowadays, first layers deposited directly onto the carbide inserts are usually TiN or TiCN [18] obtained by MT-CVD with the aid of acetonitrile as a source of both carbon and nitrogen. Acetonitrile is characterized by a relatively high reactivity at elevated temperatures and low decomposition temperature, at which the rate of carbon diffusion in the carbide substrate is low, so that the formation of the substrate decarburized near-surface layer containing η-phase can be completely eliminated. In some cases, TiN is deposited as a first layer adjacent to the carbide substrate [13]. The up-to-date coated cutting inserts usually are free or nearly free of the η-phase underlayer, which can be seen in Fig. 8.10 indicating the absence of such an η-phase underlayer under the coating.

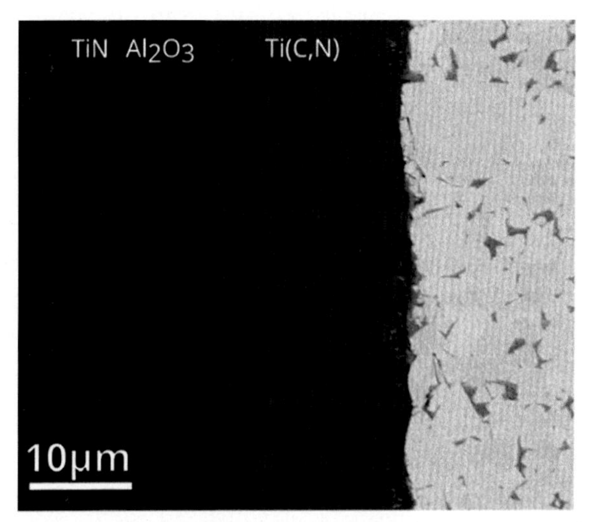

Figure 8.10 SEM image of an insert cross-section comprising a multilayer TiCN–Al_2O_3–TiN coating and microstructure of the substrate near-surface layer adjacent to the coating. *(Redrawn from Ref. [18].)*

The phase composition, structure, mean grain size, and texture of alumina layers play an important role with respect to improving performance and prolonging tool lifetime of coated cutting inserts. Stable coatings of α-Al_2O_3 can be synthesized at relatively high temperatures of nearly 1000°C. The mean grain size of such coatings is, however, relatively high leading to their reduced performance. Smoother Al_2O_3 coatings with a significantly finer mean grain size can be produced in MT-CVD processes at temperatures below 900°C; however, the metastable κ-phase forms at these deposition conditions. The mechanical and thermal properties of this phase are lower than those of α-Al_2O_3 [19]. Nevertheless, due to the significantly lower mean grain size of the κ-Al_2O_3 layers (50–100 nm), they are considerably smoother and usually ensure better performance properties of coated cutting inserts in machining.

There are approaches allowing one to obtain fine-grain α-Al_2O_3 coatings with growth textures by the use of tailored nucleation steps described in Refs. [20,21]. The nucleation process of α-Al_2O_3 can be enhanced as a result of performing such steps comprising short pulses and purges, in particular, oxidizing periods. The chemistry of the nucleation surface was found to be an important factor for predetermining the phase composition and growth texture of the Al_2O_3 layers. The optimized nucleation process can result in obtaining relatively fine-grain and defect-free textured α-Al_2O_3 layers ensuring the best wear resistance of coated cutting inserts in machining. Additions of H_2S to the gas mixtures and other technological approaches to the process of CVD of α-Al_2O_3 coatings lead to obtaining fine-grain highly textured thick layers with improved performance [22]. The approaches mentioned above represent only a limited number of examples of producing alumina coatings with improved properties and performance. It can be expected that many other technological approaches employed in the cemented carbide industry belong to "know-how" of different carbide manufacturers and are therefore not disclosed.

Later on some, special variants of CVD coatings were developed and implemented in industry, for example (Ti, Al)N, (Ti, Si)C,N, etc. Nevertheless, multilayer coatings comprising different combinations of Al_2O_3, TiCN, and TiN as well as in some cases (Ti,Al)N and ZrN with the total thickness of up to nearly 25–30 μm for turning (Fig. 8.10) and of about 6–8 μm for milling (Fig. 8.13) remain the basic CVD coatings fabricated by different carbide manufacturers.

Additionally, cutting inserts for machining nonferrous metals, for example, Al–Si alloys, abrasive composites, etc., are presently fabricated

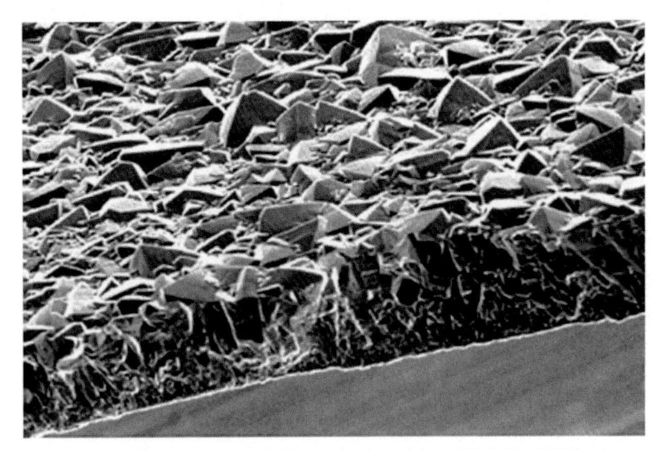

Figure 8.11 Polycrystalline diamond film obtained by PACVD. *(With the permission of Element Six.)*

with thin polycrystalline diamond coatings (Fig. 8.11); such coatings are obtained by various techniques, for example, plasma-assisted CVD or hot-filament CVD. The major problem that must be overcome in the fabrication of such coatings is related to the catalyzing effect of the Co binder on the growing diamond coatings leading to their graphitization. This problem can be solved by selective removal of the binder phase from the surface layer of carbide inserts or deposition of special barrier interlayers on the cemented carbide substrates [23,24].

Another important aspect of the CVD coatings is related to the fact that as a result of their deposition followed by cooling they are characterized by residual tensile stresses, which can lead to coatings' cracking [17]. Nowadays, the CVD coatings on carbide indexable cutting inserts are subjected to intensive mechanical posttreatment, for example, top blasting or brushing [18], to reduce the residual tensile stresses or obtain residual compressive stresses in their near-surface layer.

Figs. 8.12 and 8.13 illustrate the evolution of CVD coatings on carbide tools for turning and milling steels over the last 25 years [13]; modern coatings contain usually five to eight layers of different compositions. The first layer adjacent to the substrate is usually TiN or TiCN, since such coatings have good adhesion to the carbide substrate and prevent its decarburization during the vapor deposition process. Intermediate layers usually consist of various combinations of TiN, TiCN, and Al_2O_3. The outer layer of coatings of the first and second generations consisted of TiN and in the coatings of the third and fourth generations of Al_2O_3, although

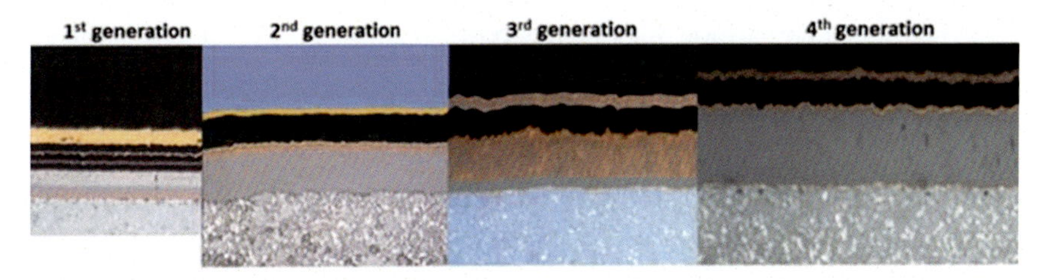

Figure 8.12 Evolution of CVD multilayer coatings on carbides substrates for turning steels: first generation: TiN—TiCN—TiC—Al_2O_3/TiCN multilayers — TiN, total thickness: 12 μm. second generation: TiN—MT—CVD TiCN—HT-CVD TiCN—Al_2O_3—TiN, total thickness: 15 μm. third generation: TiN—MT—CVD TiCN—HT—CVD TiCN—Al_2O_3—HT—CVD TiCN—Al_2O_3, total thickness: 18 μm. fourth generation: TiN—MT—CVD TiCN—HT—CVD TiCN—Al_2O_3—HT—CVD TiCN—Al_2O_3, total thickness: 22 μm. *(Redrawn from Ref. [13].)*

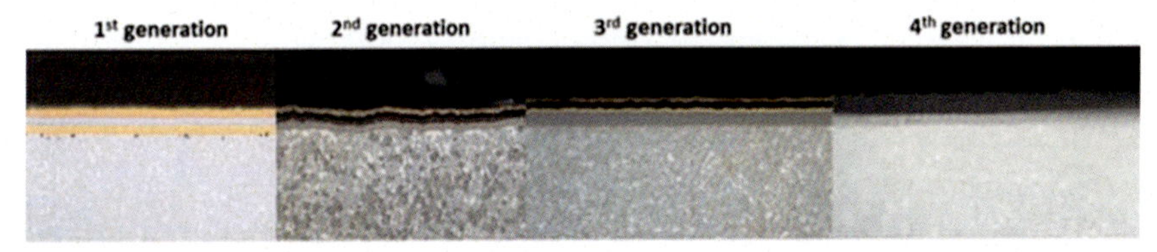

Figure 8.13 Evolution of CVD multilayer coatings on carbides substrates for milling steels: first generation: TiN—TiCN—TiC, total thickness: 6 μm. second generation: TiN—TiC—Al_2O_3—TiCN—Al_2O_3—TiN, total thickness: 6 μm. third generation: TiN—MT—CVD TiCN—Al_2O_3—TiN, total thickness: 8 μm. fourth generation: TiN—MT—CVD TiCN—(Al,Ti)N, total thickness: 8 μm. *(Redrawn from Ref. [13].)*

some companies still produce cutting inserts with multilayer coatings comprising TiN as a top layer. The architecture of multilayer coatings fabricated by different carbide manufacturers of cutting inserts can vary; the technological parameters of the coatings' deposition and their special features and properties are usually not disclosed.

8.4 Coatings obtained by physical vapor deposition (PVD)

Physical vapor deposition (PVD) methods of obtaining wear-resistant coating on carbide cutting inserts are based on creating metals' vapors, which are usually partially ionized, in a vacuum with a simultaneous supply of a reactive gas (N_2, CH_4, etc.) at a low pressure. As a result, the condensation of thin films of refractory compounds occurs on the surface of carbide articles. Typical parameters of PVD processes are low pressures (of the order of 0.01 Pa) and temperatures (nearly 500°C), at which no decarburization of carbide substrates occurs due to very low carbon diffusion rates. PVD coatings are usually under residual compressive stresses, so this very important obstacle combined with the absence of an η-phase underlayer leads to the fact that there is no or very insignificant effect of embrittlement of the coated cutting edges. This results in the high efficiency of employing PVD processes for coating carbide inserts with sharp cutting edges (for example, for grooving) and those operating under high impact loads (in milling and interrupted cutting). Employing low-pressure plasmas allows the synthesis of entirely new material systems: metastable, ternary nitrides of the MeAlN system, where Me can be almost any refractory metal, TiAlN and Al_2O_3 as well as mixed oxide coatings, such as mixed oxides of aluminum and chromium, etc. By introducing Si and B into the crystal lattice of conventional PVD coatings, e.g., TiAlN, it is possible to obtain a novel type of coatings in form of nanocomposites thus resulting in an increase of their hardness, wear resistance, thermal stability, and resistance to high-temperature oxidation [25–27]. Over the past decade, various multicomponent nanostructured PVD coatings were developed, for example, of the Ti–B–N, Ti–Si–N, Ti–Si–C–N, Ti–Al–B–N and Ti–Cr–B–N systems [25–27].

There are PVD techniques of the three major types employed for obtaining PVD coatings in the cemented carbide industry.

8.4.1 Electron beam evaporation (ion plating)

According to the technique of electron beam evaporation (ion plating), schematically illustrated in Fig. 8.14, a metal (usually a refractory metal) is

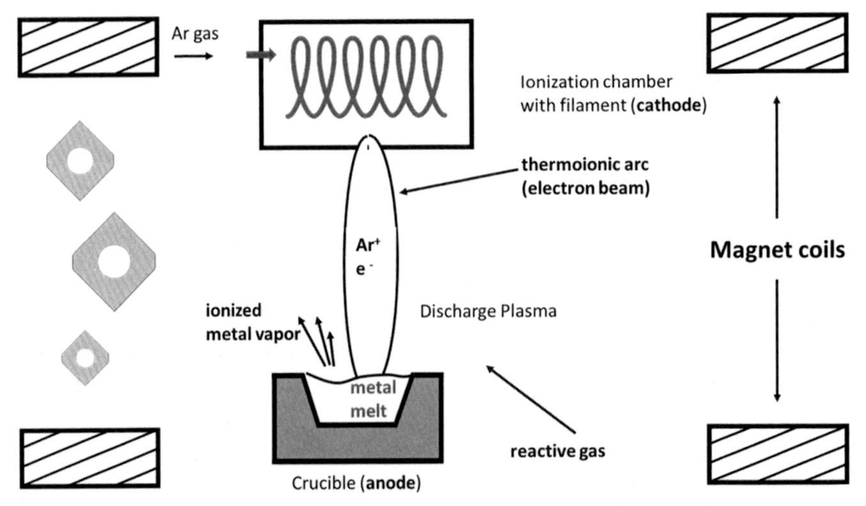

Figure 8.14 Schematic diagram illustrating the ion plating technique.

evaporated by a thermoionic arc (electron beam) comprising Ar ions by the use of the evaporation chamber as a cathode and the crucible with the liquid metal being evaporated as an anode. The discharged plasma formed in such a way is steered by magnetic coils and directed toward to the surface of the articles being coated. Bombardment of the substrate by atomic-sized energetic particles prior to deposition is used to sputter clean the substrate surface. As a result of interaction of the evaporated metals with a reactive gas fed into the chamber, a thin film of a refractory compound forms on the substrate surface. The major advantages of this method are related to high energies achievable on the substrate surface due to the bombardment resulting in high coatings' adhesion, flexibility with respect to the level of ion bombardment, and enhanced chemical reactions on the substrate surface. The major disadvantages of ion plating are increased variables and, in some cases, inconsistent uniformity of plating and excessive heating of the substrate surface.

TiN and TiCN coatings produced in such a way and intended for plating drills were introduced by Balzers under the trademark "Balinit A" and "Balinit G" in 1980. Later on, this technique was implemented for coating carbide cutting inserts [13].

8.4.2 Electric arc evaporation

According to the DC arc evaporation technique, an electric arc is created between a cathode composed of a metal that must be evaporated and an anode (Fig. 8.15); the typical process conditions are voltage of 40 V and current of 170—400 A. Special magnetic coils are located in the deposition chamber, which are intended for steering the arc, so that it steadily and fastly moves on the cathode surface resulting in partial melting spots of the cathode material, which is usually a refractory metal. As a result, the highly ionized metal vapor is created, which usually contains small droplets of the cathode material. The vapor reacts with a reaction gas fed into the chamber (nitrogen, hydrocarbons, etc.) on the substrate surface resulting in the formation of thin films of refractory compounds (nitrides, carbonitrides, carbides, etc.). The adhesion of the films to various substrates is usually very high, as the substrates are typically subjected to ion bombardment and ion etching on the initial stage without introducing reactive gases.

The problem of the presence of small droplets of the cathode material (titanium droplets in the case of Ti cathodes) can be solved by use of the so-called "flittered arc evaporation" system, comprising a special system of magnetic coils allowing the flux of ionized metal vapor to be bent at an

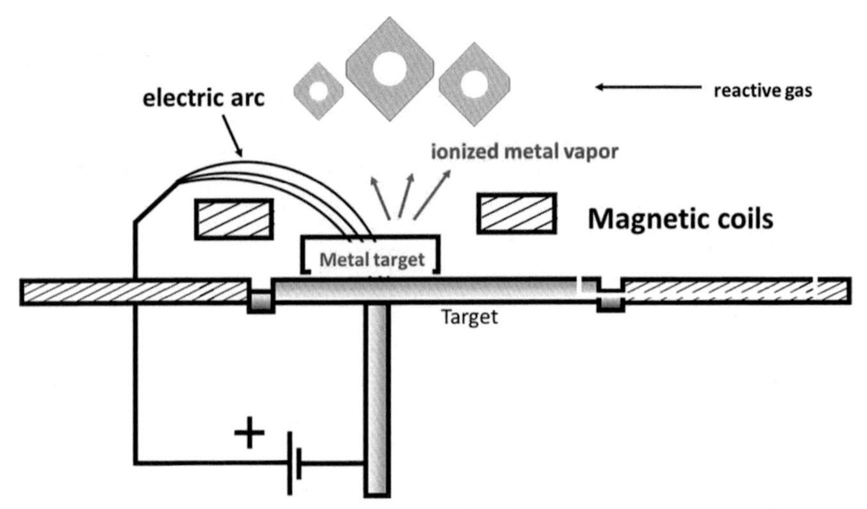

Figure 8.15 Schematic diagram illustrating the electric arc evaporation technique.

angle of 90 degrees. As a result, all the uncharged microdroplets and atoms of the cathode material fly in the direction nearly perpendicular to the cathode surface and are filtered out on the internal walls of the system. Only metal ions are bent under the influence of the magnetic field created by the magnetic coils and reach the substrate surface resulting in the possibility of obtaining very high ionization rates of the flux (up to 100%). This allows depositing extremely hard droplet-free nanostructured thin films of nitrides and other refractory compounds [28,29].

The first arc evaporation deposition units were developed and patented in the Soviet Union in the 70s and afterward widely commercialized worldwide for coating both cemented carbides and steels. One of the variants of the arc evaporation systems is the so-called "pulsed arc evaporation." In the pulsed arc, the discharge current is large compared to the conventional DC arcs. The arc evolves from rapidly created new cathode spots leading to reducing the thermal stresses at the cathode spot surface and the size and number of eroded microparticles. Cutting inserts with thin films of TiAlN obtained by pulsed arc evaporation are presently fabricated in the carbide industry on a large scale. In 2006, Balzers developed and implemented mixed Al—Cr oxide coatings by use of pulse arc evaporation, which are presently produced in combination with TiAlN coatings on a large scale for carbide indexable cutting inserts intended for milling cast iron and steels.

8.4.3 Magnetron sputtering

Another deposition technique employed for obtaining thin films of refractory compounds is magnetron sputtering. Sputtering is a plasma-based deposition process, in which energetic ions of gases (for example, argon) are accelerated toward a target, which is a metal that must be evaporated. The surface of the target is eroded by such high-energy ions within the plasma, and the evaporated atoms travel through the vacuum environment and deposit onto a substrate to form thin hard films as a result of their reaction with a reactive gas (Fig. 8.16).

In a typical sputtering deposition process, a chamber is first evacuated to high vacuum to minimize the partial pressures of all background gases and potential contaminants. After evacuation of the deposition chamber, a sputtering gas is fed into the chamber and the total pressure is adjusted (typically in the milliTorr range) using a pressure control system. Typical deposition conditions are voltage of 350–450 V and current of 10–40 A. To initiate plasma generation, high voltage is applied between the cathode (usually located directly behind the sputtering target) and the anode (commonly connected to the chamber as electrical ground). Electrons which are present in the sputtering gas are accelerated away from the cathode causing collisions with nearby atoms of the sputtering gas. These collisions lead to obtaining an electrostatic repulsion resulting in the gas ionization. The positively charged sputter gas atoms are accelerated toward the negatively charged cathode, leading to high energy collisions with the surface of the target. Each of these collisions can cause atoms at the

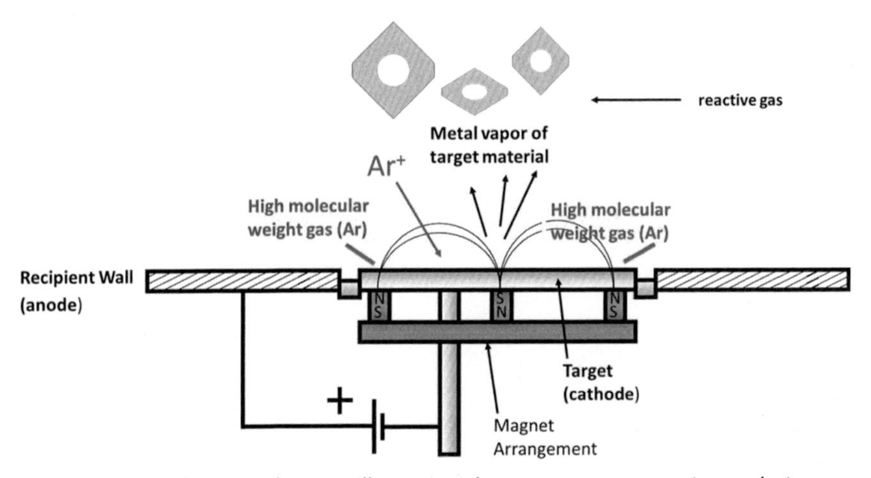

Figure 8.16 Schematic diagram illustrating the magnetron sputtering technique.

surface of the target to be ejected into the chamber volume. In order to enhance the atoms' ejections and increase the deposition rate, the sputtering gas is typically chosen to be a high molecular weight gas such as argon or xenon. If a reactive sputtering process is desired, reactive gases such as oxygen or nitrogen are introduced into the chamber during the film growth.

The major advantage of the sputtering technique compared to the arc deposition technique is the possibility to employ a great variety of target materials including insulators; however, the ionization rate of the metal vapor of the target material is significantly lower than that of the arc deposition techniques. Typical deposition rates achievable by use of magnetron sputtering are also noticeably lower than those achievable by use of arc evaporation.

There are different variants of the sputtering deposition systems including dual magnetron sputtering, high powder impulse magnetron sputtering, etc. In 2005, a technology for deposition of thin Al_2O_3 films by use of magnetron sputtering was developed and implemented.

Figs. 8.17 and 8.18 show typical architectures of up to date PVD coatings and microstructures of modern nanostructured PVD coatings [14]. As it can be seen in Fig. 8.17, there is a great variety of coating architectures—from

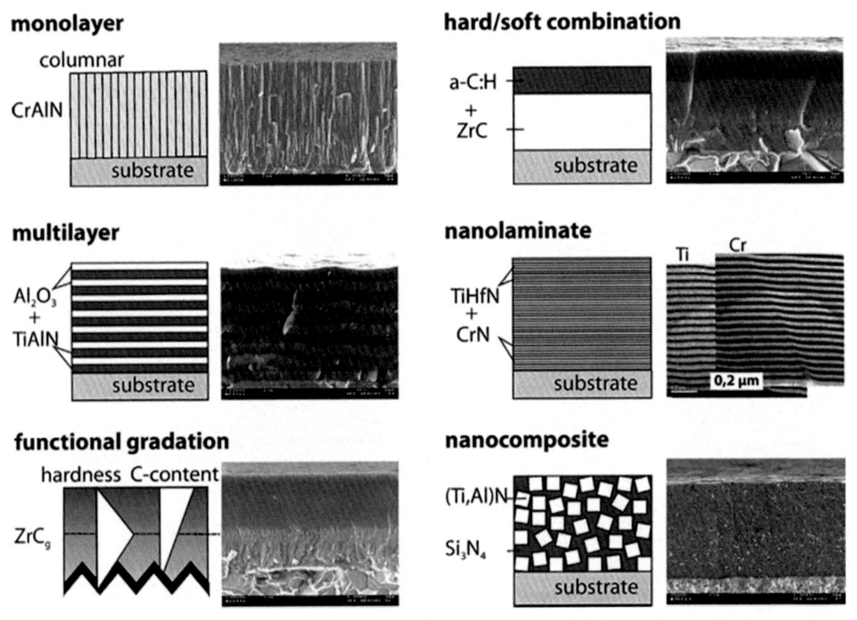

Figure 8.17 Typical architectures of up-to-date advanced PVD coatings. *(Redrawn from Ref. [14].)*

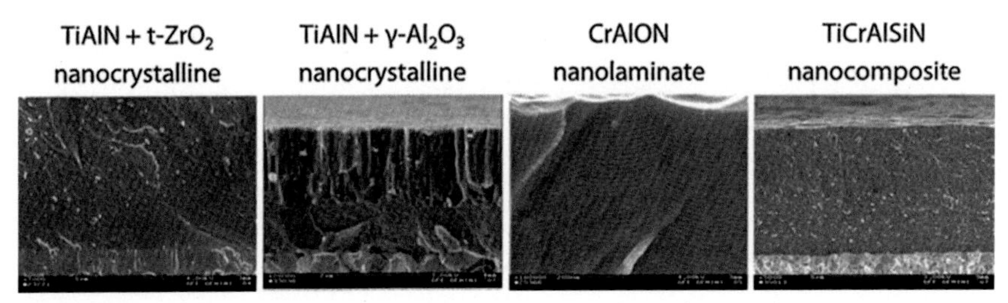

Figure 8.18 Microstructures of modern nanostructured PVD coatings. *(Redrawn from Ref. [14].)*

nanolaminates and nanocomposites to functionally graded coatings. Carbide manufacturers produce indexable cutting inserts with PVD coatings of different types and compositions varying from TiAlN and Al_2O_3 to multi-component nanostructured coatings with complicated structure.

References

[1] DIN standard 8589 T11, Fertigungsverfahren — Schleifen mit rotierendem Werkzeug, Einordnung, Unterteilung, Begriff, 1984.
[2] M. Meister, Vandemecum des Schleifens, Hanser Verlag, 2011.
[3] H.K. Tönshoff, Spanen — Grundlagen, VDI Book, 1995.
[4] Internet site https://www.zerspanungstechnik.at/detail/weg-zu-220-ms-schnittgeschwindigkeit-geoeffnet_17592. Dated April 2021.
[5] J. Yang, M. Oden, M.P. Johansson-Joesaar, L. Llanes, Grinding effects on surface integrity and mechanical strength of WC-Co cemented carbides, Procedia CIRP 13 (2014) 257—263.
[6] Internet site www.richieburnett.co.uk/indheat.html. Dated May 2021.
[7] I.E. Petrunin, Handbuch Löttechnik, Deutschsprachige Ausgabe Verlag Technik GmbH Berlin, 1991 (in German).
[8] W. Müller, J. Müller, Löttechnik: Leitfaden für die Praxis, Deutscher Verlag für Schweißtechnik, Düsseldorf, 1995.
[9] L. Dorn, Hartlöten Grundlagen und Anwendungen, Expert Verlag, Sidelfingen, 1985.
[10] Internet site https://en.wikipedia.org/wiki/Shrink-fitting datedMay 2021.
[11] L. Dorn, Vergleich der Klebetechnik mit Herkömmlichen Fügeverfahren. VDI Berichte (in German).
[12] Fügetechnik im Vergleich, Tagung Baden-Baden, 24 und 25 April 1991, VDI Verlag, Düsseldor, 883-894 (in German).
[13] R. Haubner, M. Lessiak, R. Pitonak, A. Köpf, R. Weissenbacher, Evolution of conventional hard coatings for its use on cutting tools, Int. J. Refrac. Metals Hard Mater. 62 (2017) 210—218.
[14] K. Bobzin, High-performance coatings for cutting tools, CIRP J. Manuf. Sci Technol. 18 (2017) 1—9.
[15] I. Konyashin, The mechanism of interaction between Cl-containing gaseous phase and cemented carbides in the chemical vapor deposition process, Thin Solid Films 249 (1994) 174—182.
[16] I. Konyashin, The influence of Fe-group metals on CVD of titanium carbide, Adv. Mater. CVD (2) (1996) 199—208.

[17] I. Konyahsin, CVD coated hardmetals - causes of transverse rupture strength decrease, Funct. Mater. 1 (1994) 106–110.
[18] L. Toller-Nordström, J. Östby, S. Norgren, Towards understanding plastic deformation in hardmetal turning inserts with different binders, Int. J. Refract. Metals Hard Mater. 94 (January 2021) 105309, https://doi.org/10.1016/j.ijrmhm.2020.105309.
[19] S. Ruppi, A. Larsson, A. Flink, Nanoindentation hardness, texture and microstructure of a-Al_2O_3 and k-Al_2O_3 coatings, Thin Solid Films 516/18 (2008) 5959–5966.
[20] S. Ruppi, Deposition, microstructure and properties of texture-controlled CVD α-Al_2O_3 coatings, Int. J. Refract. Metals Hard Mater. 23 (2005) 306–316.
[21] S. Ruppi, Method for Manufacture Enhanced Textured Alumina Abrasion-Resistant Coating on Cutting Tool Inserts, EP 1655392, 2004.
[22] S. Ruppi, Influence of process conditions on the growth and texture of CVD alpha-alumina, Coatings 10 (2) (2020) 158.
[23] I. Konyashin, M. Guseva, Thin films comparable with WC-Co cemented carbides as underlayers for hard and superhard coatings: the state of the art, Diam. Relat. Mater. 5 (1996) 575–579.
[24] I. Konyashin, M. Guseva, V. Babaev, V. Khvostov, G. Lopez, A. Alexenko, Diamond films deposited on WC-Co substrates by use of barrier interlayers and nano-grained diamond seeds, Thin Solid Films 300 (1997) 18–24.
[25] P.V. Kiryukhantsev-Korneev, M.I. Petrgik, A.N. Sheveiko, The influence of Al, Si and Cr on thermal stability and resistance to high-temperature oxidation of coatings on the basis of titanoim boron nitride, Metal Phys. Mater. Sci. 104 (2007) 176–183.
[26] D.V. Shtansky, E.A. Levashov, A.N. Sheveiko, et al., Synthesis and characterization of Ti-Si-C-N films, Metall. Mater. Trans. A. 30 (1999) 2439–2447.
[27] D.Y. Shtansky, P.V. Kiryukhantsev-Korneev, A.N. Sheveyko, et al., Hard Tribological Ti-Cr-B-N coatings with enhanced thermal stability, corrosion and oxidation resistance, Surf. Coating. Technol. 202 (2007) 861–865.
[28] I. Konyashin, G. Fox-Rabinovich, V. Dodonov, TiN thin films deposited by filtered arc evaporation: structure, properties and applications, J. Mater. Sci. 32 (1997) 6029–6038.
[29] I. Konyashin, G. Fox-Rabinovich, Nanograined titanium nitride thin films, Adv. Mater. 10 (12) (1998) 952–955.

CHAPTER 9

The influence of compositions and microstructural parameters on properties and applications of WC–Co cemented carbides

It is well known that the major mechanical and performance properties of WC–Co cemented carbides are hardness, fracture toughness, compressive strength, and transverse rupture strength (TRS), which are in turn influenced by the Co content and the WC mean grain size. In addition to the Co content and WC mean grain size, some other characteristics of cemented carbides (the presence and amounts of grain growth inhibitors, shape of WC grains, impurities and contaminations, etc.) also play an important role in achieving the best combination of cemented carbides' properties and performance.

9.1 Cobalt content and mean WC grain size

Most industrial grades of fine-grain, medium-grain, coarse-grain, and ultracoarse-grain cemented carbides with different combinations of properties are made by varying the Co content and WC mean grain size. The effect of the Co content and WC mean grain size on mechanical and performance properties of WC–Co cemented carbides was reported in numerous publications, the results of which were summarized in the monography [1] and review article [2] (see also Chapter 2).

It is well known that the fracture toughness and hardness, which are the main parameters determining the performance characteristics of cemented carbides, change monotonically when the Co content and mean WC grain size change. Fracture toughness increases, and hardness decreases with increasing both Co content and WC mean grain size (see Chapter 2). Except for some special cases, which are described below and in Chapter 11, the fracture toughness can only be increased at the expense of hardness and vice versa.

Cemented Carbides
ISBN 978-0-12-822820-3
https://doi.org/10.1016/B978-0-12-822820-3.00002-1

For many applications, in addition to the hardness and fracture toughness, some other physical and mechanical properties, particularly, the compressive strength and TRS, play an important role.

It is well known that the most important microstructural parameter determining both the TRS and compressive strength of WC—Co materials is the average thickness of Co interlayers among WC grains, or the so-called "Co mean free path." When producing cemented carbides with different combinations of the WC mean grain size and Co content, the value of the Co mean free path can be varied in a wide range; generally, it is one of the major parameters affecting both the TRS and compressive strength of straight WC—Co grades.

As it can be seen in Figs. 9.1 and 9.2, the dependences of the TRS and compressive strength on the Co mean free path are characterized by the presence of maxima.

According to Ref. [2], the curves indicating the dependence of the TRS on the Co mean free path for the cemented carbides with 6—25 wt.% Co and WC grain size varying from ultrafine to coarse-grain have a broad maximum at the Co mean free path in the range of nearly 0.3—0.5 μm. In particular, a carbide grade with 12 wt.% Co has the maximum TRS value at the WC mean grain size of nearly 3 μm [2]. It should be noted that at the time of publishing Ref. [2] (1970) the methods of measuring the WC mean grain and consequently Co mean free path were based on light microscopy and medium-resolution scanning electron microscopy, which did not allow all the WC/Co and WC/WC grain boundaries as well as extremely fine WC grains to be properly identified. It can therefore be assumed that if one employs modern techniques of the WC grain size measurement based on

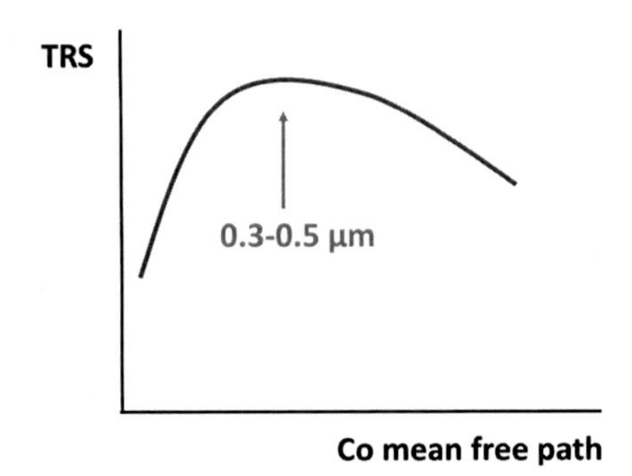

Figure 9.1 Schematical drawing illustrating the dependence of the TRS on the Co mean free path in WC—Co cemented carbides with different combinations of the WC mean grain size and Co content [1—4].

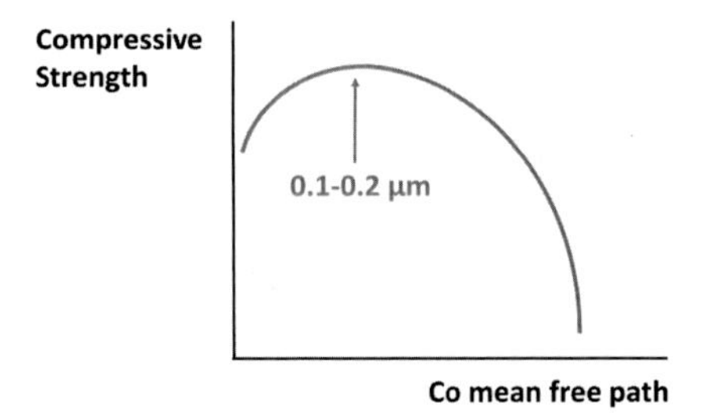

Figure 9.2 Schematical drawing illustrating the dependence of the compressive strength on the Co mean free path in WC–Co cemented carbides with different combinations of the WC mean grain size and Co content [1,2].

high-resolution transmission electron microscopy (HRSEM) and electron backscattering diffraction (EBSD) (see Chapter 7), the values of the Co mean free path corresponding to the maximum TRS on the curve shown in Fig. 9.1 would be noticeably shifted toward lower values (estimated as roughly 0.2–0.3 μm).

According to the data of Ref. [1], the dependence of the compressive strength on the Co content of WC–Co cemented carbides with a WC mean grain size varying from 1.4 to 5.3 μm is characterized by a broad maximum at about 5 - 6 wt.% Co; the maximum compressive strength is achieved at a WC mean grain size of the order of nearly 2 μm. When the grain size is reduced to a level of about 1 μm, the compressive strength decreases, but only insignificantly. According to the data of Ref. [2], the curve indicating the dependence of the compressive strength on the Co content is characterized by a maximum at about 5–6 wt.% Co for a medium-grain carbide grade with a WC mean grain size of nearly 2 μm. Summarizing the results of refs. [1,2], the dependence of the compressive strength of cemented carbides with different combinations of WC mean grain size and Co content on the Co mean free path shown in Fig. 9.2 has a relatively broad maximum at a Co mean free path of nearly 0.1–0.2 μm. Similar to the value of the Co mean free path corresponding to the highest TRS value, one can assume that the Co mean free path value corresponding to the maximum compressive strength of cemented carbides should be shifted toward lower values (estimated as roughly 0.1 μm or less) when using the up-to-date techniques of measuring the WC mean grain size and consequently Co mean free path (HRSEM and EBSD).

When taking into account the importance of choosing the optimal combination of the WC mean grain size and Co content to achieve the maximum values of either TRS or compressive strength, cemented carbides must have the optimum combination of both the WC mean grain size and Co content to achieve the best performance properties for each concrete application.

Fine-grain cemented carbides used for the manufacture of indexable cutting inserts for metal cutting and wear parts are usually not exposed to high compressive loads, but often operate under conditions of high cyclic and impact loads. Due to this, their microstructure and composition, in particular the WC mean grain size, Co content, and amounts of grain growth inhibitors (which are addressed in detail below) are selected in such a way that one can achieve the maximum hardness combined with the optimum fracture toughness for each particular application.

Inserts of medium-grain and coarse-grain carbide grades for the mining industry, in particular for percussive drilling, should have an optimum combination of compressive strength, fracture toughness, and hardness, since they operate under high compression loads, severe fatigue, and high impact loads. Such a combination is usually achieved with a Co content of 6 wt.% and WC mean grain size of 1.6−1.9 μm (Fig. 9.3A). Typical properties of such carbide grades are Vickers hardness of 14−15 GPa, TRS of above 2400 MPa, fracture toughness of nearly 12 MPa $m^{1/2}$, and compressive strength of more than 4500 MPa [3].

Inserts of coarse-grain carbide grades for roller-cone bits for rotary drilling and oil-and-gas (O&G) drilling must have an optimal combination of TRS, fracture toughness and hardness, as they are subjected to high bending loads. According to Ref. [4], the TRS of the carbide grade with 12% Co has a maximum at a WC mean grain size of about 2.5 μm. Cemented carbides for rotary drilling and O&G drilling have the maximum value of TRS at the Co content of 8−10 wt.% and WC mean grain size of about 2.5 μm. The thickness of Co interlayers in such cemented carbides, calculated using the equation given in Ref. [5], is about 0.5 μm, which corresponds to the maximum value of the TRS according to Fig.9.1. Conventional grades intended for rotary drilling and O&G drilling are characterized by the following properties: Vickers hardness of 11.5−12.5 GPa, fracture toughness of more than 14 MPa $m^{1/2}$, and TRS of over 3900 MPa. For some special applications, carbide grades with a high Co content (up to 15%) with elevated fracture toughness can be used. However, such grades have a reduced value of TRS as compared to the grades containing 8−12wt.% Co.

Cemented carbides for road-planing and mining picks must have a very high level of fracture toughness due to the presence of high thermal loads

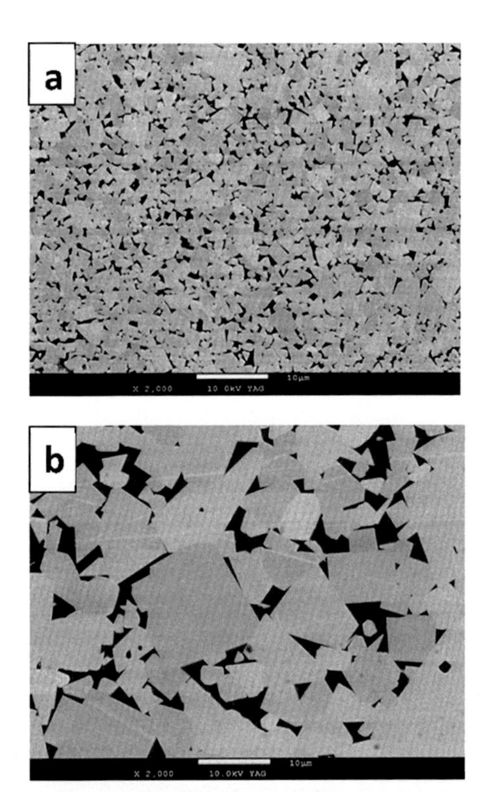

Figure 9.3 Microstructures of typical medium-grain and ultracoarse-grain carbide grades: (A)—grade for percussive drilling with 6% Co and (B)—grade for road-planing with 6.5% Co (HRSEM). *(Redrawn from Ref. [3].)*

and, consequently, thermal fatigue, as well as high impact loads. Usually, for such picks, ultracoarse-grain grades with a WC mean grain size of about 5 μm containing 6—11 wt.% Co, the microstructure of one of which is shown in Fig.9.3B, are used. The microstructure of such grades contains very thick Co interlayers between ultracoarse WC grains. This makes it possible to effectively suppress the initiation and propagation of thermal cracks during the exploitation of road-planing and mining picks (see Chapter 12). However, the high fracture toughness of such grades is achieved at the expense of TRS and compressive strength, as well as reduced wear resistance related to the presence of thick binder interlayers. The typical values of TRS for such grades are above 2300 MPa, compressive strength of at least 3500 MPa, and fracture toughness of more than 15 MPa $m^{1/2}$. It should be noted that such ultracoarse-grain grades with a mean WC grain size of about 5 μm are relatively "young" compared to other types of coarse-grain cemented carbides developed in the 50s and 60s of the last century. The development of such grades was started in the 1980s by Kennametal Inc. (USA) that developed and patented a technology for producing tungsten carbide powders, the so-called Menstruum process [6], allowing one to manufacture extremely coarse-grain WC powders (see Chapter 6).

9.2 Uniformity of microstructure

The microstructure of medium-coarse, fine-grain, submicron, ultrafine, and near-nano carbide grades for fabricating indexable cutting inserts and wear parts must be as uniform as possible with the absence of WC grains larger than the WC mean grain size by nearly 2 to 3 times. Their microstructures, typical examples of which are shown in Fig. 9.4, must ensure the

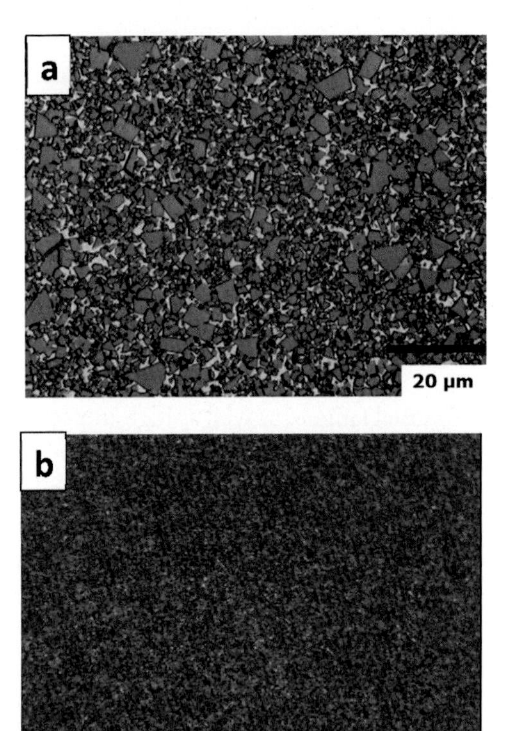

Figure 9.4 Typical microstructures of (A)—medium-coarse cemented carbide with 10 wt.% Co (optical microscopy, after etching in the Murakami reagent), and (B)—fine-grain cemented carbide with 10 wt.% Co (optical microscopy, after etching in the Murakami reagent). *(Redrawn from Refs. [3,7].)*

maximum combination of hardness, fatigue resistance, and TRS. The fracture toughness of such carbide grades can be noticeably lower than that of coarse-grain carbide grades for mining, construction, and other applications characterized by the presence of high impact loads and severe thermal and mechanical fatigue.

In literature, there is no general opinion on the optimal microstructure of medium- and coarse-grain cemented carbides in terms of their homogeneity. Most such grades are characterized by a certain degree of bimodality of the microstructure (Fig. 9.4A), so that there are both a fine-grain and a relatively coarse-grain fraction in the microstructure, which allows obtaining the optimum combination of hardness and fracture toughness.

A generally accepted viewpoint on the characteristics of the microstructure of coarse-grain cemented carbides is that the microstructure must not comprise abnormally large WC grains, whose size exceeds the average WC grain size by more than 3 times. Such abnormally large WC grains shown in Figs. 9.5 and 9.6 can initiate the formation and propagation of cracks during the operation of carbide articles under conditions of high impact loads. It is believed that the reason for the formation of such grains is associated with local deviations in the composition of cemented carbide, mainly in the carbon content and the presence of abnormally large WC grains in the initial WC powder [9,10]. However, the phenomenon of the formation of such abnormally large WC grains in the microstructure is not fully understood, as they can appear in the microstructure of cemented carbide obtained from defect-free WC powders containing much fine and nano WC fraction (see Chapter 6). In many cases, the growth of large WC

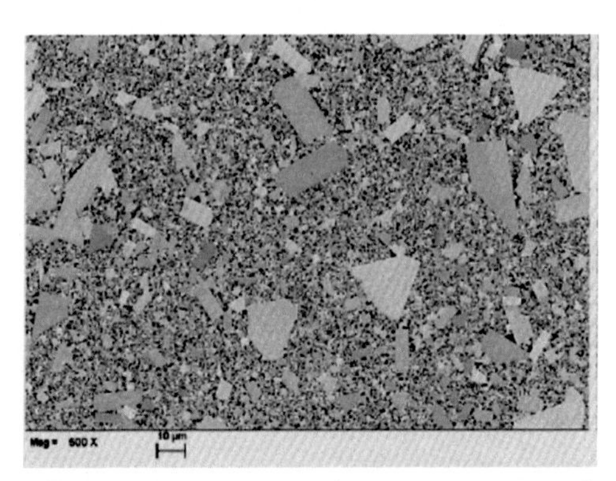

Figure 9.5 Abnormally large WC grains in the microstructure of WC–Co cemented carbide. *(Redrawn from Ref. [8].)*

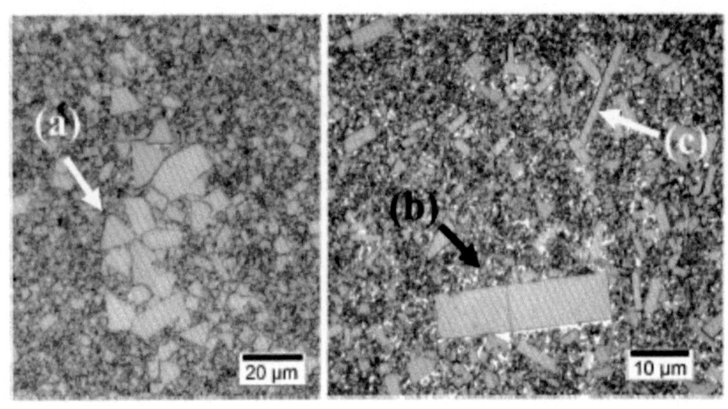

Figure 9.6 Abnormally large WC grains in the microstructure of cemented carbide according to Ref. [9]: (A)—isotropic grains formed as a result of the presence of large WC grains in the initial WC–Co mixture; (B and C)—anisotropic WC grains in form of platelets growing during sintering.

grains can be suppressed or eliminated by adding grain growth inhibitors. However, almost all the grain growth inhibitors (except for tantalum carbide and chromium carbide) are unacceptable for the fabrication of cemented carbides for mining and construction, as their additions lead to reduced fracture toughness.

Various approaches to the elimination of the growth of abnormally large WC grains in the microstructure of WC–Co cemented carbide are described in literature. In Ref. [11], the authors proposed a method of preliminary solid-phase sintering at elevated temperatures, which reduced the probability of formation of abnormally large carbide grains. A novel technology for manufacturing W and WC powders is described in Ref. [12]; it is said to completely prevent the formation of abnormally large WC particles in tungsten carbide powders thus eliminating the presence of large WC grains in the microstructure. In Ref. [13], the authors proposed preliminary sintering of carbide compacts in nitrogen, which also prevented the formation of large WC grains in the microstructure.

Another viewpoint on the microstructure of coarse-grain cemented carbides is reflected in the microstructure of the XT-type grades developed by Sandvik (Sweden) according to Ref. [14]. The microstructure of such grades is said to be characterized by a very uniform microstructure, so that only WC grains of nearly the same size are present in it, whereas the fine-grain fraction is absent. It is reported that cemented carbide of the XT48 grade has higher performance properties because of the use of new technological approaches, which are not disclosed. It can be assumed that the

XT48 grade contains less Co than conventional grades with the same hardness and fracture toughness, as it is characterized by a higher density, which means a higher WC content.

9.3 Carbon content

There are numerous publications in literature on the effect of carbon content on the mechanical properties of cemented carbide, the most important of which is the work of Suzuki and Kubota [15]. The main results of this work are presented in Fig. 9.7. As can be seen in Fig. 9.7, the TRS drastically decreases beyond the border lines of the two-phase region as a result of the appearance of inclusions of either η-phase or free carbon in the microstructure. Therefore, the presence of the η-phase or free carbon in industrial cemented carbides is unacceptable.

The maximum TRS value is found in WC–Co materials lying in the two-phase region near the border line corresponding to the formation of free carbon. The carbon content affects values of the hardness and Co-based binder lattice parameter within the two-phase region. The hardness increases with a decrease in the carbon content due to an increase in the content of tungsten dissolved in the binder, which can be evaluated on the basis of the lattice parameter of the Co-based binder phase. It was found that the microhardness of the Co-based binder phase decreases from 4.5 GPa for low-carbon alloys with 25% Co to 2.5 GPa for the corresponding high-carbon alloys [16].

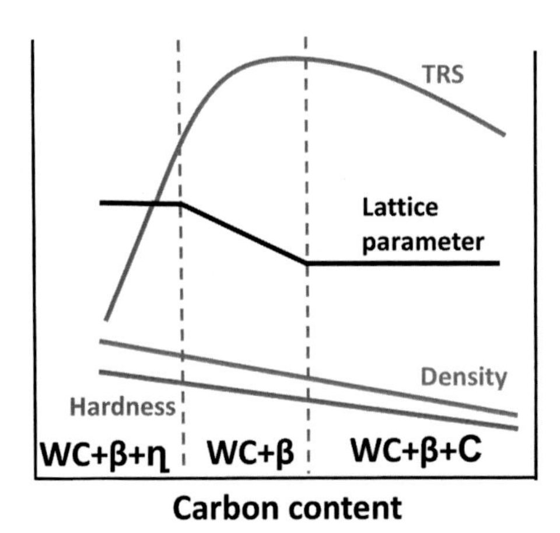

Figure 9.7 Curves indicating the TRS, density, hardness, and Co-based binder crystal lattice parameter of the cemented carbide with 10 wt.% Co cobalt binder depending on the carbon content [15].

The results mentioned above should be considered when taking into account ref. [17] devoted to the study of the dependence of the WC recrystallization rate on the carbon content during liquid-phase sintering. It was found that even within the two-phase region, the recrystallization rate of the WC fine-grain fraction in low-carbon cemented carbides is about 6 times lower than that in the high-carbon cemented carbides. The microstructures of cemented carbides with low and high carbon contents are shown in Fig. 9.8. For the abovementioned reason, coarse-grain cemented carbides with a low carbon content reported in Ref. [15] should be noticeably finer and therefore harder due to the presence of a larger amount of fine WC fraction in their microstructure in comparison with the high-carbon carbide samples.

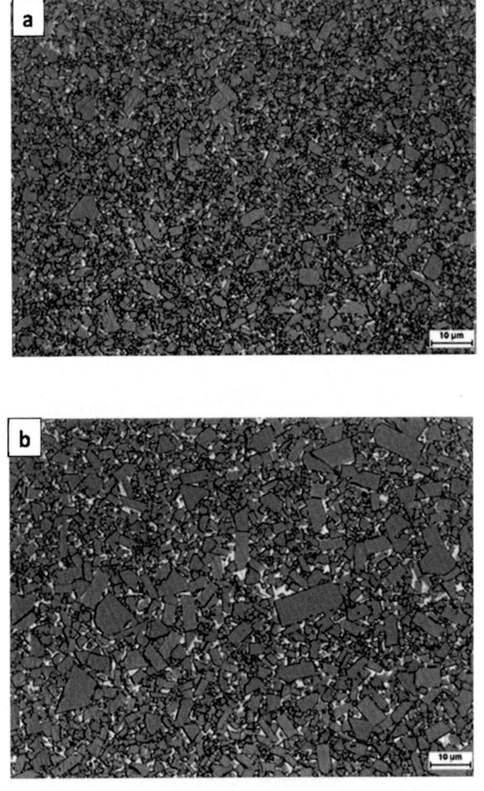

Figure 9.8 Microstructure of WC–10% Co alloys with (A) low and (B) high carbon contents after sintering at 1460°C for 60 min (optical microscopy, after etching in the Murakami reagent). *(Redrawn from Ref. [17].)*

A mechanism explaining the phenomenon mentioned above is proposed in Ref. [17]. It explains this phenomenon by inhibiting the process of dissolving fine WC grains in a liquid cobalt binder during sintering due to the absence of carbon on the surface (or in the surface layer with a thickness of two—three atomic monolayers) of the WC grains due to the high rates of dissolution of carbon atoms in the liquid binder containing very little dissolved carbon. The carbon supply from the inner part of the WC grain towards the WC/binder interface is in turn limited by the internal self-diffusion of carbon in tungsten carbide. The authors of Ref. [17] came to this conclusion on the basis of the closeness of the values of the experimentally established apparent activation energy of the WC grain growth in alloys with a low carbon content and the activation energy of carbon self-diffusion in tungsten carbide.

When taking into account the abovementioned, coarse-grain cemented carbides are usually produced with a high carbon content (near to the region of formation of free carbon in accordance with the W—Co—C phase diagram), and fine-grain cemented carbides are conventionally produced with a low carbon content.

Carbon content in cemented carbides can be controlled by adjusting the total carbon content in WC—Co graded powders and by carburizing or decarburizing green carbide articles during presintering. The carburization is usually carried out using hydrogen-based gas mixtures containing methane or other hydrocarbons [18]. The decarburization is conventionally conducted in hydrogen or CO_2 [19].

Another important aspect related to the effect of carbon on the microstructure and properties of cemented carbide is the difference in the wettability of tungsten carbide by liquid cobalt binders with different carbon contents, which is discussed in detail in Section 9.7.

9.4 Composition and state of the binder phase

The composition and state of the binder play an important role in obtaining the optimum properties of WC—Co cemented carbides, especially coarse-grain and ultracoarse-grain alloys. As it was mentioned above, the thickness of the binder interlayers in such cemented carbides is comparable to the size of carbide grains, so that the properties of the binder phase, especially its hardness, play a decisive role in many applications. A conventional Co-based binder in WC—Co cemented carbides has a relatively low hardness, which leads to its rapid wear during the operation of carbide tools, leading to intensive wear of the carbide insert or carbide tool as a whole. As it is shown above, the hardness of the cobalt-based binder of cemented carbides can be slightly increased by reducing the carbon content within the two-phase region and, accordingly, increasing the content of tungsten dissolved in the binder (see Fig. 9.7). In connection with this, numerous attempts were made to increase the hardness and wear resistance of the binder due to its dispersion hardening or nanostructuring achieved by various processes of heat treatment [20–24]. Ref. [15] describes the results of experiments on heat treatment (aging) of cemented carbides with different carbon contents; results of experiments on the dispersion hardening of cemented carbide binders due to heat treatments are described in Refs. [20–24] (see also Chapter 5).

Some other methods for hardening the binder phase of cemented carbides are disclosed in literature, in particular by introducing nanoparticles of alumina [25], carbides or carbonitrides of refractory metals [26,27], diamond or cubic boron nitride [28], carbon nanotubes [29], etc. Such technological approaches have not found application in the carbide industry because of the strong undesirable interaction of such nanoparticles with the liquid cobalt binder during sintering leading to their fast dissolution. The only type of industrial cemented carbides with nanograin reinforced binders is presently the carbides grades produced by Element Six GmbH under the trademark MasterGrade™ [30] (see Chapter 11).

9.5 Origin of tungsten carbide powders

The mean grain size and homogeneity of WC powders strongly depend on the parameters of tungsten reduction and carburization of metallic tungsten, which is described in detail in Chapter 6. Nevertheless, one of the parameters of the production technology, namely the influence of the WC carburization temperature on the mechanical properties of cemented carbides, is briefly described below.

As a result of studying the properties of tungsten carbide obtained by carburization at various temperatures, it was found that high temperatures of carburization significantly affect the plastic characteristics of coarse-grain cemented carbides. A high-temperature carburization technology for producing WC powders was developed in the Soviet Union in the 60s of the last century. It includes the fabrication of tungsten metal powders by high-temperature hydrogen reduction of WO_3 at 900°C–1200°C and high-temperature carburization of such tungsten powders at 2000°C–2200°C; this technology was subsequently patented worldwide [31].

The fact of much higher plastic deformation of WC–Co cemented carbides, obtained from high-temperature WC powders, is well known and described in detail in Ref. [32]. For example, samples of WC–20%Co, obtained from WC powders after carburization at a low temperature (1450°C) and high temperature (2200°C), are characterized by the plastic deformation before failure in compression of, respectively, 2.9%–3.1% and 5.5%–6.8% (Fig. 9.9). The plastic deformation before failure of the carbide samples with 6% Co obtained from the high-temperature carburized WC powder is approximately 2.3 times higher than that of the samples obtained from the low-temperature carburized WC powder. This leads to higher fracture toughness values and improved performance toughness of cemented carbides produced from such high-temperature carburized WC powders.

The reason for the higher plasticity of cemented carbides, obtained from the high-temperature carburized WC, remains a subject of discussion. Some authors believe that a reason for the higher plasticity is the higher degree of perfection and special features of the WC crystal lattice obtained at higher temperatures of carburization [33]. Other authors believe that this is a result of obtaining a cleaner WC powder carburized at higher temperatures, since the content of certain impurities (Mg, Na, As, Ca, and Si) is much lower in the WC powders carburized at high

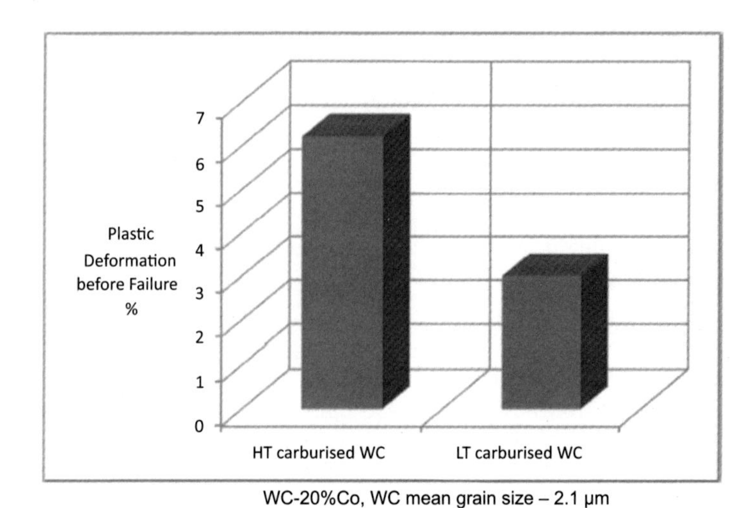

Figure 9.9 Residual plastic deformation before failure of the WC–20% Co grade with a WC mean grain size of 2.1 μm, according to Ref. [32].

temperatures [32]. The higher plasticity of cemented carbides obtained from high-temperature carburized WC powders can be a result of the larger size of WC grains and the special grain size distribution of WC grains in the carbide microstructure [1].

9.6 Inhibitors of WC grain growth

Numerous publications on the influence of various inhibitors of the WC grain growth on the microstructure and properties of cemented carbides are summarized in a review article [34].

All submicron, fine-grain, ultrafine-grain, and near-nano cemented carbides are produced using grain growth inhibitors consisting of transition metals' carbides to suppress the WC grain growth during liquid-phase sintering.

Typically, grain growth inhibitors are added to WC–Co powder mixtures in amounts not exceeding their solubility limits in the cobalt binder phase (except for TaC, TiC, and NbC, the solubility of which in the binder at room temperature is very insignificant). The values of the solubility limits of the most commonly used inhibitors of grain growth in various metallic binders based on different iron group metals are presented in Table 9.1.

Curves characterizing the effect of different grain growth inhibitors on the WC mean grain size are shown in Fig. 9.10. In Fig. 9.10, it can be seen that the grain growth inhibitors can be ranked in terms of their effect on the inhibition of grain growth in the following order: VC > NbC > TaC/TiC > Mo_2C/Cr_3C_2 > ZrC/HfC [36]. The results obtained in Ref. [36] were confirmed in a similar study on the influence of various grain growth inhibitors on the characteristics of the microstructure of the WC–6%Co grades in Ref. [37].

It should be noted that almost all the curves shown in Fig.9.10 are characterized by the saturation region achieved with increasing the content of the grain growth inhibitors to the level of their solubility limit in the

Table 9.1 The solubility limits (wt.%) of grain growth inhibitors (in comparison with WC) in Co, Ni, and Fe-based binders at 1250°C according to Ref. [35].

Binder metal	WC	TiC	VC	NbC	TaC	Cr_3C_2	Mo_2C
Co	22	1	6	5	3	12	13
Ni	15	5	7	3	5	12	8
Fe	7	0.5	3	1	0.5	8	5

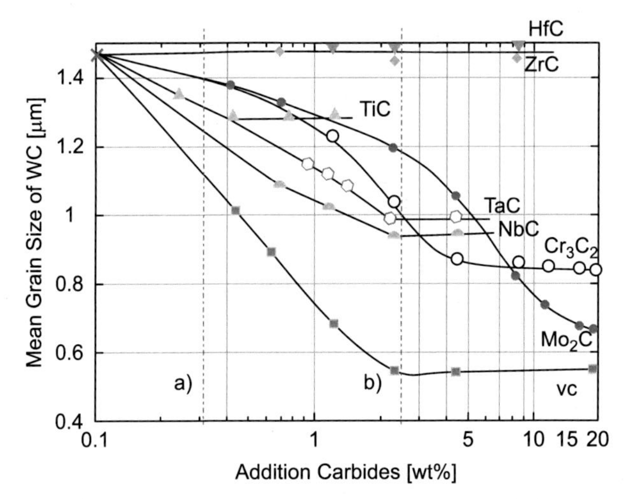

Figure 9.10 The effect of additions of various carbides on the WC mean grain size of the fine-grain cemented carbide with 20 wt.% Co [36]. *(Redrawn from Ref. [34].)*

Co-based binder. Mo_2C is the only exception due to the fact that it dissolves both in the binder and in the grains of tungsten carbide, so it is characterized by the very high solubility in cemented carbides.

As it can be seen from Fig. 9.10, vanadium carbide (VC) has the strongest effect on the grain growth inhibition, so that it is widely used in industrial ultrafine and near-nano cemented carbides. It should be noted that even small additions of VC have a very detrimental effect on the fracture toughness of cemented carbides; therefore, it is employed in relatively small amounts. As one can see in Fig. 9.10, chromium carbide is a less effective inhibitor with respect to the WC grain growth in comparison with titanium carbide, tantalum carbide, and niobium carbide. However, when taking into account its significantly higher solubility in the Co-based binders compared to the other grain growth inhibitors, it can be added to WC−Co mixtures in larger amounts, at which it becomes a more effective grain growth inhibitor than the abovementioned carbides. In addition, chromium carbide additives have a much less pronounced effect on fracture toughness of cemented carbides than VC. Due to this fact, it is widely used in submicron and ultrafine cemented carbides, for which it is important to achieve the maximum combination of hardness and fracture toughness. According to Ref. [38], chromium carbide can be employed in medium-coarse grain mining grades subjected to high impact loads; tantalum carbide can be also used for such mining grades [3].

Modern concepts with respect the mechanisms of the influence of grain growth inhibitors on the microstructure of cemented carbides are related to the experimentally established fact of the segregation of grain growth inhibitors on the surface of WC grains (see also Chapter 14). They form a thin carbide layer consisting of (W,Me)C, where Me is a metal forming the carbide phase of the grain growth inhibitor [39,40]. Fig. 9.11 shows the WC/Co interface in the cemented carbide doped with VC with atomic resolution. It is characterized by the presence of a thin layer of cubic carbide (V,W)C (of the order of several atomic monolayers). The presence of such layers was established at the grain boundaries of cemented carbides containing additions of practically all the grain growth inhibitors (except for molybdenum carbide) according to Ref. [40]. In Ref. [40], it was also found that almost all the grain growth inhibitors except for TaC segregate at WC/WC grain boundaries. It is believed that thin mixed carbide layers comprising metallic components of the grain growth inhibitors are present on the surface of fine grains of tungsten carbide during liquid-phase sintering thus suppressing their dissolution in the liquid phase during sintering and stabilizing the fine-grain microstructure of submicron, ultrafine, and near-nano cemented carbides.

Chromium carbide and VC are mainly employed for submicron and ultrafine carbide grades having high hardness values for applications in metal cutting and wear protection [34], whereas TaC is used for coarse-grain and medium-grain grades having a lower hardness and high fracture toughness, for example, carbide grades for mining applications [3,12].

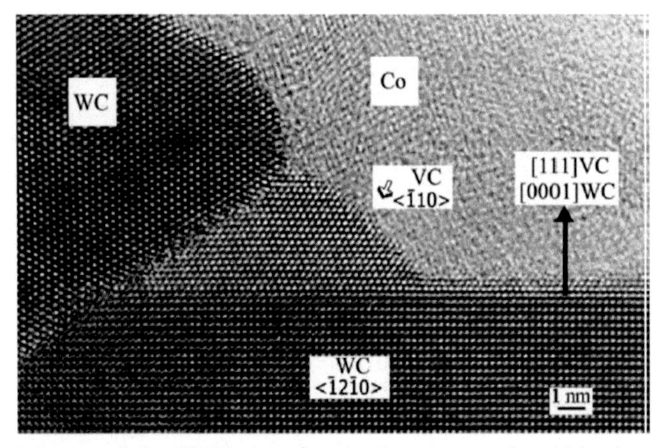

Figure 9.11 Image of the WC/Co interface in the cemented carbide doped with VC (high-resolution transmission electron microscopy). *(Redrawn from Ref. [39].)*

9.7 Shape and contiguity of WC grains

As a rule, WC—Co cemented carbides consist of WC single crystals, which have a shape of truncated trigonal prisms with sharp edges. It is believed that the sharp edges of adjacent WC grains lead to the stress concentration in cemented carbides during operation, thereby contributing to the initiation and propagation of cracks. According to the results of ref. [42], cemented carbides with rounded WC grains that do not contain sharp edges have an increased fracture toughness. According to Ref. [43], cemented carbides with rounded WC grains are characterized by a high thermal conductivity, which has a positive effect on the performance of carbide tools operating under conditions of severe thermal fatigue and elevated temperatures. Cemented carbides with rounded carbide grains can be obtained by cladding coarse-grain WC powders with cobalt. This makes it possible to eliminate the formation of a fine WC fraction during milling of WC—Co graded powders. As a result, the recrystallization of such a fine-grain fraction on large WC grains during sintering can be suppressed, thus leading to the elimination of the process of faceting such large grains of tungsten carbide during sintering. It is well known the initial WC single crystals forming as a result of carburization of tungsten metal powders are characterized by the original round shape, so that if they are not intensively milled, their rounded shape remains in the absence of the fine-grain WC fraction during sintering. Cemented carbides with rounded WC grains can therefore be obtained by replacing the milling process of WC—Co mixtures by the deagglomeration of tungsten carbide polycrystals obtained as a result of carburization followed by mixing the WC single crystals obtained in such a way with Co.

Another approach to the fabrication of cemented carbides with unusual WC grain shape was developed and patented by Toshiba Tungaloy Co. Ltd. in Japan in 1996 [44]. In this case, WC—Co cemented carbides are produced with WC grains in from of thin platelets or discs. The so-called "Disc Reinforced Technology" is said to allow obtaining a better combination of hardness and fracture toughness. This is believed to be achieved because the WC platelets have a higher percentage of the harder WC crystal planes, and the crack nucleation and propagation can be suppressed in such cemented carbides, which is illustrated in Fig. 9.12 [45]. The technology for fabrication of such cemented carbides includes the production of tungsten powders in form of flat platelet-like particles and their carburization by adding a carbon-source material, preferably graphite,

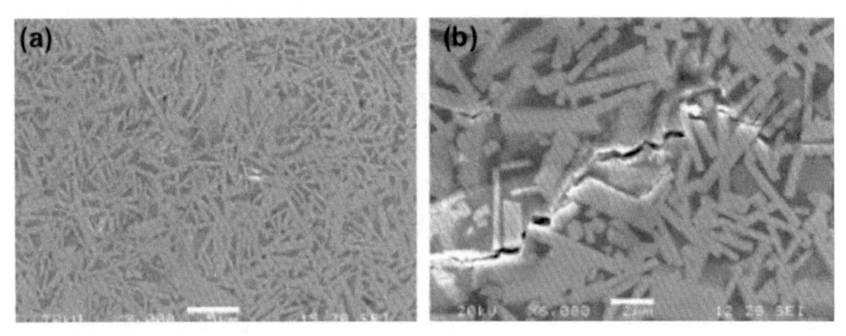

Figure 9.12 SEM image of the microstructure (A) and the path of crack propagation (B) in cemented carbide containing carbide grains in form of platelets. *(Redrawn from Ref. [45].)*

during liquid-phase sintering [44]. Nam et al. [45] suggested to produce platelet-reinforced cemented carbides by decomposition of (Ti,W)C oversaturated with tungsten during liquid-phase sintering. This is said to result in the formation of WC platelets with high aspect ratios shown in Fig. 9.12 leading to an improved combination of hardness and fracture toughness.

A technological approach based on the "rounding" the WC grains in WC–Co cemented carbides mentioned above can also affect the contiguity of the WC grains in the microstructure, which is indirectly related to the presence or absence of a continuous carbide skeleton in cemented carbides. According to Ref. [2], the hardness increases and TRS decreases with increasing a rate of the contiguity of the WC grains.

Nevertheless, in light of the results on the composition and structure of the WC/WC grain boundaries in cemented carbides [40,46,47], it is now clearly established that the most WC grains are separated from each other by thin Co interlayers or Co segregations of different thicknesses varying from the order of about one atomic monolayer or a half of atomic monolayer up to several nanometers. In this connection, the contiguity of carbide grains at the WC/WC grain boundaries, visible in optical microscopes or SEMs as direct contacts between adjacent WC grains, appears not to be an important parameter affecting mechanical and performance properties of cemented carbides. Also, the existence of a continuous carbide skeleton in the cemented carbide structure should be considered when taking into account the presence or absence of the Co interlayers and their thickness.

The results of Ref. [47] on the study of the wetting of tungsten carbide by liquid cobalt alloys with different carbon contents and WC/WC grain boundaries in cemented carbides with various carbon contents shed new light on the issue regarding the presence or absence of a carbide skeleton in

WC–Co materials. Fig. 9.13 shows droplets of the Co-based binder model alloys with different carbon contents after their melting and spreading on the surface of WC substrates. As it can be seen in Fig. 9.13A, the binder alloy with a low carbon content completely spreads over the surface of tungsten carbide; the wetting contact angle in this case is equal to 0. The binder alloy with a higher carbon content did not undergo complete

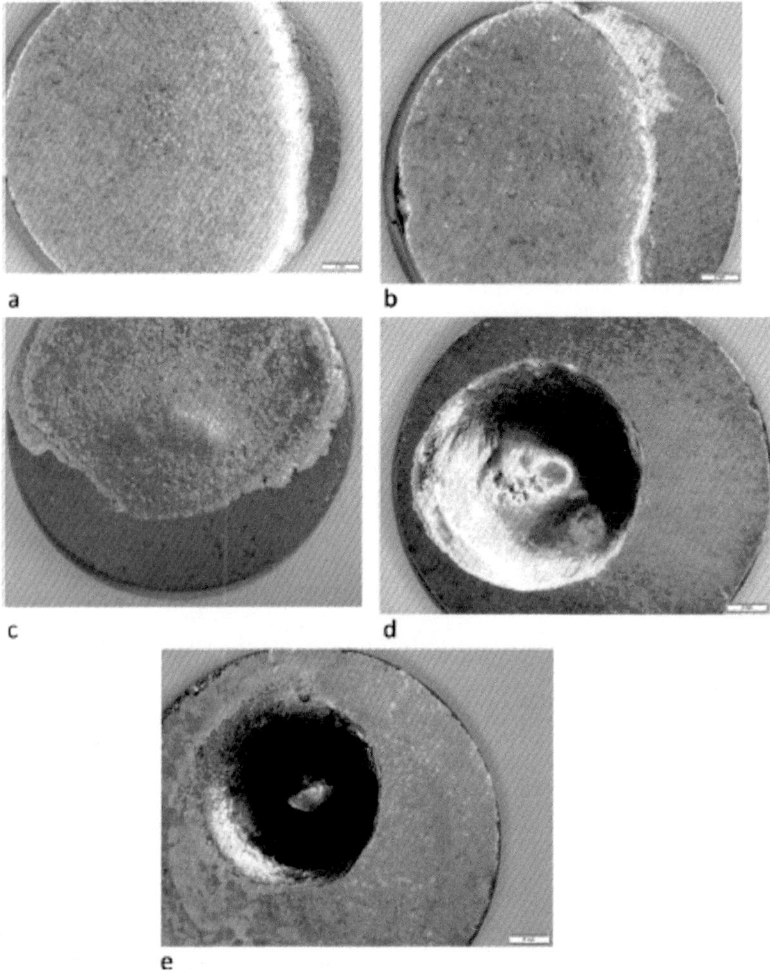

Figure 9.13 Appearance of the droplets of the binder model alloys with different carbon contents after their melting and spreading on the surface of WC substrates for different times: (A)—low carbon content, 35 s; (B)—medium-low carbon content, 35 s; (C)—medium-high carbon content, 35 s; (D)—high carbon content corresponding to the free carbon formation, 35 s; (E)—same as (D), 120 s. *(Redrawn from Ref. [47].)*

spreading on the WC surface (Fig. 9.13C). The binder alloy with a very high carbon content is characterized by relatively poor wettability with respect to WC (Fig. 9.13D and E); in this case, the wetting contact angle is equal to about 15 degrees indicating the incomplete wetting.

The different degree of wetting tungsten carbide by Co-based liquid alloys with various carbon contents has an impact on the structure of the WC/WC grain boundaries in cemented carbides with different carbon contents. As one can see in Fig. 9.14, the majority of WC/WC grain boundaries in the carbide sample with a low carbon content comprise relatively thick Co interlayers (of the order of several nanometers), which indicates the absence of the carbide skeleton in this case. In the sample with a high carbon content, the presence of nm-thick cobalt interlayers at the WC/WC grain boundaries is not established in the majority of WC/WC grain boundaries (Fig. 9.15), although the presence of Co segregations of the order of one atomic monolayer or less at the grain boundaries cannot be excluded. Thus, in the cemented carbides with the high carbon contents (close to those corresponding to the border line with the free carbon formation in the W—C—Co phase diagram), the presence of a more developed carbide skeleton can be assumed. Nevertheless, the concept of the "carbide skeleton" itself in WC—Co cemented carbides must undergo significant changes and be considered in terms of thickness of the Co interlayers or segregations present at WC/WC grains boundaries [48].

9.8 Impurities and contaminations

Among numerous publications describing the influence of trace elements on the microstructure and properties of cemented carbides, one should consider refs. [41,49,50], in which the influence of a great number of impurities and contaminations on the characteristics of W and WC powders as well as WC—Co cemented carbides was systematically examined.

In ref. [49], trace elements in the concentration of 10—200 ppm added to tungsten blue oxide prior to the hydrogen reduction were found to have a considerable effect on the reduction kinetics and properties of the resulting W powders. The particle size, particle shape, and Scott density of the W powders were affected by the trace elements even in cases when they were partially or fully volatilized during the reduction process. The concentration of the trace elements present in the W powders after reduction was further decreased as a result of carburization. The final concentration of the impurities remaining in the WC powder after carburization was found

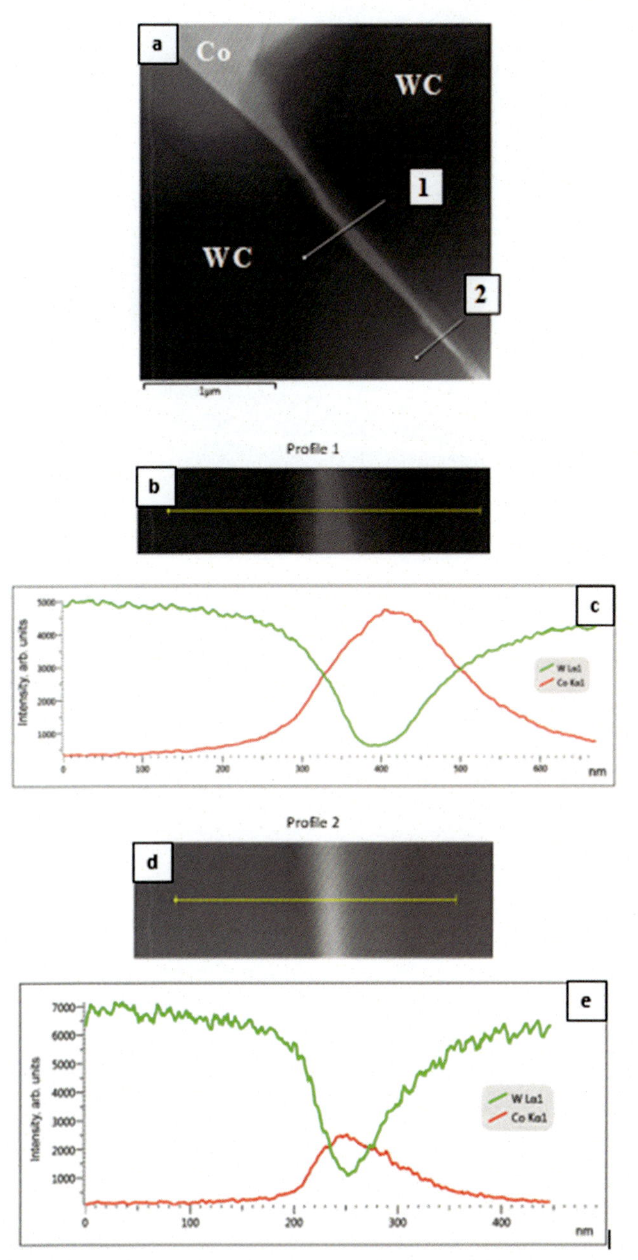

Figure 9.14 WC—WC grain boundary in the cemented carbide with a low carbon content: (A)—the image of the WC/WC interface with lines showing where the element distribution profiles were recorded (transmission electron microscopy), (B, D) - the WC/WC interfaces with the scanlines of the profile 1 and profile 2 at a higher magnification, (C, E) —curves indicating the distributions of chemical elements [47].

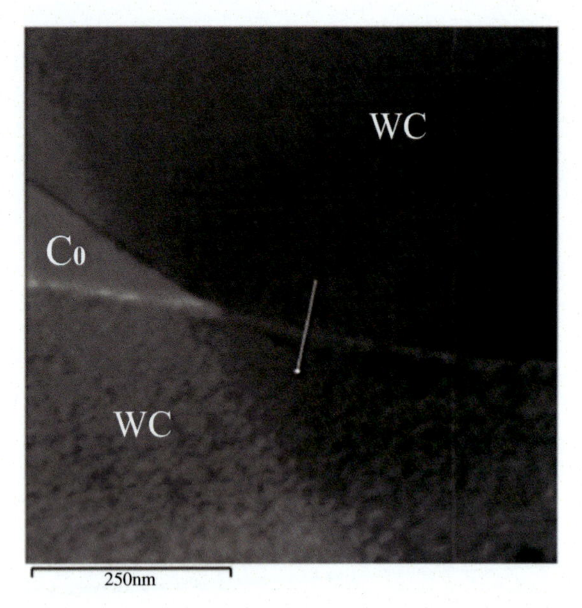

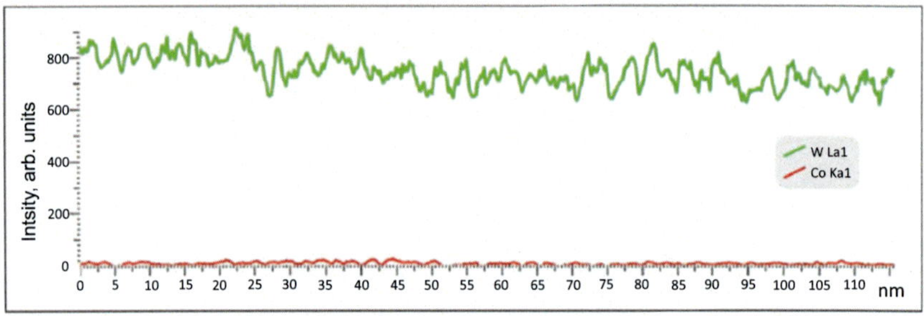

Figure 9.15 WC—WC grain boundary in cemented carbide with a high carbon content: the image of the grain boundary above comprises a line showing where the element distribution profiles were recorded (transmission electron microscopy); and curves indicating the distributions of tungsten and cobalt are shown below. *(Redrawn from Ref. [47].)*

to depend on the nature of the chemical element, its initial content in the tungsten oxide powder, duration, and temperature of the reduction and carburization processes. The impurities were found to have a significant influence on the WC mean grain size and grain size distribution in the WC powders after carburization. All the chemical elements added to the initial tungsten oxide powders can be divided into the following groups:

1. Elements totally removed during hydrogen reduction independently on the reduction conditions: zinc, sulfur, chlorine, and fluorine.

2. Elements partially evaporated during hydrogen reduction depending on the reduction conditions: sodium, potassium, lithium, boron, and phosphorous.
3. Elements remaining in the tungsten metal powder after hydrogen reduction: tin, nickel, copper, aluminum, magnesium, calcium, and barium.

It was found that impurities of the alkali metals enhance the grain growth during the fabrication of W powders by hydrogen reduction of tungsten oxides.

During the carburization process, the concentration of the most trace elements decreases; the final concentration of the trace elements depends on the carburization temperature. Nevertheless, the following chemical elements are retained in the WC powder completely during carburization: aluminum, uranium, nickel, rare earth, antimony, bismuth, phosphorous, and silicon. Elements such as copper, calcium, and barium are partially removed as a result of carburization. The following elements are completely removed during the carburization process: lithium, sodium, potassium, magnesium, boron, tin, and arsenic. The concentration of all other alkali metals is reduced to the level that is below the analytical detection limit (10 ppm).

It was found that some trace elements have an inhibiting effect of the WC grain growth when carburizing fine-grain W powders; these elements are as follows: uranium, calcium, boron, aluminum, silicon, and particularly alkali metals [50].

Very important information on the influence of trace elements on the microstructure and properties of cemented carbides was reported and summarized in Ref. [41]. The most unfavourable contaminations present in the initial powders negatively affecting the structure of fractured surfaces of sintered carbide articles are the following:

- Aluminum leading to the formation of pores and inclusions of aluminum oxide of 10–60 μm in size;
- Calcium leading to the formation of large inclusions of Ca and CaS varying from about 5 μm to nearly 50 μm;
- Phosphorous leading to the formation of large binder pools in the microstructure (although no P was detected in these pools);
- Silicon leading to the formation of large oxide inclusions + calcium sulfite inclusions comprising much dissolved Si and having sizes of up to 50 μm.

The following traces elements have the major negative influence on the TRS of sintered carbide articles:

- If only one among the following elements—Ca, Si, P and Al—in small amounts is present as a dopant, the TRS value is almost not affected, although the scattering of the results of the TRS measurements was slightly increased;
- For the samples simultaneously doped with Ca and S, a significant decrease of the TRS was established as a result of the presence of large CaS inclusions in the microstructure.

As for the performance properties of cemented carbides in metal cutting, the samples simultaneously doped with Ca + S and P + S were found to exhibit reduced performance characterized by numerous early premature failures of cutting inserts in metal-cutting tests. It was established that alkali metals impurities inhibit the WC grain growth during sintering of WC−10% Co samples. Doping the carbide samples with Mg, Zn, P and Sn leads to enhancing the WC grain growth during liquid-phase sintering.

References

[1] G.S. Kreimer, Strength of Hard Alloys, Consultants Bureau, New York, USA, 1968.
[2] H. Exner, J. Gurland, A review of parameters influencing some mechanical properties of tungsten carbide-cobalt alloy, Powder Metall. 13 (1970) 13−31.
[3] I. Konyashin, Cemented Carbides for Mining, Construction and Wear Parts, Comprehensive Hard Materials, Editor-in-Chief V.Sarin, Elsevier Science and Technology, 2014, 425-251.
[4] J. Gurland, P. Bardzil, Relation of strength, composition, and grain size of sintered tungsten-carbide-cobalt alloys, JOM 7 (1955) 311−315.
[5] W. Schatt, K.P. Wieters, B. Kiebach, Pulvermetallugie, Springer, Würzburg, Germany, 2007.
[6] C.J. Terry, Makrocrystalline Tungsten Monocarbide Powder and Process for Production, 1989. US Patent 4,834,963.
[7] I. Konyashin, S. Farag, B. Ries, B. Roebuck, WC-Co-Re cemented carbides: structure, properties and potential applications, Int. J. Refract. Metals Hard Mater. 78 (2019) 247−253.
[8] K. Mannesson, I. Borgh, et al., Abnormal grain growth in cemented carbides − experiments and simulations, Int. J. Refract. Metals Hard Mater. 29 (2011) 488−494.
[9] I. Konyashin, A. Anikeev, V. Senchihin, V. Glushkov, Development, Production and Application of Novel Grades of Coated Hardmetals in Russia, Int. J. Refract. Metals Hard Mater. 14 (1996) 41−48.
[10] M. Schreiner, T. Schmitt, E. Lassner, B. Lux, On the origin of discontinuous grain growth during liquid phase sintering of WC-Co cemented carbides, Powder Metall. Int. 16 (1984) 180−183.
[11] D.-Y. Yang, S.-J. Kang, Suppression of abnormal grain growth in WC-Co via pre-sintering treatment, Int. J. Refract. Metals Hard Mater. 27 (2009) 90−94.
[12] I. Konyashin, T. Eschner, F. Aldinger, V. Senchihin, Cemented carbides with uniform microstructure, Z. Metallkd. 90 (1999) 403−406.

[13] P. Gusafson, M. Wandenstroem, S. Norgren, Manufacture of Fine-Grained WC-Co Alloy by Sintering Under Nitrogen Before Pore Closure, 2004. European Patent Application EP1500713.
[14] Internet site https://www.rocktechnology.sandvik/en/campaigns/powercarbide/. Dated April 2021.
[15] H. Suzuki, H. Kubota, The influence of binder phase composition on the properties of tungsten carbide-cobalt cemented carbides, Planseeberichte fuer Pulvermetallurgie 14 (2) (1966) 96–109.
[16] I. Konyashin, B. Ries, F. Lachmann, R. Cooper, A. Mazilkin, B. Straumal, A. Aretz, Hardmetals with nano-grain reinforced binder: binder fine structure and hardness, Int. J. Refract. Metals Hard Mater. 26 (2008) 583–588.
[17] I. Konyashin, S. Hlawatschek, B. Ries, F. Lachmann, T. Weirich, F. Dorn, A. Sologubenko, On the mechanism of WC coarsening in WC-Co hardmetals with various carbon contents, Int. J. Refract. Metals Hard Mater. 27 (2009) 234–243.
[18] R. Lueth, Method for Carbon Control of Carbide Performs, 1986. US Patent 4,579,713.
[19] Y. Taniguchi, H. Sasaki, M. Ueki, K. Kobopri, Manufacture of Surface-Refined Sintered Alloy, 1986. Japan Patent Application 1986-314817.
[20] H. Jonsson, Studies of the binder phase in WC-Co cemented carbides heat-treated at 650°C, Powder Metall. 15 (1972) 1–10.
[21] H. Jonsson, Studies of the binder phase in WC-Co cemented carbides heat-treated at 950°C, Planseeber. für Pulvermetall. 23 (1975) 37–55.
[22] C. Wirmark, G.L. Dunlop, Phase transformation in the binder phase of Co-W-C cemented carbides, in: R.K. Viswandham, D. Rouclihle, J. Gurland (Eds.), Proc. Int. Conf. Sci. Hard Mater., Plenum, New York, 1983, pp. 311–327.
[23] D.L. Tillwick, I. Joffe, Precipitation and magnetic hardening in sintered WC-Co composite materials, J. Phys. D 6 (1973) 1585.
[24] H. Grewe, J. Kolaska, Gezielte Einstellungen von Lösungszuständen in der Binderphase technischer Hartmetalle und Folgerungen daraus, Metall 7 (1981) 563.
[25] L. Sun, J. Xiong, Z. Guo, Effects of nano-Al_2O_3 additions on microstructures and properties of WC-8Co hard metals, Adv. Mater. Res. 97–101 (2009) 1649–1652.
[26] W. Bryant, Schneideinsatz zum Fräsen von Titan und Titanlegierungen, 1981. German Patent Application DE19810533A1.
[27] S. Kang, Solid-Solution Powder, Method to Prepare the Solid-Solution Powder, Cermet Powder Including the Solid-Solution Powder, Method to Prepare the Cermet Powder and Method to Prepare the Cermet, 2009. US Patent Application US2009-0133534.
[28] Y. Zhang, G. Zhan, X. Sheng, A. Locksetdt, A. Griffo, Y. Shen, H. Deng, M. Keshavan, Nano-Reinforced WC-Co for Improved Properties, 2008. US Patent Application US2008-179104A1.
[29] J.D. Belnap, G. Zhan, X. Sheng, et al., Polycrystalline Composites Reinforced with Elongated Nanostructures, 2007. US Patent Application 2007-939350A1.
[30] I. Konyashin, B. Ries, F. Lachmann, Cemented Carbide Material, WO2012/130851, 2012.
[31] V.A. Ivensen, O.N. Eiduk, V.A. Fal'kovskii, N.M. Lukashova, L.A. Chumakova, Z.A. Gol'dberg, Process for Preparing Cemented Tungsten Carbides, 1977. Great Britain Patent 1483527.
[32] V.I. Tretyakov, Bases of Materials Science and Technology of Sintered Cemented Carbides, Metallurgiya, Moscow, 1977.
[33] V.A. Ivensen, O.N. Eiduk, S.I. Artemev, The dependence of plasticity of WC-Co cemented carbide on the carburisation temperature of tungsten und tungsten carbide, in: Hardmetals, Metallurgiya, Moscow, 1970.

[34] S. Farag, I. Konyashin, B. Ries, The influence of grain growth inhibition on the microstructure and properties of submicron, ultrafine and nano-structured hardmetals - a review, Int. J. Refract. Metals Hard Mater. 7 (2018) 12—30.
[35] H. Exner, Physical and chemical nature of cemented carbides, Int. Met. Rev. 24 (1) (1979) 149—173.
[36] K. Hayashi, Y. Fuke, H. Suzuki, Effects of addition carbides on the grain size of WC-Co alloy, J. Jpn. Soc. Powder Powder Metall. 19 (2) (1972) 67—71.
[37] H. Grewe, H. Exner, P. Walter, Behinderung des kornwachstums in hartmetall-legierungen vom iso-k10-typ durch zusatzkarbide, Z. Metallkde 64 (1973) 85—95.
[38] M. Martensson, I. Arvanitidis, K. Turba, Rock Dril Insert, 2020. US Patent Application 2020/030886A1.
[39] S. Lay, S. Hamar-Thibault, A.V.C. Lackner, Cr_3C_2 codoped WC-Co cermets by HRTEM and EELS, Int. J. Refract. Metals Hard Mater. 20 (2002) 61—69.
[40] J. Weidow, H.-O. Andrén, Grain and phase boundary segregation in WC—Co with TiC, ZrC, NbC or TaC additions, Int. J. Refract. Metals Hard Mater. 29 (2011) 38—43.
[41] H.M. Ortner, in: T. Valente (Ed.), The Influence of Trace Elements on the Properties of Hard Metals, European Commission, Brussels, 1997.
[42] R.-P. Herber, W.-D. Schubert, B. Lux, Hardmetals with "rounded" WC grains, Int. J. Refract. Metals Hard Mater. 24 (2006) 360—364.
[43] J. Akerman, T. Ericson, Cemented Carbide Body with Improved High Temperatures and Thermomechanical Properties, 1997. US Patent 6,126,709.
[44] M. Kobayashi, S. Kinoshita, Grain-Oriented WC-Base Hard Alloy Having High Fracture Toughness and Wear-Resistance and Its Manufacture, 1995. Japan Patent JP08199285.
[45] H. Nam, J. Lim, S. Kang, Microstructure of (W,Ti)C—Co system containing platelet WC, Mater. Sci. Eng., A 527 (2010) 7163—7167.
[46] A. Henjered, M. Hellsing, G. Nouet, A. Dubon, J. Laval, Quantitative microanalysis of carbide/carbide interfaces in WC-Co base cemented carbides, Mater. Sci. Technol. 2 (1994) 847—855.
[47] I. Konyashin, A.A. Zaitsev, D. Sidorenko, et al., Wettability of tungsten carbide by liquid binders in WC—Co cemented carbides: is it complete for all carbon contents? Int. J. Refract. Metals Hard Mater. 62 (2017) 134—148.
[48] B.B. Straumal, I. Konyashin, B. Ries, K.I. Kolesnikova, A.A. Mazilkin, A.B. Straumal, A.M. Gusak, B. Baretzky, Pseudopartial wetting of WC/WC grain boundaries in cemented carbides, Mater.Lett. 147 (2015) 105—108.
[49] E. Lassner, M. Schreiner, B. Lux, Influence of trace elements in cemented carbide production. Part 1, Int. J. Refract. Metals Hard Mater. 1 (2) (1982) 54—60.
[50] E. Lassner, M. Schreiner, B. Lux, Influence of trace elements in cemented carbide production. Part 2, Int. J. Refract. Metals Hard Mater. 1 (3) (1982) 97—102.

CHAPTER 10

Conventional industrial grades of cemented carbides

Different industrial grades of cemented carbides for various applications and their microstructures are presented in Refs. [1–7].

Microstructures of typical industrial WC–Co coarse-grain, medium-grain, ultrafine, and submicron grades are shown in Fig. 10.1. As one can see, they are characterized by a high rate of uniformity and do not comprise large or abnormally large WC grains.

Typical industrial cemented carbide grades for different applications are presented in Tables 10.1–10.4. The coarse-grain grades for mining, construction, and other applications characterized by the presence of high impact loads usually do not contain grain growth inhibitors. Table 10.1 show typical industrial medium-coarse, coarse, and ultra-coarse grades for such applications. Properties of a wide range of medium-coarse-grain, coarse-grain, and ultracoarse grain cemented carbides for the mining industry, construction, tools for cold-forming, etc., and other applications that are fabricated by Element Six GmbH (Germany) are shown in Table 10.1. These cemented carbide grades contain 6%–15% Co and are characterized by the Vickers hardness of 10.5–14.5 GPa and transverse rupture strength (TRS) of 3500–4500 MPa. There are also special carbide grades with higher values of fracture toughness manufactured by Element Six for extremely harsh applications.

WC–Co cemented carbides for wear parts and indexable cutting inserts are mainly medium-grain, ultrafine, or submicron grades. The medium-grain grades typically contain small amounts of grain growth inhibitors, and ultrafine, and submicron grades have to contain relatively high amounts of grain growth inhibitors (usually vanadium and chromium carbides) to prevent the WC grain growth during sintering. Table 10.2 shows properties of such carbide grades fabricated by Element Six.

Cemented Carbides
ISBN 978-0-12-822820-3
https://doi.org/10.1016/B978-0-12-822820-3.00010-0

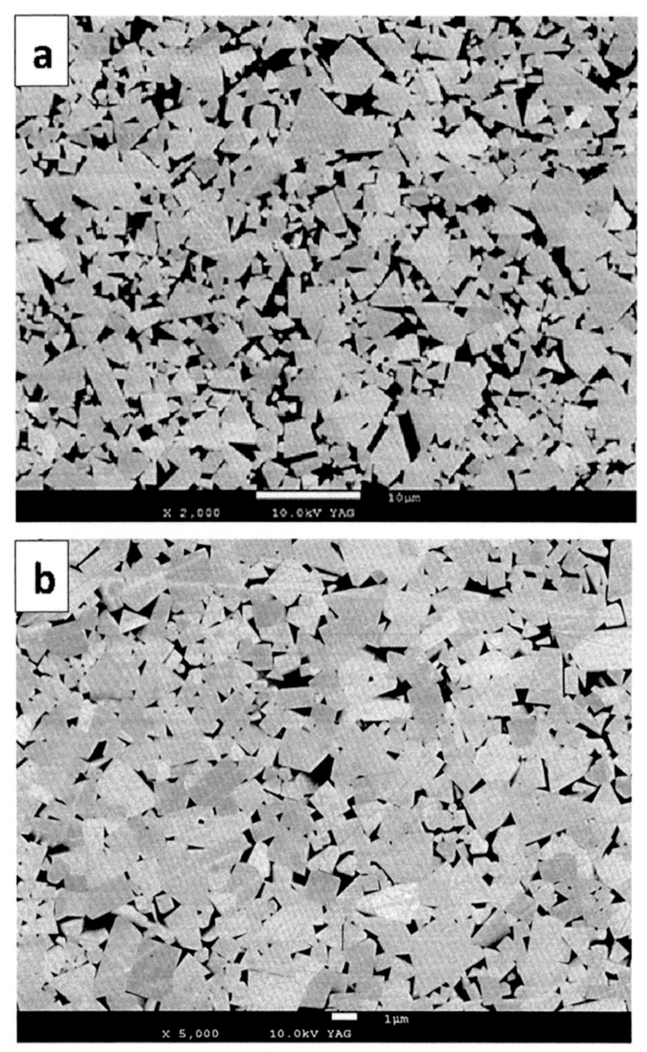

Figure 10.1 Typical microstructures of different industrial WC—Co grades: (A) - coarse-grain grade with 10% Co (B25, Element Six); (B) - medium-grain grade with 6%Co (G10, Element Six); (C) - submicron grade with 10% Co (K010, Element Six), and (D) - submicron grade with 7% Co/Ni (NK07, Element Six). Note that the magnification of the micrograph (A) is different from that of the micrographs (B—D). *(Redrawn from Ref. [1].)*

When wear parts are subjected to medium-high impact loads, medium-grain cemented carbides are employed. Such grades can also be used for metal-cutting, so that many companies do not distinguish grades for these applications. As it can be seen in Table 10.2, the medium-grain grades contain 6%—26% Co and comprise insignificant amounts of grain growth

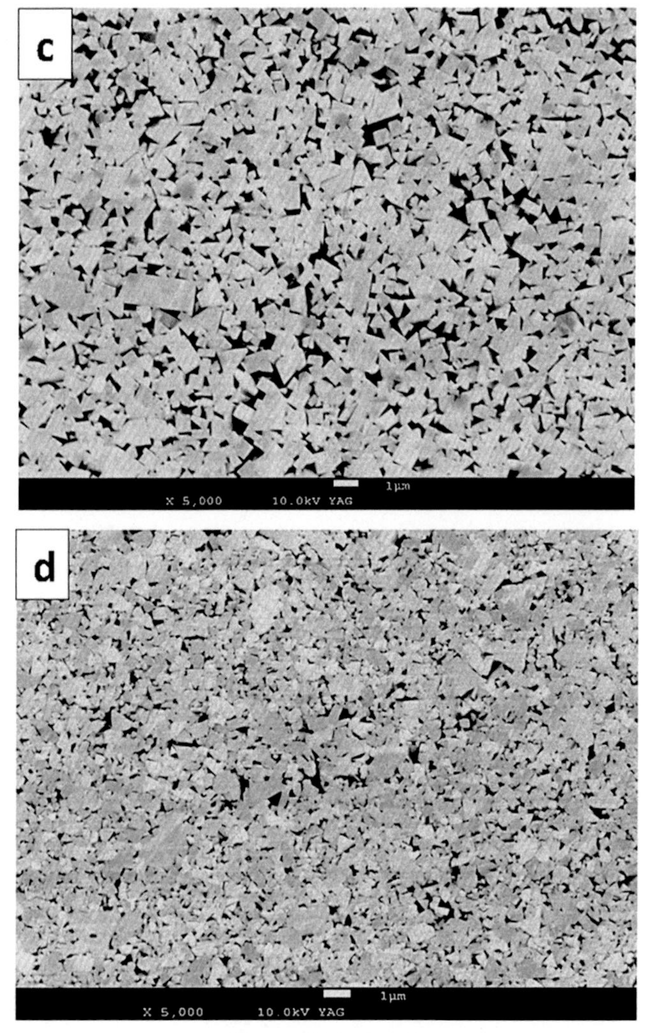

Figure 10.1 Cont'd

inhibitors to prevent the formation of abnormally large WC grains during sintering. The hardness of such grades varies from nearly 900 to 1600 Vickers units. The grades with high Co contents, such as G40 and G55 produced by Element Six, are employed in applications where extremely high fracture toughness is needed, such as stamping tools. In addition, submicron WC–Co grades with the cobalt content of 5–15 wt.% as well as grades with Ni–Co and nickel binders are fabricated; they are

Table 10.1 Industrial cemented carbide WC-Co grades mainly for mining and construction applications fabricated at Element Six GmbH.

Grade	Composition		Hardness $[HV_{20}] \pm 50$	Hardness [HRA] ± 0.3	TRS [N/mm^2]	Compressive strength [N/mm^2]
	Co [%] ± 0.2	Grain size				
T6	6.0	Medium-Coarse	1450	90.6	2400	4500
B20	8.0	Coarse	1250	88.7	2700	4000
B25	10.0	Coarse	1200	88.2	2800	4000
B30	11.0	Coarse	1150	87.7	2700	3900
B40	15.0	Coarse	1050	86.5	2800	3600
B15N	6.5	Ultracoarse	1100	87.2	2200	3600
B20N	8.6	Ultracoarse	1050	86.5	2300	3500
B25SN	9.5	Ultracoarse	1050	86.5	2300	3500

According to Ref. [1].

characterized by various combinations of hardness and fracture toughness. The carbide grades with the Ni–Co and Ni binders are intended for the manufacturing wear-resistant parts that work in corrosive environments. Table 10.2 indicates only basic grades for wear parts, that can be also employed for metal-cutting, produced by Element Six, which are supplemented by many other special grades tailored for special applications.

Table 10.3 shows properties of carbide grades for mill rolls fabricated by Element Six [2]. The assortment of such grades includes coarse-grain grades with Co and Co–Ni–Cr binders characterized by high resistance to thermal shock and the process of formation and propagation of thermal cracks during hot rolling of steels. Some of these grades have a relatively low or medium-low hardness due to the high binder contents. The grades with the Co–Ni–Cr binders are characterized by the extremely high oxidation and corrosion resistance at elevated temperatures.

The assortment of industrial cemented carbides produced by the Ceratizit Group for fabrication of wear parts and cutting inserts includes the following types of grades [3]:

P-LINE:

- ultrafine-grain grades with the Co content of 4.2%–12% (CTU08L, TSF22, and TSF44), Vickers hardness of 22–17.3 GPa, TRS of 3700–4600 MPa, and fracture toughness of 4.4 up to 9.8 MPa m$^{1/2}$.

Table 10.2 Industrial cemented carbide grades for tool and wear applications fabricated at Element Six GmbH.

	Composition								
Grade	Co [%] ± 0.2	Ni [%] ±0.2	VC [%] ± 0.2	Cr_3C_2 [%] ± 0.2	Grain size	Hardness [HV_{20}] ±50	Hardness [HRA] ±0.3	TRS [N/mm^2]	Compressive strength [N/mm^2]
G10	6.0		0.10		Medium	1600	91.9	2800	5000
G15	8.0		0.15		Medium	1480	90.9	3000	4700
G20	11.0		0.15		Medium	1320	89.4	3200	4300
G30	15.0		0.15		Medium	1200	88.2	3000	3900
G40	20.0		0.15		Medium	1050	86.5	2900	3500
G55	26.0		0.15		Medium	870	84.4	2800	3000
K05	5.0		0.2		Submicron	1800	93.0	2800	5200
K06	6.0		0.2		Submicron	1750	92.7	3100	5200
K07	7.0		0.2	0.1	Submicron	1700	92.5	3100	5400
K08	8.0		0.2	0.2	Submicron	1650	92.2	3200	5400
K010	10.0		0.3	0.3	Submicron	1620	92.0	3600	5000
K015	15.0		0.4	0.5	Submicron	1400	90.2	3600	4500
N6		6.0		0.5	Submicron	1600	91.9	2400	4800
N12		12.0		0.5	Submicron	1200	88.2	2800	3900
NK07	4.8	2.0	0.3	0.3	Submicron	1850	93.4	3000	5400

According to Ref. [1].

Table 10.3 Industrial cemented carbide grades for mill rolls fabricated at Element Six GmbH [2].

Grade	WC grain size	Co [%] ±0.2	Ni [%] ±0.2	Cr_3C_2 [%] ±0.2	Density [g/cm^3] ±0.1	Hardness [HV_{20}] ±50	Hardness [HRA] ±0.3	TRS [MPa]
J6	Coarse	6.00	–	–	14.93	1270	89.0	2700
J8	Coarse	8.00	–	–	14.70	1220	88.3	2800
J10	Coarse	10.00	–	–	14.50	1130	87.2	2800
J13	Coarse	13.00	–	–	14.20	1020	85.8	2900
J15	Coarse	15.00	–	–	14.00	990	85.7	3100
J20	Coarse	20.00	–	–	13.55	850	83.8	3000
J25	Coarse	25.00	–	–	13.10	750	82.7	3000
J30	Coarse	30.00	–	–	12.70	675	81.7	2900
P8	Coarse	3.60	4.00	0.40	14.70	1220	88.5	2750
P10	Coarse	4.50	5.00	0.50	14.52	1150	87.7	2900
P12	Coarse	5.00	6.50	0.60	14.17	980	85.4	3000
P15	Coarse	6.80	7.50	0.80	13.90	875	84.1	3100
P20	Coarse	9.20	10.60	1.20	13.45	820	83.4	3100
P25	Coarse	11.00	12.60	1.40	13.05	700	81.9	3000
P30	Coarse	14.00	15.00	1.50	12.65	605	80.8	2900

With the permission of Element Six.

- submicron grades with the Co content of 6%–15% (CTS12D, CTS15D, CCTS18D, CTS20D, CTS25D, and CTS30D), Vickers hardness of 18.2–14 GPa, TRS of 3600–4300 MPa, and fracture toughness of 9.3–13.2 MPa $m^{1/2}$.
- fine-grain grades with the Co content of 6%–12.5% (CTF12E and CTF25E), Vickers hardness of 16.2–13 GPa, TRS of 2200–3500 MPa, and fracture toughness of 9.9–15 MPa $m^{1/2}$.

E-LINE:

- submicron and fine-grain grades with the Co content of 7%–12% (WF05, WF15, and WF25), Vickers hardness of 15.8–17.9 GPa, TRS of 3500–3700 MPa, and fracture toughness of 7.6–9.1 MPa $m^{1/2}$.

S-LINE:

- fine-grain grade with the Co content of 10% and the cubic carbide content of about 2% (TMG30), Vickers hardness of 15.7 GPa, TRS of 3600 MPa, and fracture toughness of 9.3 MPa $m^{1/2}$.

In addition to the grades mentioned above, Ceratizit manufactures a number of carbide grades containing cubic carbides for steel-cutting.

Typical microstructures of an industrial grade for steel-cutting containing grains of cubic carbide (Ti,Ta,W)C or cubic carbide (Ti,Ta,Nb)C are shown in Fig. 10.2.

According to Ref. [4], Sandvik (Sweden) produces numerous fine-grain grades (GC1025, GC1105, H10, H10A, H10F, etc.), submicron grades (GC1115, GC1125, etc.), and medium-grain grades (GC3210, GC3215, etc.) mainly for metal-cutting. A significant part of these grades is intended for cutting steels and must contain cubic carbides; however, the compositions of the grades are not disclosed.

The assortment of cemented carbide grades containing cubic carbides, which are employed mainly for steel-cutting, produced by Böhlerit GmbH & Co.KG (Austria) according ref. [5] is presented in Table 10.4. Three grades of cemented carbides for finish and semifinish turning and milling steels (LW225, SB10, SB20, SB29, and SB30) contain large amounts of cubic carbides (17.3–33 wt.%) and 9–10 wt.% Co. They are characterized by the Vickers hardness of 15–15.75 GPa, TRS of 2000–2300 MPa, and fracture toughness of 9.3–10 MPa $m^{1/2}$. The grade SH40 for rough machining steels contains 12 wt.% cubic carbide and 11% Co and is characterized by the hardness of 14 GPa and fracture toughness of 12 MPa $m^{1/2}$.

There are other three carbide grades for turning and milling stainless steels manufactured by Böhlerit (EV10, EB15, and EB40) containing slightly lower amounts of cubic carbides (5–10 wt.%) and different

Figure 10.2 Above - microstructure of a typical industrial WC-(Ti,Ta,W)C—Co grade for steel-cutting containing grains of cubic (Ti,Ta,W)C carbide in amount of 7 wt.% (visible as dark inclusions after etching according to the procedure described in Chapter 7) with 12 wt.% Co (optical microscopy image); and below - SEM BackScattered Electron image of a WC-(Ti,Ta,Nb,W)C—Co grade indicating the different morphology of the WC phase (facetted grains) and cubic carbide phase (rounded grains) according to Ref. [7].

amounts of Co (5—10.7 wt.%). They are characterized by the Vickers hardness of 13.5—16 GPa, TRS of 2200—2600 MPa, and fracture toughness of 9.5—12.4 MPa $m^{1/2}$. Böhlerit fabricates also carbide grades containing insignificant amounts of cubic carbide (0.25—2 wt.%) and different amounts of cobalt (6—26wt.%) that are intended for cutting cast iron; for example, the grades HB01F, HB20—HB50, GB10—GB30, and GB56.

Table 10.4 Industrial cemented carbide grades manufactured by Böhlerit for metal-cutting [5].

Grade	ISO appl. Code ISO 513	WC [wt.%]	TiC +TaC [wt.%]	Co [wt.%]	WC grain size [μm]	Density ISO 3369 [g/cm^3]	Hardness ISO 3878 [HV_{30}]	Comp. strength ISO 4506 [N/mm^2]	TRS ISO 3327 [N/mm^2]	Fracture toughness K_{1C} [MPa·m$^{1/2}$]	Modulus of elasticity ISO 3312 [kN/mm^2]
SB10	P05 −P15	57.5	33	9.5	2.5	10.3	1575	5300	2000	9.3	530
SB20	P15 −P25	69	22	9	2.5	11.20	1550	5200	2000	9.3	540
LW225	P20 −P40	72.7	17.3	10	1.25	12.55	1525	5100	2300	9.8	550
SB29	P20 −P40	74.1	15.9	10	1.25	12.35	1525	5100	2200	9.8	550
SB30	P25 −P30	69	21	10	2.5	11.60	1500	5100	2200	10.0	550
SB40	P35 −P45	77	12	11	5.3	13.15	1400	5000	2400	12.0	560

Cemented carbides obtained by the full or partial substitution of tungsten carbide as a hard, ceramic phase by other carbides or carbonitrides are designated in literature as "cermets." Although the materials science and technology of cermets are special topics, which are not in the scope of this book, as cermets do not formally belong to the materials family of cemented carbides, below there is general information on industrial grades of cermets.

Fig. 10.3 shows typical microstructures of two different grades of cermets [7]. The hard phase of cermets is mainly Ti-based cubic carbonitride. Conventional cermets usually contain molybdenum carbide to improve wettability of Ti(C,N) by liquid Ni- or Co-based binders. Modern cermets exhibit multicomponent microstructures based on titanium carbonitride and various amounts of Mo, W, Ta, Nb, V, and other chemical elements. The binder is a Co or Co—Ni alloy and contains considerable amounts (20—40 wt.%) of dissolved Ti, Mo, W, V, and other metals. Modern grades of cermets contain also other carbides (WC, TaC, NbC, etc.) to improve their mechanical and performance properties. Grains of the hard phases normally exhibit a core-rim structure clearly seen in Fig. 10.3. The core areas comprise undissolved TiCN, whereas the rim forms as a result of the dissolution—precipitation of the carbides in the liquid binder during sintering and consists of a cubic (Ti,Me) (C,N) phase, where Me = Mo/W/Ta/Nb [8]. The rim can comprise two parts, the inner rim and the outer rim, characterized by different ratios between titanium and the other carbide-forming chemical elements [9].

Generally, cermets are characterized by a combination or hardness and fracture toughness inferior to conventional WC—Co cemented carbides. Nevertheless, they have improved performance properties in some operations of metal-cutting, i.e., in finish turning of steels, allowing one to achieve a better surface quality of workpieces as result of finish machining. Cermets are therefore employed almost exceptionally for the fabrication of indexable cutting inserts. Indexable cutting inserts made from cermets with Co-based binders can be coated by conventional CVD and PCD coatings, whereas conventional CVD coatings cannot be employed for cermets with Ni-based binders [10]; in this case, tailored PVD coatings or special wear-resistant layers (see, e.g., Ref. [11]) should be used.

Nearly 2 decades ago, functionally graded cermets obtained by sintering of TiCN-based cermets in a nitrogen atmosphere were developed [12].

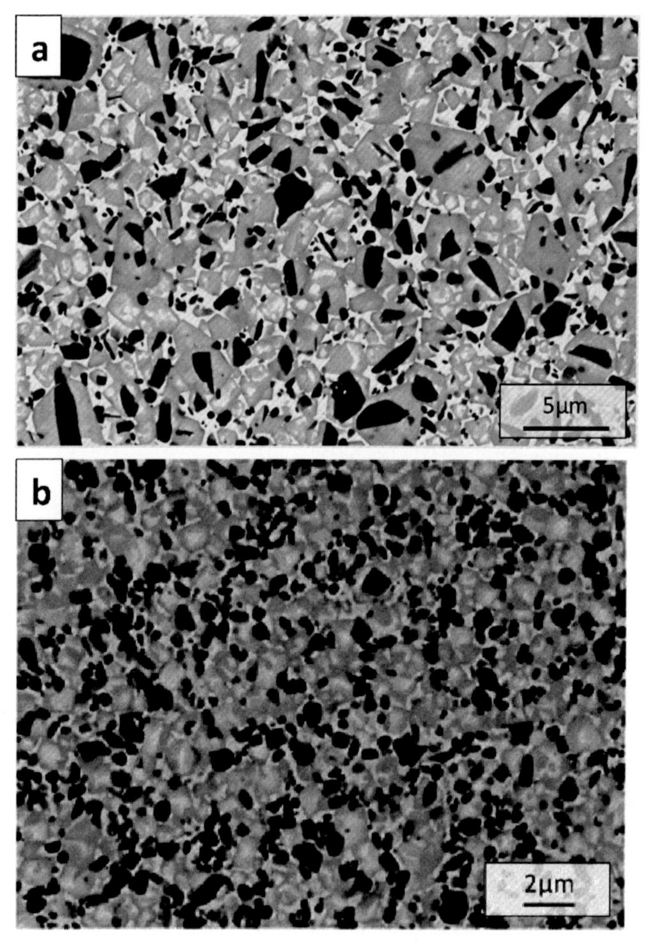

Figure 10.3 (A) - typical microstructure of a cermet grade based on a hard phase of Ti-based cubic carbide or carbonitride (the *black areas* correspond to the core regions of the hard phase grains, the *dark-grey areas* correspond to the rim, and the *bright-grey areas* represent the binder phase); (B) - microstructure of the Ti(C,N)—WC—NbC—Co cermet (the core areas of the hard phase grains are composed of Ti(C,N); the rim areas are composed of (Ti,Nb,W) (C,N) and the binder is composed of mainly Co. *(Redrawn from Ref. [7].)*

Such functionally graded cermets consist of a hard wear-resistant cubic-carbide—enriched surface layer, a tough core containing WC, cubic carbonitrides, Co—Ni binder phase and an intermediate zone with graded composition. It is also possible to produce functionally graded cermets containing wear resistant nitiride surface layers by their nitriding.

Most of the companies producing cemented carbide cutting inserts fabricate also cermet indexable cutting inserts (see, for example, refs. [3,4]); however, their share in the total production of metal-cutting inserts is relatively insignificant.

References

[1] I. Konyashin, Cemented carbides for mining, construction and wear parts, in: V. Sarin (Ed.), Comprehensive Hard Materials, Elsevier Science and Technology, 2014, 425-251.
[2] Internet site, 2021. www.e6.com.
[3] Internet site, 2021. https://www.ceratizit.com/de/produkte/staebe-formteile/hart-metallsorten/.
[4] Internet site, 2021. https://www.home.sandvik/de/produkte/.
[5] Internet site, 2021. http://www.boehlerit.com/.
[6] J. Prakash, Fundamental and general applications of hardmetals, in: V. Sarin (Ed.), Comprehensive Hard Materials, Elsevier Science and Technology, 2014, pp. 29–90.
[7] J. García, V. Collado, C.A. Blomqvist, B. Kaplan, Cemented carbide microstructures: a review, Int. J. Refract. Metals Hard Mater. 80 (2019) 40–68.
[8] W. Wan, J. Xiong, M. Liang, Effects of secondary carbides on the microstructure, mechanical properties and erosive wear of Ti(C,N)-based cermets, Ceram. Int. 43 (1) (2017) 944–952.
[9] D. Park, Y. Lee, Effect of carbides on the microstructure and properties of Ti(C,N)-based Ceramics, J. Am. Ceram. Soc. 82 (11) (1999) 3150–3154.
[10] I. Konyashin, The influence of Fe-group metals on CVD of titanium carbide, Adv. Mater. CVD (2) (1996) 199–208.
[11] I. Konyashin, Wear-resistant coatings for cermet cutting tools, Surf. Coat. Technol. 71 (1995) 284–291.
[12] K. Tsuda, A. Ikegaya, K. Isobe, N. Kitagawa, T. Nomura, Development of functionally graded sintered hard materials, Powder Metall. 39 (4) (1996) 296–300.

CHAPTER 11

Current range of advanced uncommon industrial grades of cemented carbides

11.1 Functionally graded cemented carbides

Hardness and fracture toughness of cemented carbides and other hard materials are traditionally considered to be contradictory and incompatible properties. As it is shown in Chapter 10, for conventional WC—Co materials, hardness can be increased only at the expense of fracture toughness. The fabrication of WC—Co cemented carbides in the form of functionally graded materials is a unique approach that allows both the hardness and fracture toughness to be increased and a tailored combination of microstructure and properties in the near-surface layer and core region to be achieved.

11.1.1 Functionally graded WC—Co cemented carbides

First attempts to produce cemented carbides with various WC grain sizes and Co contents were based on pressing articles from two different graded WC—Co powders. However, this approach had a limited practical applicability, as liquid Co drifted very fast from a part with a higher Co content into a part with a lower Co during liquid phase sintering. Also, it is very difficult or impossible to produce carbide articles of complicated shape by use of this approach.

The fabrication of functionally graded carbide articles or bilayer articles having a significant Co gradient is possible due to pressing two WC—Co graded powders with various WC mean grain sizes followed by liquid-phase sintering. It is well known that, if two WC—Co parts with different WC mean grain sizes are sintered together, the liquid Co drifts from the part with the larger WC grain size into the part with the finer WC grain size due to different capillary forces in these parts. The binder drifts in couples of alloys with the same binder content and carbon content,

Cemented Carbides
ISBN 978-0-12-822820-3
https://doi.org/10.1016/B978-0-12-822820-3.00007-0

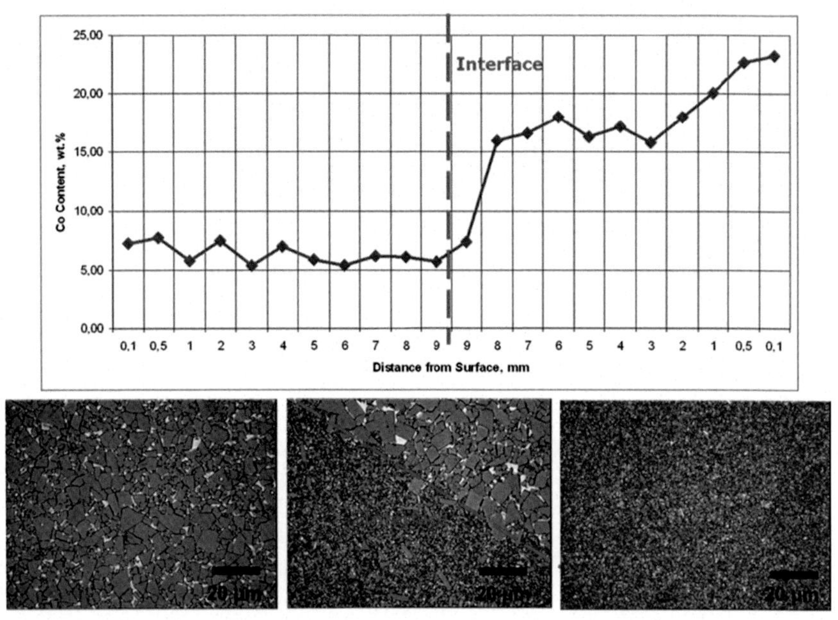

Figure 11.1 Curve indicating the Co distribution after sintering of a couple of cemented carbides with various WC mean grain sizes of about 4.8 μm and 0.2 μm; and microstructures of these cemented carbides and an interface between them after pressing the WC—Co graded powders together followed by liquid phase sintering [2].

but various WC mean grain sizes during liquid phase sintering are reported in Refs. [1,2]. Fig. 11.1 shows a curve indicating the Co distribution between the cemented carbide with the coarse microstructure (the WC mean grain size of 4.8 μm) and that with the extremely fine structure (the WC mean grain size of about 0.2 μm) initially containing 10 wt.%Co as well as their microstructures. It can be seen that Co migrates from the coarse-grain part into the fine-grain part as a result of capillary forces attracting the liquid binder into the fine-grain part during liquid-phase sintering. As a result, the Co content in the coarse-grain part becomes equal to about 5 wt.% and that the in the fine-grain part is equal to nearly 15 wt.%, so the difference in Co contents between these two parts is roughly 3 times.

Bilayer carbide articles can also be produced by pressing together and sintering of graded powders with different compositions of the binder phase, for example, WC—Co and WC—FeNi cemented carbides shown in Fig. 11.2 [1]. This allows obtaining various WC mean grain sizes in the parts

Figure 11.2 SEM image of bilayer WC–Co/WC–FeNi cemented carbide (above) and corresponding EBSD image showing two microstructures characterized by different WC grain sizes and a sharp interface (below). *(Redrawn from Ref. [1].)*

with the Co and FeNi binders due to the different behavior of the liquid Co and FeNi binders with respect to the WC recrystallization and grain growth during sintering.

Although it is difficult to produce carbide articles of complicated shape by pressing together two graded WC–Co powders and shape distortions may arise due to differences in shrinkage of the consistent parts of such bilayer carbide articles, the bilayer articles produced by this approach were implemented in industry [5].

Nevertheless, if one can employ just one WC–Co graded powder and create drifts of liquid binders either from the surface toward the core or from the core region toward the surface layer, the problems mentioned above can be solved and obtaining functionally graded cemented carbides can be simplified.

Three major technological approaches to the fabrication of industrial grades of functionally graded cemented carbides from a single WC—Co powder are described in literature. All of them are based on selective carburization of surface layers of WC—Co articles leading to the formation of a carbon gradient between the surface and core regions, which results in obtaining Co and hardness gradients. The phenomenon of the Co drift is based on the well-known fact that liquid Co-based binders migrate from a carbide part with a high carbon content into a carbide part with a low carbon content during liquid-phase sintering [6—9].

Among these three approaches, the technique for the fabrication of industrial functionally graded cemented carbides of the first type, which are designated in literature as "DP" (dual properties) grades, was developed at Sandvik about 40 years ago [3,4]. Such grades are employed for rock drilling and mineral cutting. Functionally graded cemented carbides that are similar to the Sandvik DP grades were produced and examined in detail by Zhang et al. [10], Liu et al. [11,12], and others. Microstructures of different areas of a carbide article made of the DP grade are shown in Fig. 11.3. The DP grades are produced in a two-stage technological process comprising the following stages: (1) the fabrication of fully sintered carbide articles deficient with respect to carbon and therefore containing η-phase and (2) carburization of the articles in a carburizing atmosphere at temperatures above 1300°C in the presence of liquid phase. As a result of this fabrication process, carbon diffuses from the surface toward the core leading to the carburization of η-phase and its transformation into WC + Co in the near-surface layer. This causes a Co drift from the surface toward the interface with the η-phase—containing region. As a result, a graded structure, which generally comprises the three following areas shown in Fig. 11.3, forms. The functionally graded structure comprises the following parts: (1) a near-surface layer characterized by a significantly lower Co content and higher hardness than on overage; (2) an interlayer adjacent to the near-surface layer comprising no η-phase and having a noticeably higher Co content and lower hardness than on average; and (3) a core region comprising η-phase. Despite the presence of the brittle core containing η-phase, the functionally graded cemented carbides can be employed in mining due to the presence of the near-surface layer with high hardness and wear-resistant characterized by the presence of residual compressive stresses suppressing the initiation and propagation of cracks. Nevertheless, the applicability of the DP grades in

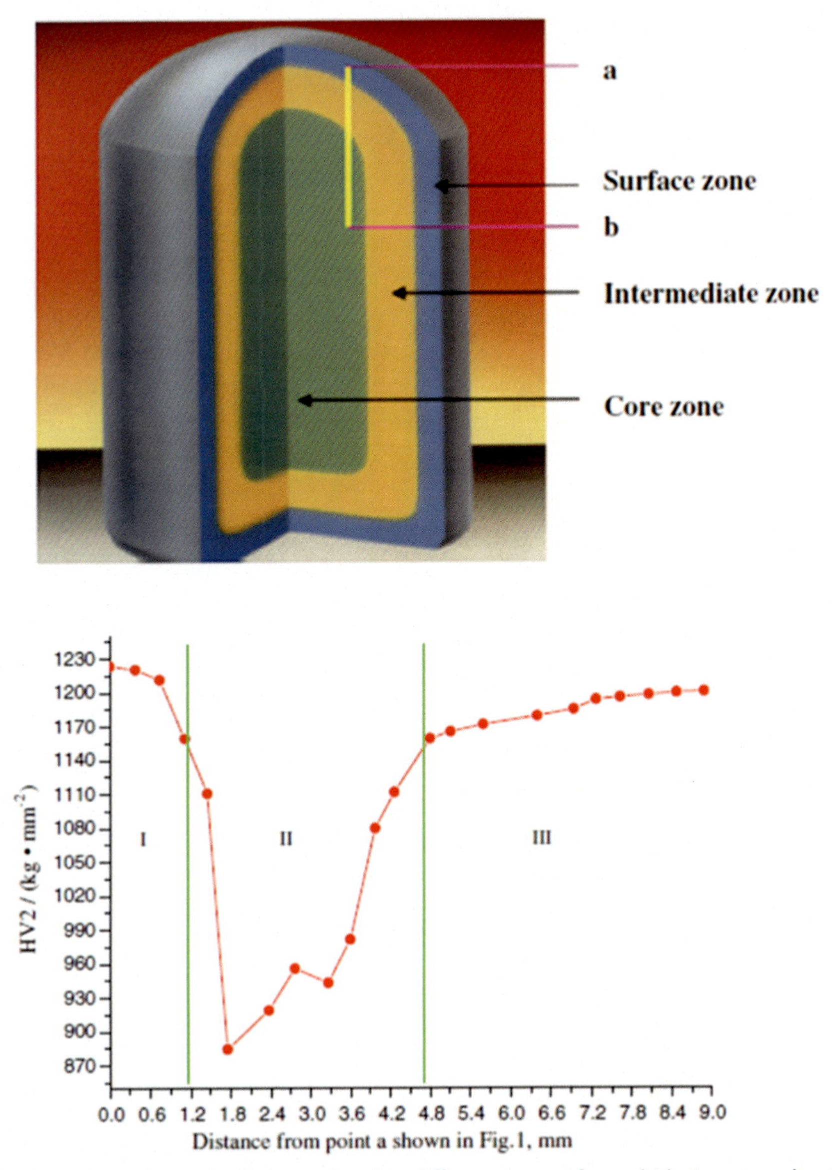

Figure 11.3 Schematic diagram showing different parts of a carbide insert made of the DP grade (above) and a curve indicating the hardness distribution in the carbide insert (below). *(Redrawn from Ref. [10].)*

mining applications, particularly in percussive drilling, is limited because of the presence of the brittle core region comprising η-phase. Usually, worn carbide inserts for percussive drilling bits are subjected to several regrinding processes, which cannot be conducted in the case of inserts made of the DP grades.

As it was mentioned above, a significant disadvantage of the DP functionally graded cemented carbides is the presence of the brittle core region comprising η-phase. The production of functionally graded WC–Co cemented carbides not containing η-phase was therefore the objective of intensive research resulting in the development and implementation of three types of η-phase-free functionally graded cemented carbides.

One type of these functionally graded cemented carbides developed and presently produced by Sandvik is designated as the GC80 grade, which appears not to contain η-phase [13]. It is stated that an innovative manufacturing process, details of which are not disclosed, makes it possible to produce carbide buttons that are wear resistant on the outside and tough in the center. It is claimed that the use of such functionally graded cemented carbide buttons "... further extends service life and grinding intervals ..." in percussive drilling [13].

The second type of the functionally graded cemented carbides not containing η-phase and a technology for their fabrication were developed and implemented at the University of Utah and Heavystone Laboratory, LLC [14–16] (Fig. 11.4).

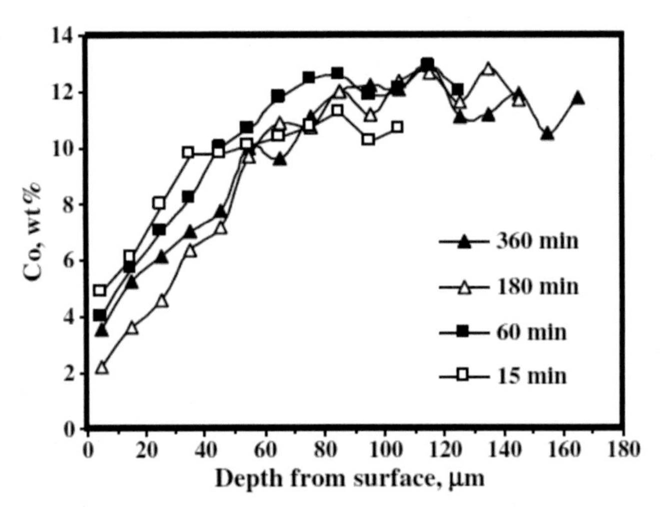

Figure 11.4 Curves indicating the distribution of cobalt in WC–10 wt.%Co specimens after carburization heat treatment of the fully sintered samples by use of the technology described in Refs. [14–16]. *(Redrawn from Ref. [16].)*

According to this technology, fully sintered WC–Co articles with initial low or medium carbon contents are subjected to carburization at a temperature at which WC, liquid cobalt, and solid Co coexist. Although there is neither η-phase nor free carbon in the functionally graded cemented carbides obtained in such a way, this technology allows the formation of only relatively thin hard surface layers having a thickness of the order of 0.5 mm. It appears that such thin hard layers might have a limited application in mining and construction, as the wear of carbide inserts in these applications is typically of the order of 1 mm or more. Nevertheless, the functionally graded carbide inserts obtained by use of this technology can be reground several times, which is a big advantage in comparison with the functionally graded carbide articles made of the DP grades.

The third type of the functionally graded cemented carbides not containing η-phase also known as "gradient carbides" and a technology for their fabrication were developed and implemented at the Element Six Group [17–19]. It includes the following stages: (1) presintering of green carbides articles with original low carbon contents to obtain a certain residual open porosity and consequently gas permeability; (2) their carburization in a carburizing atmosphere to selectively increase the carbon content in the surface layer; and (3) final liquid phase sintering at tailored conditions to obtain the full densification and Co drift from the surface toward the core.

Fig. 11.5 shows an insert for percussive drilling of the functionally graded WC–Co cemented carbides obtained by use of the technology described in Refs. [17–19] and its microstructures in the surface layer and core region, and Fig. 11.6 shows curves indicating the Co and hardness gradient in the insert. Neither the surface layer not the core comprises η-phase or free carbon. There are noticeable Co and hardness gradients between the surface and core region of the insert, so that the Co content in the core is higher compared to the surface layer by a factor of nearly two leading to the hardness gradient of about 250 Vickers units. Note that the initial Co content in the WC–Co graded powder employed for obtaining the gradient carbide insert was 10 wt.%.

Fig. 11.7 shows curves indicating the fracture toughness versus hardness for conventional WC–Co materials and the point indicating the combination of toughness and hardness for the surface layer of the gradient carbide. The surface layer was found to be characterized by high residual compression stresses of above −400 MPa in both the WC phase and the binder phase resulting in a high combination of fractures toughness and

Figure 11.5 Insert for percussive drilling (right) of the functionally graded WC–Co cemented carbides obtained by use of the technology described in Refs. [17–19] and its microstructures (left) in the surface layer and core region (the tungsten carbide grains are gray and binder areas are red). *(Redrawn from Ref. [21].)*

hardness of the surface layer. Results of numerous laboratory and field tests of mining tools with inserts of gradient carbides produced by the Element Six Group indicated their significantly improved performance and prolonged lifetime (by a factor of two to three).

In spite of the fact, that the formation of Co drifts from the surface toward the core of carbide articles as a result of the carburization of cemented carbides with low carbon contents is employed in industry on a large scale, its mechanism was not fully understood up to recent times.

In the case of the DP functionally graded cemented carbides, the transformation of η-phase into WC + Co as a result of its carburization should play an important role in the formation of the Co drift, as this process is accomplished by a noticeable volume change due to the different densities of η-phase on the one hand and WC + Co on the other hand. Nevertheless, drifts of liquid Co-based binders occur from a carbide part with a high carbon content into a part with a low carbon content not

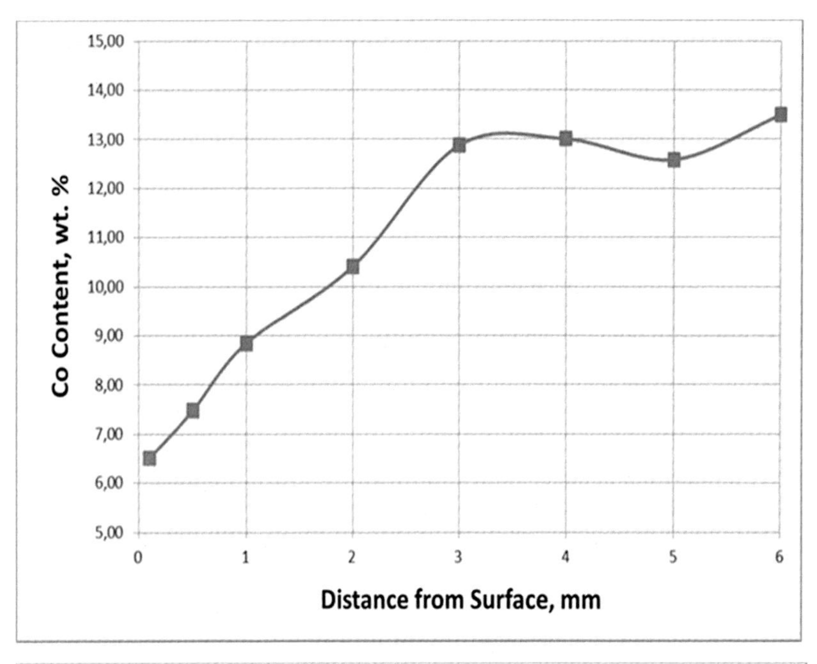

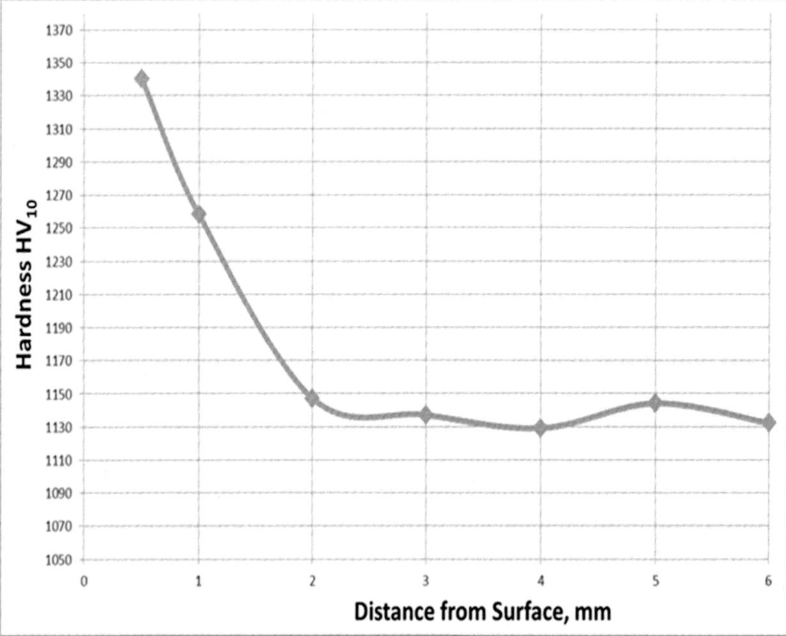

Figure 11.6 Curves indicating Co and hardness gradients in the functionally graded carbide articles produced according to the Element Six technology [18].

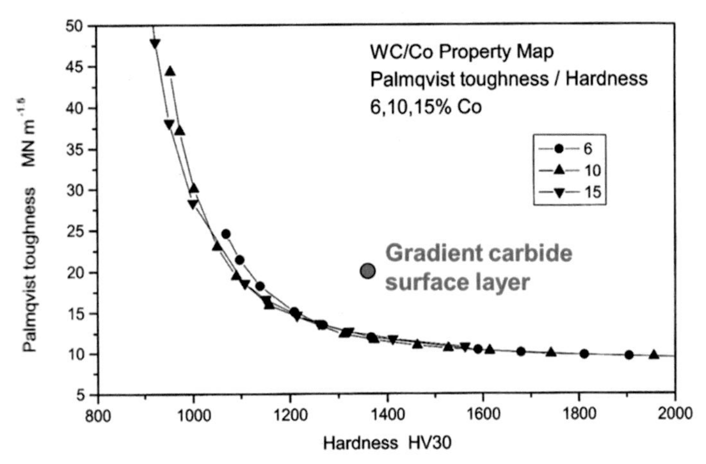

Figure 11.7 Curves indicating the fracture toughness versus hardness for conventional WC–Co materials according to Ref. [20] and the point representing the combination of toughness and hardness for the gradient carbide surface layer obtained according to the Element Six technology [18].

comprising η-phase, so that the mechanism of the binder drifts appears to be universal and independent on the presence of η-phase.

In order to establish the mechanism of obtaining the functionally graded cemented carbides as a result of carburization of carbide articles with the initial low carbon content, the wettability of tungsten carbide with model binder alloys was examined in Ref. [21]. Samples of two model WC–Co binder alloys containing 65 wt.% Co with low and high carbon contents were prepared, placed on a surface of porous-free WC discs and heated up to temperatures slightly above the melting point. Afterward, the kinetics of the model alloys' spreading over the WC surface were examined and recorded by a high-velocity camera. Fig. 11.8 shows pictures indicating the process of melting the model alloys and their appearance after solidification. The melted alloy with the low carbon content is completely spread over the WC disc after 17 s and the wetting angle measured after the solidification of this alloy is equal to nearly zero indicating complete wetting in this case. The shape of the melted droplet of the model alloy with the high carbon content reaches the equilibrium shape after 15 s of melting and does not change further up to the end of the melting process (35 s). The melt of this alloy remains as a droplet and is not spread over the WC surface; the wetting angle in this case is found to be nearly 15 degrees providing evidence for incomplete wetting.

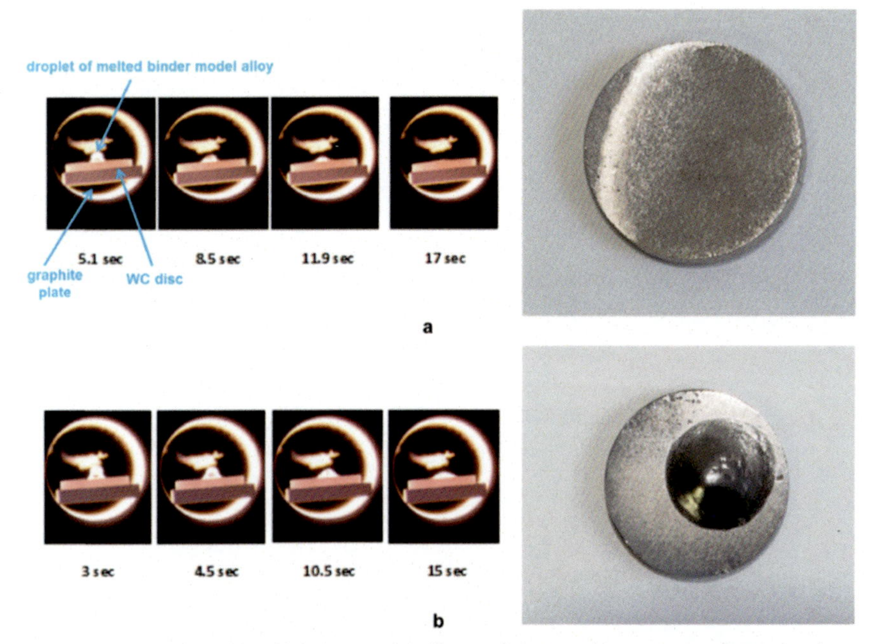

Figure 11.8 (A) Droplet of binder model alloy with low carbon content during melting and spreading over WC at various times (on the left hand side) and that after melting for 35 s followed by solidification (on the right hand side); and (B) droplet of binder model alloy with high carbon content during melting and spreading over WC at various times (on the left hand side) and that after melting for 35 s followed by solidification (on the right hand side). *(Redrawn from Ref. [21].)*

Thus, the wettability of tungsten carbide by liquid Co-based binders is complete at the low carbon content and incomplete at the high carbon content, so that the binder drifts mentioned above are likely to be caused by capillary forces related to various wettability rates of WC by liquid binders having various carbon contents. These capillary forces were found to increase due to different structures of WC—WC grain boundaries in cemented carbides with high and low carbon contents shown in Fig. 11.9. The majority of WC—WC grains boundaries in the carbide sample with the low carbon content comprise nm-thick Co interlayers, whereas those in the carbide sample with the high carbon content do not comprise such nm-thick Co interlayers (see also Chapter 9). Thus, the mechanism explaining the drifts of liquid binders during fabrication of functionally graded cemented carbides is thought to be explained by creating capillary forces attracting the liquid binders toward the core due to various

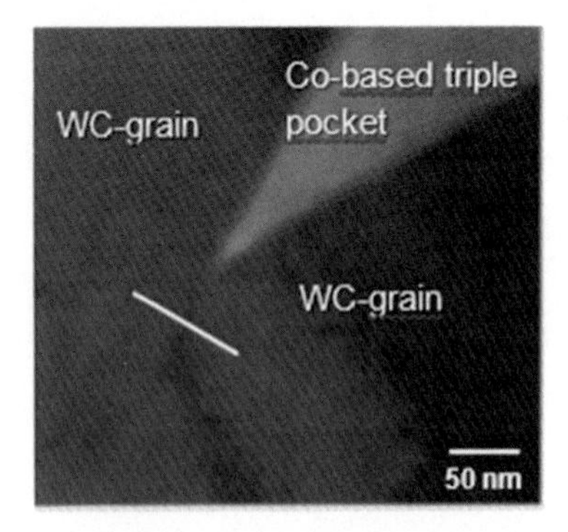

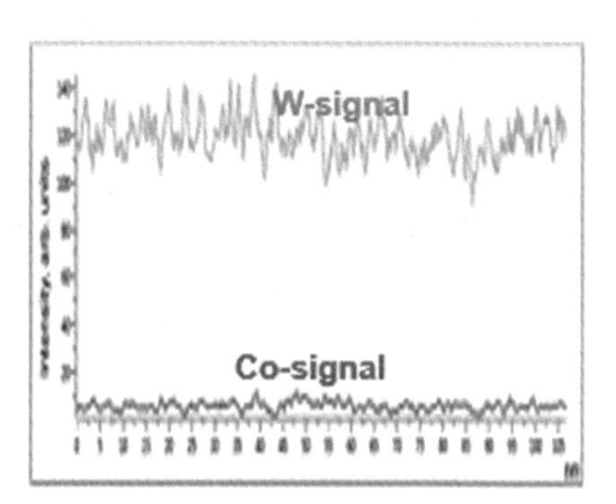

a

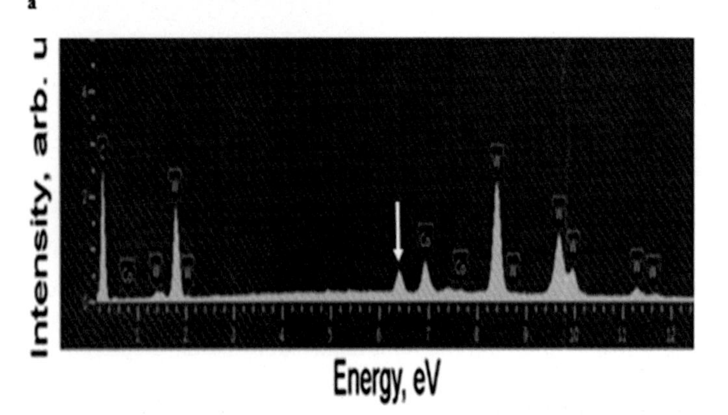

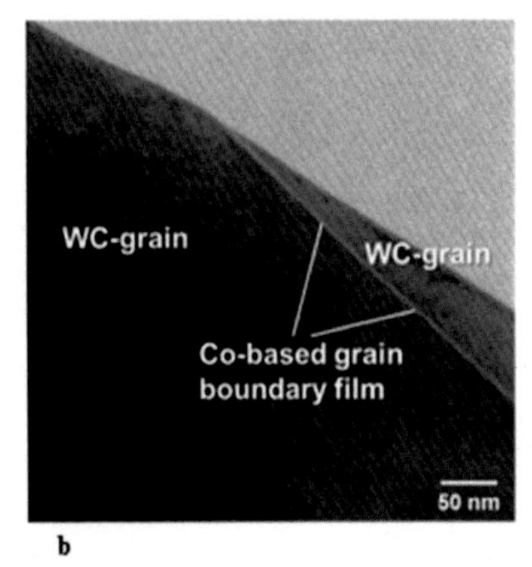

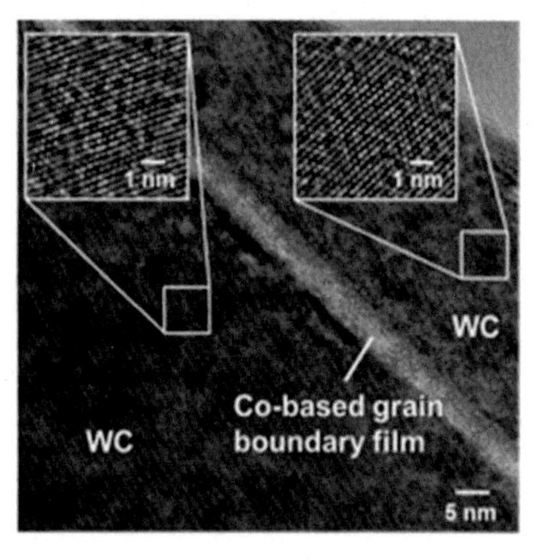

b

Figure 11.9 (A) Carbide sample with a high carbon content: left - STEM image of two WC single crystals and EDX element profile indicating distributions of W and Co at the WC—WC grain boundary; (B) carbide sample with a low carbon content: above - the EDX spectrum recorded from the region of the WC—WC grain boundary (the Fe Kα is indicated by an arrow); below - bright-field TEM image of the WC—WC grain boundary (left) and HRTEM image of the WC—WC grain boundary with insets, which are Fourier-filtered images from the small boxed areas (right) revealing the high crystallinity of the WC grains. *(Redrawn from Ref. [21].)*

wettability rates of WC by liquid binders in the near-surface layer and core region. The capillary forces are strongly enhanced as a result of the presence of numerous nm-thick channels at WC—WC grain boundaries in the low-carbon core region and the limited number of such channels in the high-carbon near-surface layer. As it was shown in Ref. [22], the difference in the wettability of tungsten carbide by liquid Co-based binders with different carbon contents is not related to the segregation of carbon or other chemical elements at WC—Co interfaces. The main reasons for the reduced wettability of WC by liquid binders with high carbon contents are thought to be special features of such liquid binders oversaturated with carbon at sintering temperatures or the presence of carbon terminated surfaces on tungsten carbide grains at WC/Co grain boundaries. According to Ref. [23], this should negatively affect the wettability of WC by the liquid binders at high carbon contents.

11.1.2 Functionally graded cemented carbides with cubic carbide depleted near-surface layers

Another class of industrial functionally graded cemented carbides is represented by carbide grades for steel-cutting containing grains of Ti-based cubic carbonitrides such as (Ti,Ta,Nb,W)C,N. Such functionally graded cemented carbides are exceptionally produced in the form of coated indexable cutting inserts comprising chemically vapor deposited (CVD) thin wear-resistant coatings. The near-surface layer adjacent to the coatings of such cutting inserts is subjected to severe thermal and mechanical fatigue in processes of steel turning and milling, so its fracture toughness plays a very important role. It would therefore be desirable to form a WC—Co near-surface layer of the cutting inserts adjacent to the wear-resistant coatings, as such a layer is characterized by a noticeably higher fracture toughness and fatigue resistance compared to those of cemented carbides containing grains of cubic carbides or carbonitrides. This would lead to an increased resistance of the surface layer to the formation and propagation of cracks during metal-cutting thus resulting in improved performance and reliability of the coated cutting inserts (see Chapter 12).

A phenomenon of forming cubic carbide depleted near-surface layers as a result of sintering articles of cemented carbides initially containing Ti-based cubic carbonitrides, for example Ti(C,N) or (Ti,Ta,Nb,W)C,N, illustrated in Fig. 11.10 was first reported in Ref. [24] and, afterward, examined in detail in Refs. [25—27]. By addition of such carbonitrides to

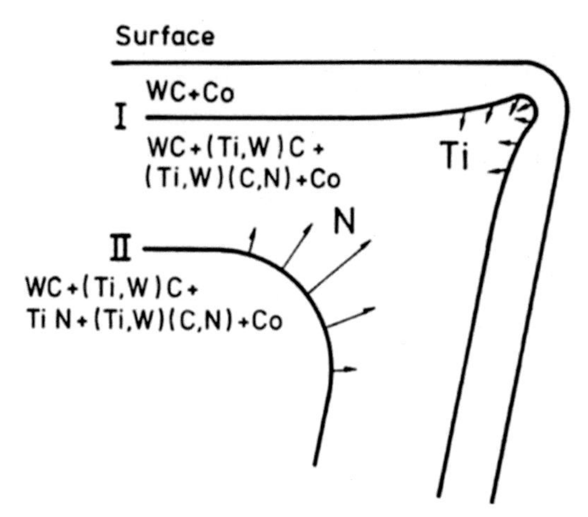

Figure 11.10 Diffusional phenomena occurring at the (cutting) edge of a cemented carbide article containing Ti-based cubic nitride and carbonitride as a result of sintering in vacuum. *(Redrawn from Ref. [25])*

WC—Co materials, nitrided or denitrided gradient zones can form as a result of nitriding in nitrogen-containing atmospheres or denitriding near-surface layers of cemented carbides due to sintering in a vacuum. These phenomena can be employed to tailor the structure and composition of near-surface areas of carbide articles, thus obtaining functionally graded cemented carbides with microstructures and properties smoothly changing from their surface toward the core.

The process of dissolving the grains of the Ti-based cubic carbonitrides depends on the nitrogen pressure in the gas atmosphere during liquid-phase sintering. As one can see in Fig. 11.11, the content of such cubic carbonitrides in the near-surface layer of the carbide articles can remain unchanged or even increase when the gas atmosphere contains relatively much nitrogen during liquid-phase sintering. When the gas atmosphere contains no or little nitrogen, the grains of cubic carbonitrides completely dissolve in the near-surface layer, so that it comprises only WC grains and a Co-based binder.

When the carbide article is nitrided at a high nitrogen partial pressures during sintering leading to its enrichment with the carbonitride grains, titanium diffuses from the core region toward the surface. When the carbonitride depleted near-surface zone of WC—Co forms as a result of sintering in vacuum or nitrogen-free atmospheres, titanium diffuses from the surface toward the core, which is schematically shown in Fig. 11.12. The Co content in the WC—Co surface layer, that does not contain the carbonitride grains, noticeably

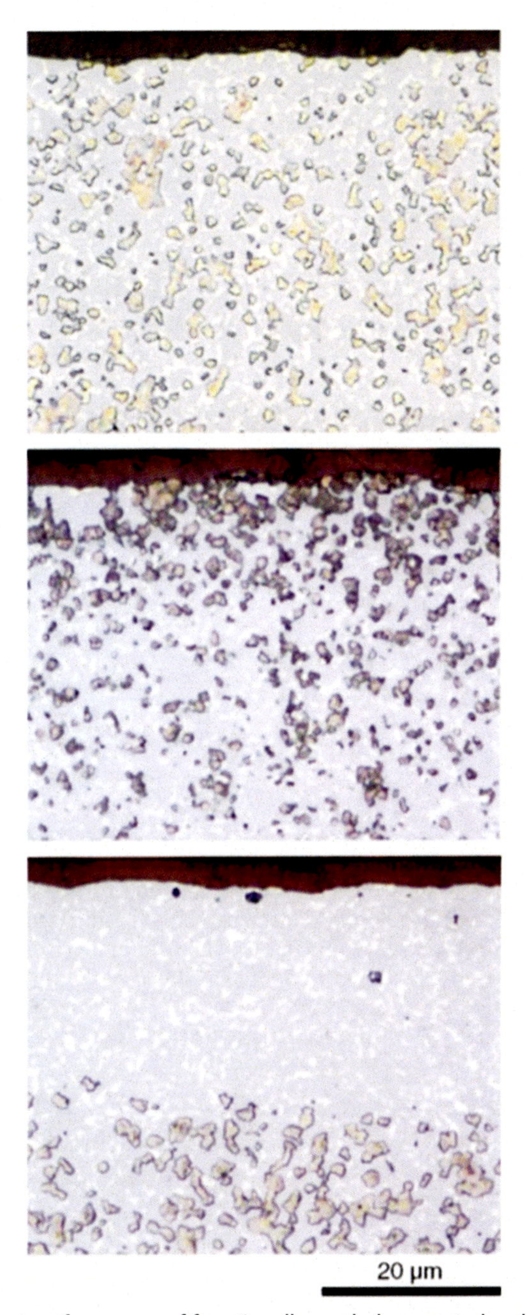

Figure 11.11 Near-surface areas of functionally graded cemented carbides sintered at a nitrogen equilibrium partial pressure (above), at a high nitrogen partial pressure (middle) and at a lower than equilibrium nitrogen pressure (below). The near-surface areas are enriched with grains of carbonitrides having a yellow-brown color (middle) or depleted with such grains (below) [28].

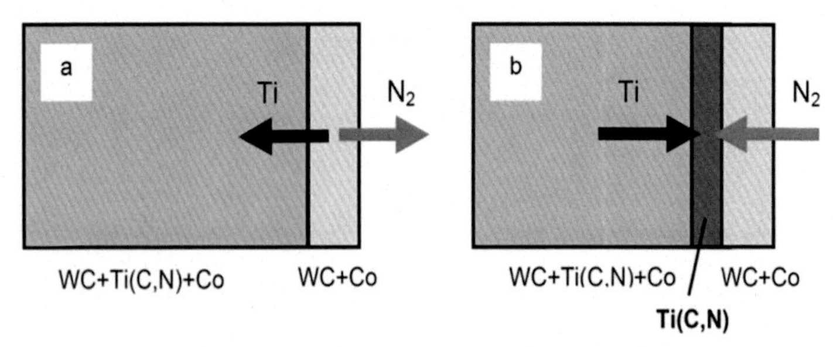

Figure 11.12 Schematic diagram indicating (A) processes of diffusion of titanium and nitrogen leading to the formation of the WC—Co carbonitride-deleted layer as a result of sintering in vacuum or nitrogen-free atmospheres, and (B) diffusion processes leading to the formation of the Ti(C,N)-enriched zone between the core region and the WC—Co carbonitride-deleted surface layer. *(Redrawn from Ref. [29].)*

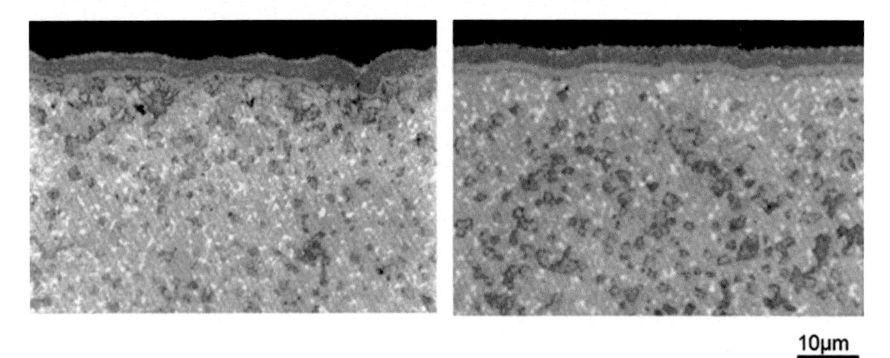

Figure 11.13 Coated functionally graded cemented carbides with carbonitride/nitride-enriched surface (left) and carbonitride-depleted surface (right). The coating consists of a layer of nitrogen-rich Ti(C,N) adjacent to the carbide substrate (yellow), an interlayer of carbon-rich Ti(C,N) second (brown-yellow) and thin alumina surface layers, which are not clearly visible. *(Redrawn from Ref. [30].)*

increases ensuring its high resistance to the spread of cracks from CVD coatings during metal-cutting.

Thus, it is possible to obtain tailored functionally graded carbide substrates for wear-resistant coatings containing either higher amounts of cubic carbonitrides in the surface zone adjacent to the coating or no grains of cubic carbonitrides and increased amounts of Co; the both cases are illustrated in Fig. 11.13.

The coated cutting inserts for machining steels and similar workpiece materials comprising the WC—Co layer not containing grains of cubic carbonitrides adjacent to CVD coatings are presently the state of the art in the manufacture of coated indexable cutting inserts for a wide range of applications. Such cutting inserts provide the best performance in metal-cutting operations characterized by interrupted cutting conditions at high impact loads due the high toughness and stability of cutting edges. For special meat-cutting applications, coated indexable cutting inserts having tailored gradients of microstructure and composition of the near-surface layer adjacent to the wear-resistant coating can be employed.

11.2 Nanostructured cemented carbides

There is a general trend in the carbide industry to produce WC–Co materials with WC mean grain size as small as possible, having a grain size in the region of nanocrystalline materials.

Recently, many works devoted to obtaining WC–Co nanostructured cemented carbides, which are produced from WC nanopowders, were published (see, for example, [31–33]). Fig. 11.14 shows images of WC nanopowders, which can potentially be used for fabricating nanostructured cemented carbides. Nevertheless, all the attempts to obtain nanograined cemented carbides with the WC mean grain size of below 100 nm described in the literature were unsuccessful so far because of the very intensive growth of WC nanopowders during sintering. This growth starts at relatively low temperatures during solid state sintering, which is illustrated in Fig. 11.15, followed by the very fast grain growth of separate WC grains during liquid-phase sintering illustrated in Fig. 11.16 and cannot be eliminated by adding large amounts of grain growth inhibitors. Fig. 11.16

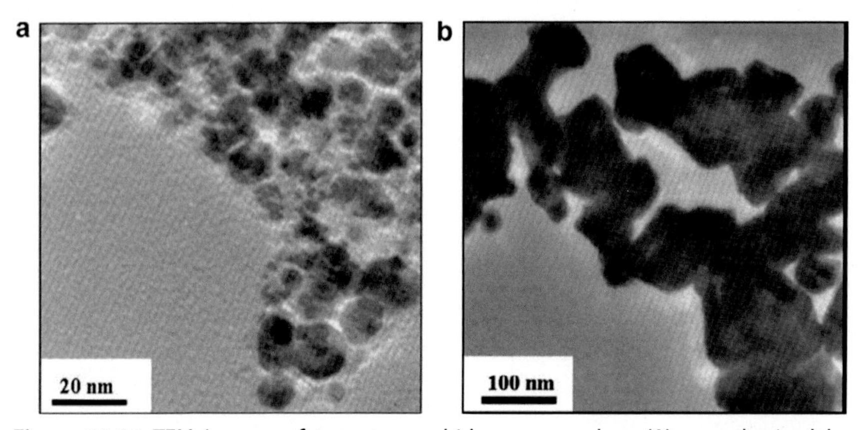

Figure 11.14 TEM images of tungsten carbide nanopowders: (A) - synthesized by plasma-assisted CVS from WCl6, and (B) - obtained after posttreatment of WC1-x powder by hydrogen at 900°C for 5 h. *(redrawn from Ref. [35].)*

Figure 11.15 Microstructure of cemented carbide obtained from a WC nanopowder as result of solid-state sintering at 1200°C for 1 min. *(Redrawn from Ref. [34].)*

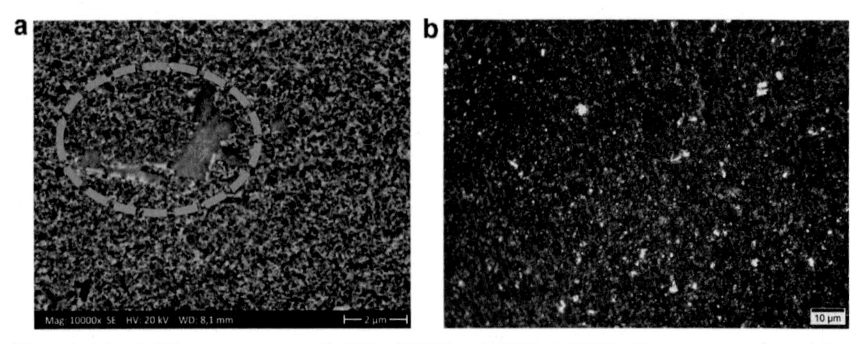

Figure 11.16 Microstructure of WC–10%Co–1%VC–0.5%Cr_3C_2 cemented carbide with a low carbon content made from a WC powder with a mean grain size of about 80 nm: (A) - HRSEM image (large WC grains are indicated by a *dashed circle*), and (B) - overview image obtained by light microscopy. *(With the permission of Element Six.)*

shows the microstructure of WC–10%Co–1%VC–0.5%Cr_3C_2 cemented carbide with a low carbon content made from a WC powder with a mean grain size of about 80 nm. The high amounts of added grain growth inhibitors (vanadium and chromium carbides) are close to their solubility limit in the binder phase, so that their contents correspond to the maximum amounts of the grain growth inhibitors that can be added without the formation of the second carbide phase. The carbon content, which also affects the WC grains growth, is very low and close to the border line with the η-phase formation according to the W–Co–C phase diagram, so the grain growth in this case should be strongly suppressed. Nevertheless, as one can see in Fig. 11.16, the microstructure of the cemented carbide sample is very inhomogeneous and comprises numerous large WC grains (up to nearly 5 μm) embedded in a fine-grain WC–Co matrix, which is unacceptable for most applications. Another problem with respect to the fabrication of nanostructured cemented carbides is related to the fact that their microstructure is very sensitive to small deviations in carbon contents, so that if the carbon content is high (close to that corresponding to the free carbon formation), the microstructure becomes significantly coarser in comparison with that obtained at low carbon contents. Also, there is a problem related to the very high susceptibility of nanostructured WC–Co green articles with respect to deviations of carbon content in gas atmospheres during sintering or presintering. If the carbon potential in the sintering furnace is slightly above a certain level, free carbon forms in the microstructure of nanostructured cemented carbides. If the carbon potential in the sintering furnace is below a certain level, the decarburization of nanostructured cemented carbides can easily occur leading to the formation of η-phases (Co_3W_3C or Co_6W_6C) in their microstructure [41].

Recently, nanostructured or so-called "near-nano" WC powders shown in Fig. 11.17 (see also Fig. 6.5), which are characterized by the relatively low activity with respect to the WC grain growth and coarsening during solid-state and liquid-phase sintering, were developed and implemented in industry on a large scale [36–39].

Both problems regarding the growth of abnormally large WC grains and coarsening WC grains at high carbon contents in nanostructured WC–Co cemented carbides mentioned above can be solved in the manufacture of nanostructured or so-called "near-nano" cemented carbides with the WC mean grain size of about 150 nm by the use of their tailored composition and production technology [39].

The microstructure of the near-nano cemented carbide with 10 wt.% Co is shown in Fig. 11.18. One can see that the microstructure of the near-nano cemented carbide is extremely fine and uniform; the majority of WC grains are below 300 nm in size. The WC mean grain size of the near-nano carbide grade obtained on the basis of the EBSD image shown in Fig. 11.19C is found to be about 150 nm, so that according to the ISO standard (ISO 4499-2:2010), it can be considered as a nanostructured carbide grade (Fig. 11.20).

Fig. 11.19 shows diagrams indicating hardness, wear, and fracture toughness of the near-nano grade with 10% Co and conventional ultrafine carbide grade with 10% Co. The Vickers hardness of the near-nano

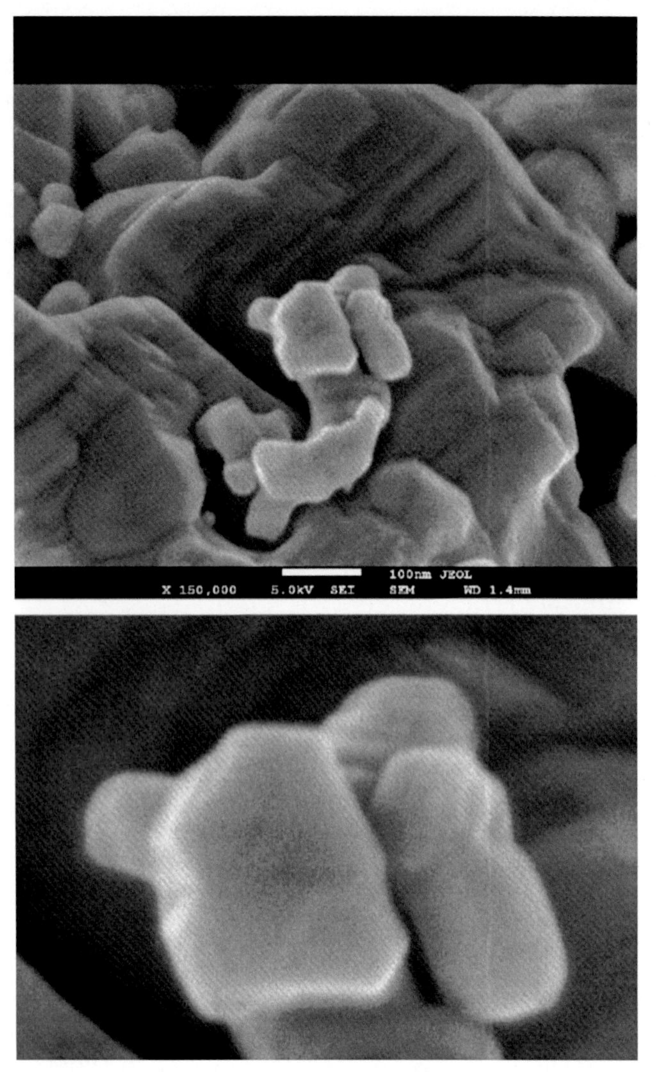

Figure 11.17 HRSEM images of the near-nano powder with the WC mean grain size of about 150 nm (4NP0, H.C. Starck, Germany) obtained at a high magnification. *(Redrawn from Ref. [40].)*

grade with 10% Co is quite high (nearly 19 GPa or higher) ensuring its exceptionally high wear resistance. The wear resistance of the near-nano cemented carbide is greater compared to that of the conventional ultra-fine grade by a factor of up to four. As one can see in Fig. 11.19, the hardness and wear resistance of the near-nano cemented carbide is dramatically increased in comparison with the corresponding ultrafine grade

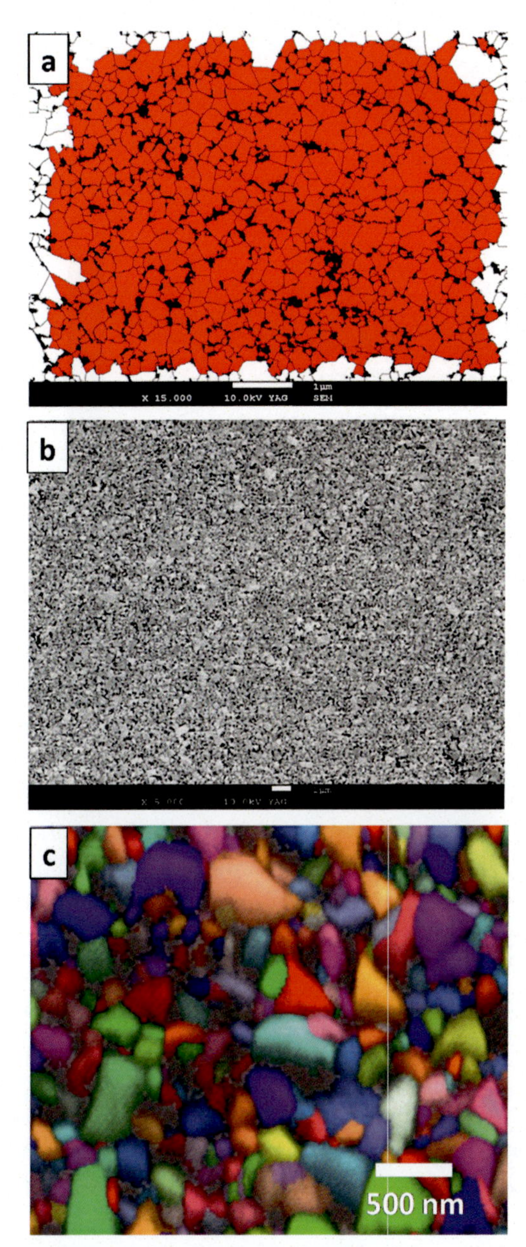

Figure 11.18 Microstructure of the near-nanostructured cemented carbide with 10% Co: (A) - and (B) - HRSEM images at different magnifications, (C) - EBSD image. *(Redrawn from Ref. [41])*

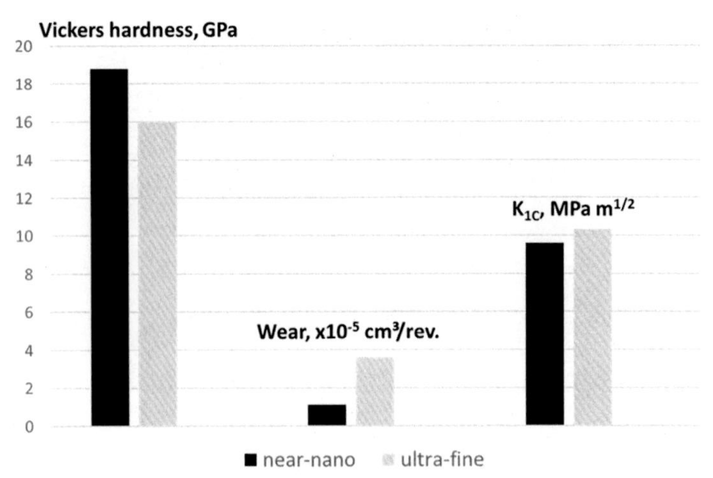

Figure 11.19 Vickers hardness, wear in the ASTM B611 test and fracture toughness of the industrial near-nano carbide grade with 10% Co in comparison with the corresponding ultrafine grade with 10% Co. *(Redrawn from Ref. [41].)*

with 10% Co almost without sacrificing the fracture toughness. Thus, the near-nano cemented carbide is characterized by a noticeably better combination of hardness, wear resistance, and fracture toughness in comparison with the conventional ultrafine grade. Due to the high hardness and extremely low values of the WC mean grain size leading a very low Co mean free path, the wear resistance of the near-nano cemented carbides in applications of sliding wear and abrasion is exceptionally high [42]. Fig. 11.20 shows wear surfaces of the near-nano and ultrafine grades after performing the ASTM B611 test. As one can see, the wear surface of the near-nano grade is smooth and almost does not comprise traces of the predominant wear of Co interlayers due to their low thickness. In contrast to that, the wear surface of the ultrafine grader is characterized by the predominant wear of the thicker Co interlayers leaving the WC grains unsupported. The near-nano grade NA10 with 10 wt.% Co having a hardness of 1900±50 Vickers units and TRS of at least 3000 MPa is fabricated on an industrial scale at Element Six GmbH (Germany). As a result of its high hardness and extremely fine and uniform microstructure, lifetime of various tools and wear parts made of this nanostructured cemented carbide is significantly prolonged in comparison with that of articles made of conventional ultrafine and submicron carbide grades.

To obtain the fine and homogeneous microstructure of near-nano cemented carbides, they must contain significant amounts of grain

Figure 11.20 Wear surfaces of different cemented carbides with 10% Co after performing the ASTM B611 test: above - near-nano, below - ultrafine. *(Redrawn from Ref. [41].)*

growth inhibitors (usually vanadium and chromium carbides). This reduces their fracture toughness in comparison with straight medium-coarse, coarse, and ultracoarse WC—Co grades with the same hardness, which is shown in Fig. 11.21. Therefore, the use of near-nano cemented carbides in the mining and construction industry is not practiced, as in these applications characterized by high impact loads, and severe thermal and mechanical fatigue, they have a reduced performance in comparison with the coarse and ultracoarse carbide grades not containing grain growth inhibitors.

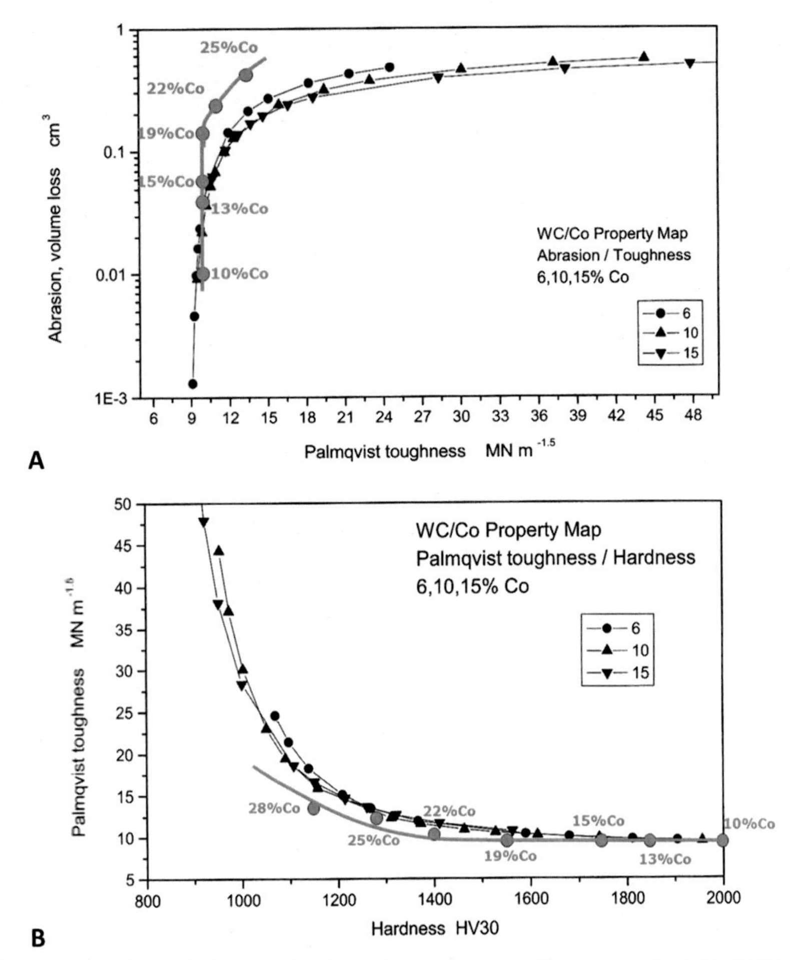

Figure 11.21 Curves indicating the dependences between the wear in the ASTM B611 test and fracture toughness (A) and the fracture toughness and hardness (B) for conventional WC–Co cemented carbides (shown in black), and the corresponding curves for near-nano cemented carbides with different Co contents (shown in red [gray in print]). *(Redrawn from Ref. [43].)*

As it is shown above, nanostructured cemented carbides with WC mean grain size significantly smaller than 100 nm cannot be produced from WC nanopowders by powder metallurgy techniques, so only unconventional methods allows one to fabricate hard materials on the basis of tungsten carbides with carbide mean grain sizes as low as 10–20 nm. One of the examples of such industrial nanostructured hard materials manufactured presently on large scale in form of coatings is the materials with the brand name "Hardide" [44] produced by CVD (Fig. 11.22). Although these

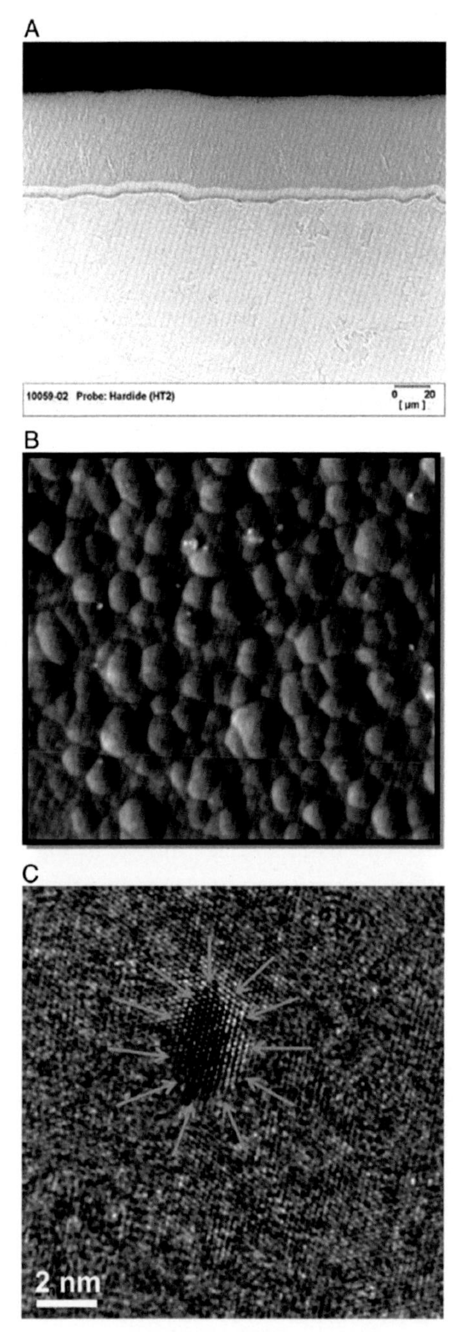

Figure 11.22 Nanostructured W_2C—W material Hardide-T: (A) the microstructure, (B) surface morphology, and (C) nanostructure with atomic resolution (HRTEM). *(Redrawn from Ref. [43].)*

materials do not directly belong to cemented carbides, their microstructure on the nano level is similar to that of cemented carbides comprising nanograins of W_2C embedded in a tungsten metal matrix, which is shown in Fig. 11.22C. According to Ref. [43], the wear resistance of the Hardide-T grade with the hardness of roughly 14.5 GPa is comparable with that of the WC–6 wt.%Co grade having nearly the same hardness.

11.3 Industrial cemented carbides with nanograin reinforced binder

As it is shown in Chapter 5, most attempts to use heat-treatments for improving properties of Co-based binders in WC-Co cemented carbides failed because of a limited increase of hardness accomplished by a significant decrease of transverse rupture strength (TRS) and fracture toughness.

The only industrial carbide grades with nanograin reinforced binders obtained by use of a special heat treatment technology are the cemented carbides produced under the brand name MasterGrade™ by Element Six GmbH [45–48]. A range of such cemented carbides with nano/micro-hierarchical structure are presently widely employed in the construction industry as well as in some other applications. The microstructure of the MasterGrade™ material for road-planinig picks consists of coarse rounded tungsten carbide grains with the WC mean grain size of about 5 μm (Fig. 11.23) embedded in a Co-based binder phase reinforced and hardened by hard nanoparticles. The nanoparticles shown in Fig. 11.24 consist of a metastable phase with the primitive Cu_3Au-type cubic lattice and have an average size of about 3 nm. The crystal lattice of the nanoparticles is coherent with that of the Co matrix, which can be clearly seen in Fig. 11.24C.

The unusual combination of the ultracoarse-grain microstructure structured on the μm-level and the binder phase structured on the nm-level was found to provide an extraordinary combination of both TRS/fracture toughness and hardness/wear resistance, which is illustrated in Figs. 11.25 and 11.26. For example, the cemented carbide grade for the construction industry is characterized by a TRS of about 2500 MPa, which is nearly 20% higher than that of a conventional carbide grade, significantly higher binder nanohardness and wear resistance increased by a factor of two to three (Figs. 11.25 and 11.26). This is achieved without sacrificing its fracture toughness resulting in revolutionary improved performance properties and noticeably prolonged lifetime of tools for mining and construction.

Figure 11.23 Microstructure of the MasterGrade™ for road-planing picks (EBSD image). *(Redrawn from Ref. [43].)*

The MasterGrade™ materials are produced by a novel technology including the fabrication of WC–Co articles with the Co-based binder oversaturated with dissolved tungsten followed by a special heat treatment process [48].

Another example of industrial cemented carbides with nanograin reinforced binders is a new WC–Fe–Cr–Si carbide grade fabricated in form of hard-facings [49]. Hard-facing materials obtained by thermal spraying, plasma spraying, and welding are not a subject of this book, as their microstructure and properties are usually considerably different from those of cemented carbides [50]. Nevertheless, the carbide hard-facing material mentioned above is characterized by the microstructure, which is similar to that of conventional cemented carbides. It is well known that ordinary hard-facings obtained by plasma spraying and welding are characterized by a combination of hardness, wear resistance, and fracture toughness significantly inferior to that of conventional WC–Co cemented carbides. One of the major reasons for that is that conventional hard-facings on steel articles must contain a relatively high proportion of the binder phase allowing them to be sintered to a full or nearly full density being applied to steel substrates for a short time. Therefore, in the majority of

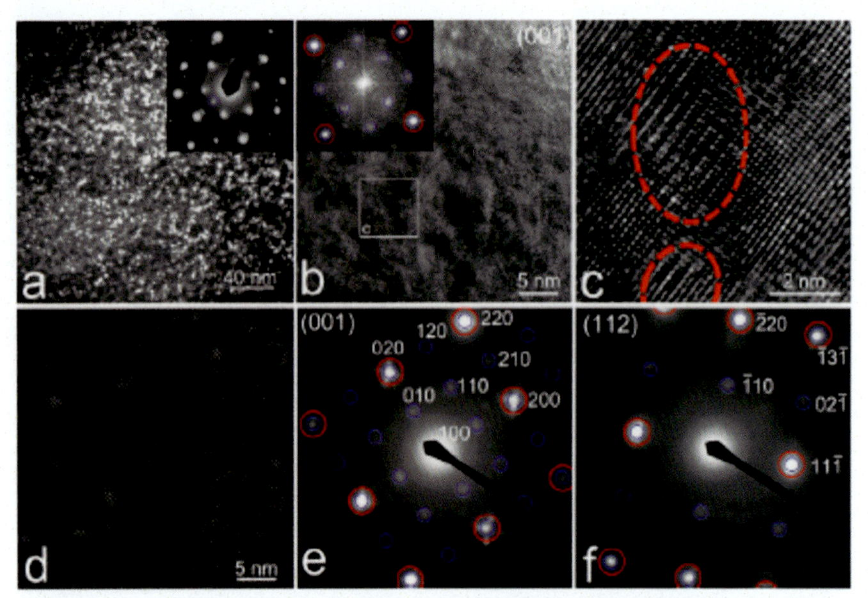

Figure 11.24 Structure of the binder phase in the MasterGrade™ material: (A) dark field TEM with the appropriate electron diffraction pattern (inset). The satellite reflection corresponding to the image in (A) arises from the nanoparticles and is marked by blue; (B) and (C) HRTEM images of nanoparticles embedded in the fcc Co matrix and the corresponding FFT pattern (inset in (B)), where the reflections from the nanoparticles are marked by blue and those from the fcc Co matrix are marked by red; two nanoparticles in (C) are marked by *red circles*; (D) the same as in (B) but after the FFT filtering using the reflections from the nanoparticles to enhance their contrast; (E) and (F) electron diffraction patterns from the fcc Co matrix containing the nanoparticles obtained by tilting toward two different crystal zones, i.e., (001) and (112), confirming that the nanoparticles have the cubic Cu_3Au lattice type. The red and blue circles indicate the location of the reflections in the simulated diffraction patterns from fcc Co and from the phase with the cubic Cu_3Au lattice type correspondingly. *(Redrawn from Ref. [45].)*

applications, their performance properties are incomparable with those of WC−Co materials. In contrast, the uncommon hard-facing material of the WC−Fe−Cr−Si system obtained by plasma transfer arc welding (PTA welding) has a combination of performance properties comparable with that of WC−Co cemented carbides, which is achieved due to the binder nanograin reinforcement and hardening. Fig. 11.27 shows the structure of the nanograin reinforced binder comprising nanoparticles of η-phase and mixed Cr−Fe carbides, which are embedded in the nanostructured binder matrix. The presence of hard nanoparticles of nearly 10−20 nm in the binder phase allows one to significantly increase its nanohardness up to the

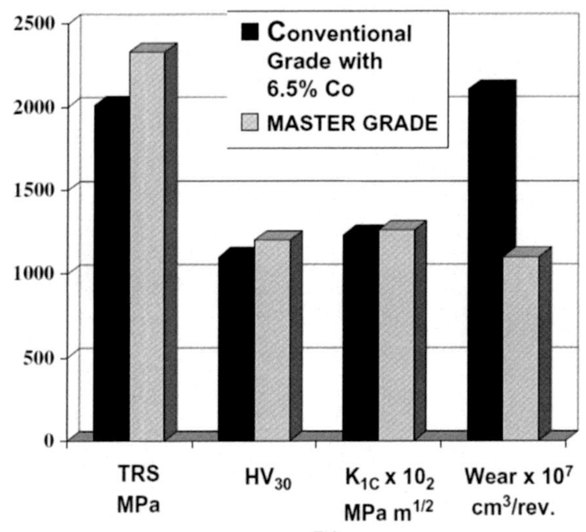

Figure 11.25 Properties of the MasterGrade™ material for road-planing in comparison with the standard grade with the same Co content and similar WC mean grain size. *(Redrawn from Ref. [46].)*

level typical for high-Co cemented carbides (12 GPa), which is achieved without sacrificing its fracture toughness. The Vickers hardness of the hard-facing varies from nearly 9.5 to 11.5 GPa depending on their composition and welding parameters at the fracture toughness of roughly 12–14 MPa $m^{1/2}$. In spite of relatively moderate values of macrohardness, the wear resistance of the hard-facing is comparable with that of conventional WC–Co mining grades with 6–10 wt.% Co, which is a result of the binder nanograin reinforcement and hardening (Fig. 11.28). This ensures the novel cemented carbides in form of hard-facing to be employed for the fabrication of articles having a complicated geometry or large sizes, which cannot be produced by conventional powder metallurgy routes. Presently, these cemented carbides in form of hard-facings are fabricated in industry on a large scale for different applications varying from agricultural tools to rock-cutting rings for tunneling machines.

11.4 Cemented carbides with alloyed binder phases

In spite of numerous publications on alloying WC–Co cemented carbides with different chemical elements, only few alloying elements are employed for industrial cemented carbide grades (besides additions of grain growth inhibitors described in detail in Chapter 9).

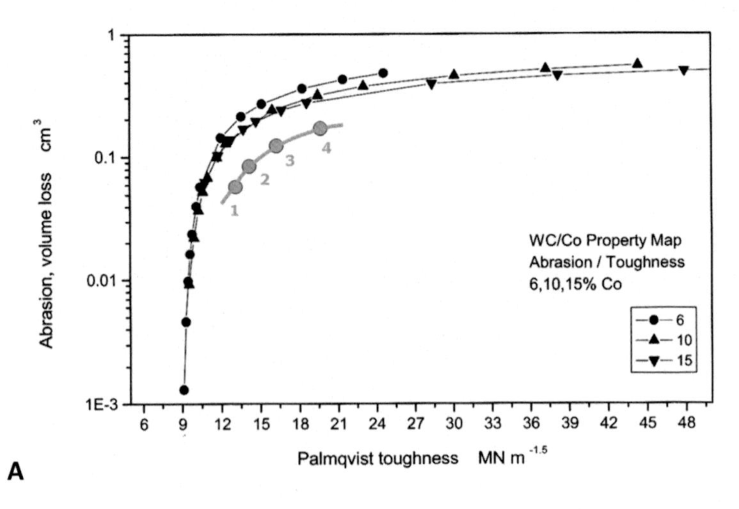

A

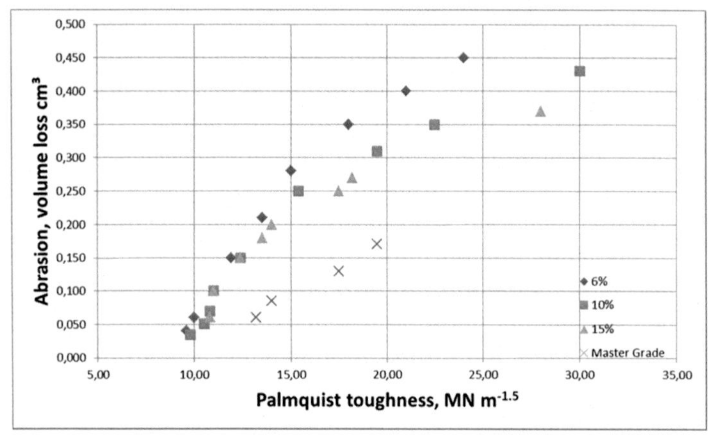

B

Figure 11.26 The MasterGrade^{TM} materials versus conventional WC–Co grades: (A) the curve indicating the dependence of wear of the Master Grades with different combinations of WC mean grain size and Co content indicated by numbers: 1–1.9 μm, 6%Co, 2–2.5 μm, 10%Co, 3–4.8 μm, 6.5%Co, 4–4.9 μm, 9.5%Co, in comparison with the baselines for conventional WC–Co cemented carbides (the ASTM B611 test, the logarithmic coordinates); (B) the same as (A) but reconstructed in the straight co-ordinates. *(Redrawn from Ref. [43].)*

One of such chemical elements is ruthenium, the positive influence of which on the binder hardness is well known (Fig. 11.29) (see, for example, Refs. [51,52]).

Cemented carbides containing a binder phase alloyed with Ru are fabricated on a large scale under the trade name X-Grade by Kennametal Inc [53]. It is stated that the combination of ruthenium with cobalt leads to

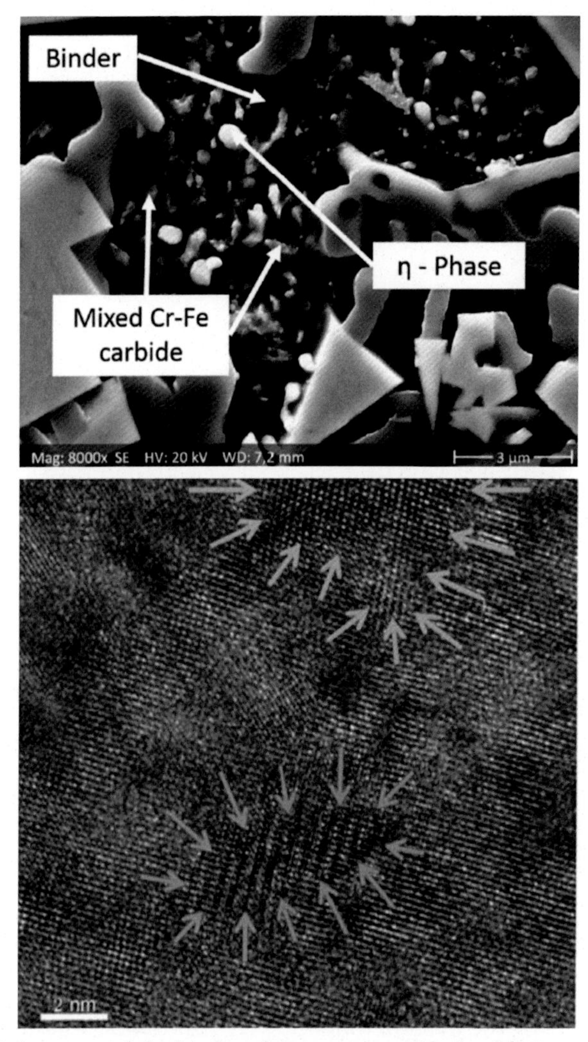

Figure 11.27 Structure of the binder of the novel carbide hard-facing reinforced and hardened by nanoparticles of η-phase and mixed Cr–Fe carbides: above - an SEM image, and below - an HRTEM image obtained with atomic resolution (the carbide nanoparticles are indicated by *arrows*). *(Redrawn from Ref. [50].)*

"... an exclusive binder ..." for "... cutting tool solutions that deliver industry leading metal removal rates."

Another approach to fabricating cemented carbides with alloyed binder phases is realized in the Dyanite grade containing boron, which was developed by MultiMetals [54]. In particular, it is stated that "... Dyanite makes it possible to obtain cemented carbides with increased hardness

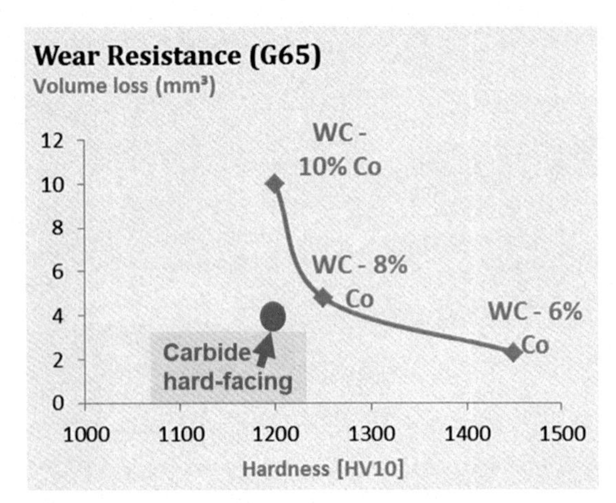

Figure 11.28 The dependence of the wear resistance in the ASTM G65 test on hardness for conventional mining WC—Co grades with 6%, 8%, and 10% Co in comparison with that of the novel carbide hard-facing. *(With the permission of Element Six.)*

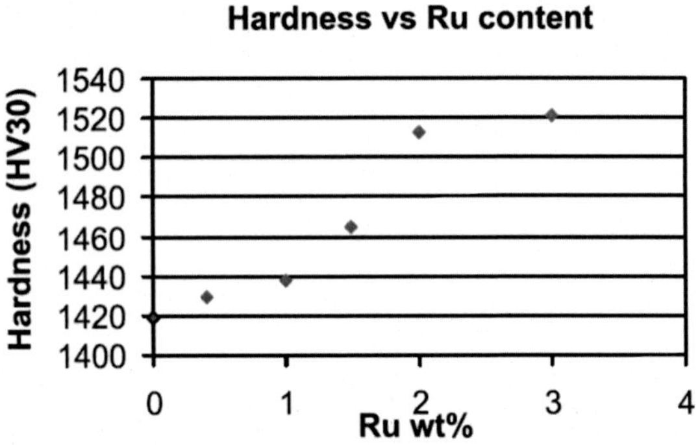

Hardness vs Ru content

Hardness (HV30)

1540
1520
1500
1480
1460
1440
1420
1400

0 1 2 3 4

Ru wt%

Figure 11.29 The influence of Ru additions on the Vickers hardness of WC—10 wt.%Co cemented carbide. *(Redrawn from Ref. [52].)*

without decreasing the fracture toughness ... ." It can be assumed that boron is introduced into a binder using a special technology, when taking into account the negative influence of even insignificant boron additions on the homogeneity of the microstructure of cemented carbides (see Chapter 9).

When considering industrial carbide grades with alloyed binders, the publication of Lian Sha [55] on the production of rare earth doped cemented carbides in China should be mentioned. It is reported that small additions (below 1 wt.%) of rare earth (Y, La, Ce, etc.) to cemented carbides lead to increasing the strength and toughness of such alloyed WC—Co materials, stabilizing the fcc Co modification and inhibiting the WC grain growth.

Another alloying element that can significantly improve high-temperature properties of WC–Co cemented carbides is known to be rhenium [56,57]. Industrial WC–Co–Re grades VRK-10 and VRK-15 were fabricated on a large scale in the Soviet Union in the 80s. It was established that cemented carbides of the WC–Co–Re system represent a new class of hard materials having a significantly increased Young's modulus, hot hardness, and high temperature creep resistance. Rhenium being dissolved in the Co-based binder is found to be a very strong grain growth inhibitor with respect to WC coarsening during liquid-phase sintering, which is illustrated in Fig. 11.30. As one can see in Fig. 11.31,

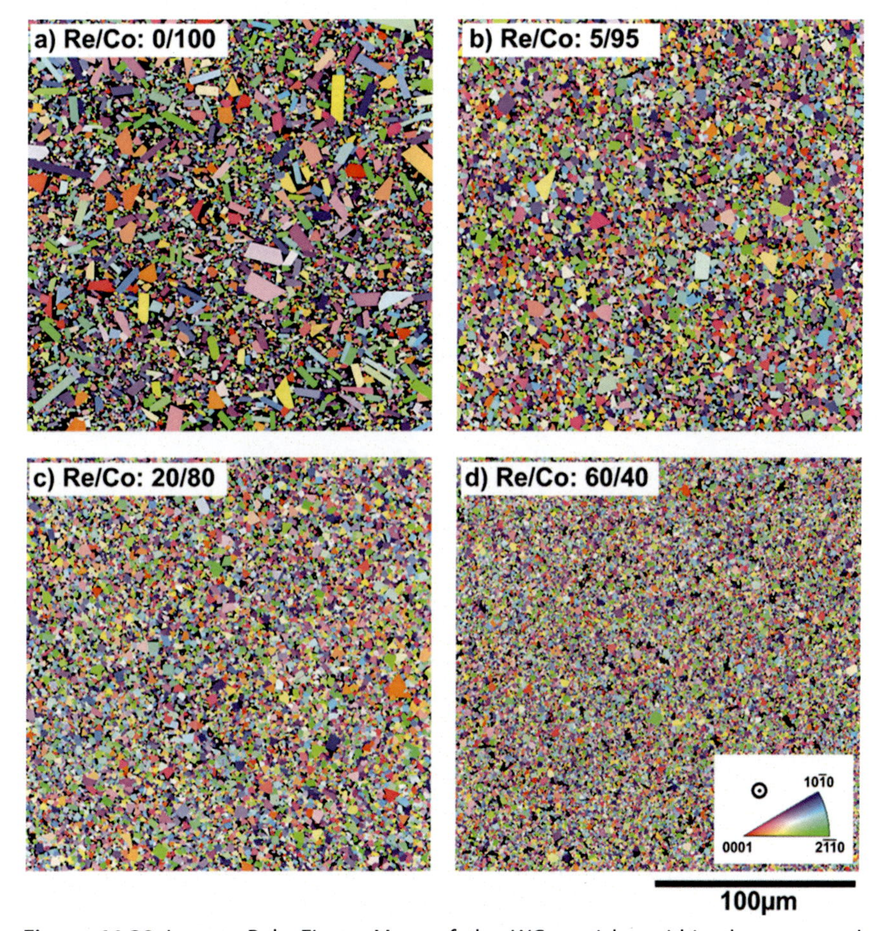

Figure 11.30 Inverse Pole Figure Maps of the WC particles within the cemented carbides with different Re:Co ratios (wt.%) in comparison with a control WC–Co sample containing no rhenium. *(Redrawn from Ref. [57].)*

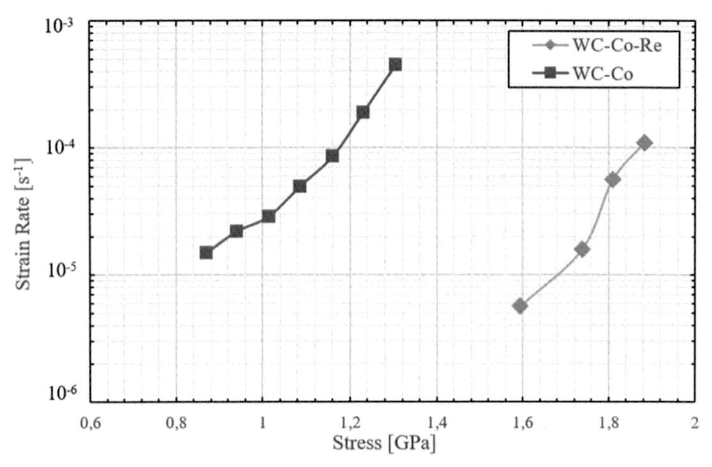

Figure 11.31 Dependences of strain rates for the WC–Co–Re and WC–Co cemented carbides on the compression stress at a temperature of 800°C. *(Redrawn from Ref. [56].)*

high temperature creep resistance of the WC–Co–Re cemented carbides is dramatically improved in comparison with conventional WC–Co materials. Although Re is very expensive, due to their unique properties, the WC–Co–Re cemented carbides can find applications in use in high-pressure high-temperature components for synthesis of diamond and c-BN as well as cutting Ni-based superalloys and other heat-generating workpiece materials.

Besides special carbides grades comprising Co–Ni–Cr binders with high Cr contents for mill rolls described in Chapter 10, WC–Co mining grades containing Cr alloyed binders were recently patented by Sandvik [58]. The Cr alloy binders are stated to have an improved performance and wear resistance in mining applications, particularly, in percussive drilling, as they are characterized by the so-called "self-hardening" effect being subjected to severe fatigue during percussive drilling. It is stated that the Sandvik SH70 grade is characterized by such a "self-hardening" effect, and "... it has the ability to become both more wear resistant and tougher as drilling progresses." It is also stated by Sandvik that "The hardness of the insert surface continuously increases during drilling ..." [13].

11.5 Cemented carbide substrates in combination with layers of polycrystalline diamond

At the present time, tools for oil-and-gas (O&G) drilling, mining, and construction with inserts comprising cemented carbide substrates and layers of polycrystalline diamond (PCD), which is a composite material containing diamond grains embedded in a metallic binder matrix, have become very popular and partially substituted conventional tricone bits for O&G drilling with cemented carbide inserts [59,60]. Despite relatively high production costs and consequently high prices of such cemented carbide/PCD inserts fabricated at ultrahigh pressures, their employment in many applications is economically reasonable due to their dramatically prolonged tool life. According to the forecast of Bellin et al. [60], drilling bits with PCD cutters will substitute conventional roller-cone bits with cemented carbide inserts for O&G drilling by at least 80%.

Up to recent times, conventional WC–Co cemented carbides were used as substrates for PCD, which caused the following problems: (1) the erosion resistance of carbide substrates during O&G drilling is insufficient because of employing very erosive liquid–solid slurries for cooling PCD inserts and removing finely dispersed products of rock-drilling, and (2) in many cases, there is intensive WC grain growth in carbide substrates during the fabrication of PCD inserts at ultrahigh pressures and high temperatures as a result of carbon diffusion from the PCD table into the substrates. To solve the problems mentioned above, novel special WC–Co–Ni–Cr carbide grades were developed and implemented in industry on a large scale [61,62]. Such grades are characterized by dramatically improved erosion resistance. They are very stable with respect to the WC grain growth during PCD press sintering due to the tailored composition, WC grain size distribution, and microstructure. Fig. 11.32 shows microstructures of the new cemented carbides before and after the PCD press sintering at ultrahigh pressures. As one can see in Fig. 11.32, the carbide microstructure almost does not change as a result of the PCD press sintering process. Fig. 11.33 shows microstructures of the substrate/PCD interface and PCD table obtained by use of the new carbide substrates indicating the absence of abnormally large WC grains at the substrate/PCD interface.

11.6 Cemented carbides with Co-enriched surface layers

In many cases, the surface of sintered carbide articles subjected to brazing contains very little cobalt leading to its poor wetting by braze alloys. Sometimes such layers are present on the surface of carbide articles after sintering; however, in many cases when articles of the same carbide grade are repeatedly sintered in the same production sinter-HIP furnace, there are

Figure 11.32 Microstructures of the new WC—Co—Ni—Cr grade for substrates for PCD (optical microscopy): above - before PCD press sintering, and below - after PCD press sintering [61]. *(With the permission of Element Six.)*

no such layers on the carbide surface. Fig. 11.34 illustrates the above-mentioned showing carbide articles of different grades and geometries sintered in various furnaces. Some of these articles have a shiny surface indicating the presence of Co-enriched surface layers, some of them are dull indicating only little Co on the surface, and some of them are dark-grey or black providing evidence for the absence of Co on the surface.

A reason for the presence or absence of Co thin layers or Co-enriched layers on a surface of WC—Co articles during sintering has been a riddle for

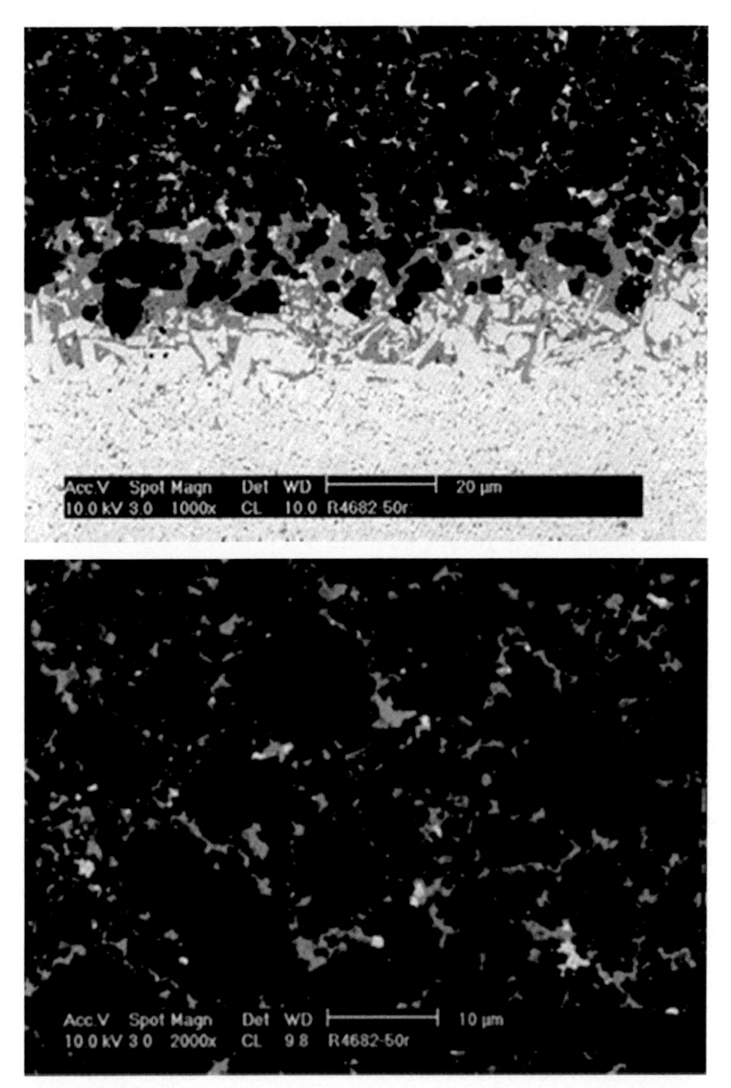

Figure 11.33 SEM images of the PCD table on the substrate of the new WC–Co–Ni–Cr cemented carbide grade: above - microstructure of the substrate/PCD interface, and below - microstructure of the PCD table [61]. *(With the permission of Element Six.)*

a long time. As a result of sintering, such thin shiny Co layers are sometimes present on the surface of WC–Co articles and their formation is referred to as "Co capping" in literature. A mechanism explaining the presence or absence of such Co layers after sintering was proposed in Refs. [63,64]. The proposed mechanism is based on considering wetting phenomena of WC by liquid Co on the surface of carbide articles and capillarity phenomena

Figure 11.34 Carbide articles of different grades and geometries sintered in various industrial sinter-HIP furnaces. *(With the permission of Element Six.)*

acting on liquid Co in narrow channels between WC grains in the carbide near-surface layer. The mechanism explains all the phenomena of the Co layer formation during sintering of various carbide grades followed by either fast or slow cooling.

The understanding of the Co capping phenomenon allowed the development of a new technology for the fabrication of functionally graded cemented carbides having a surface layer enriched with Co [65], which is designated as the "Co capping technology." The technology comprising stepwise cooling of carbide articles after liquid-phase sintering at tailored cooling rates [66] was implemented at Element Six GmbH. The fabrication of such cemented carbides with Co-enriched surface layers leads to significant improvements of the TRS, surface fracture toughness, and carbide surface wettability by braze alloys.

Fig. 11.35 shows the microstructure and morphology of the Co-coated cemented carbides obtained by the Co capping technology [66].

The surface of nonground carbide articles always comprises microcracks and other defects after sintering [67], which can lead to a significant reduction of their TRS and consequently performance properties. The Co films can completely "heal" such surface cracks and defects. The typical Palmqvist cracks forming near the Vickers indentation on the carbide surface shown in Fig. 11.36A completely disappear after the formation of the Co-enriched layer, which is clearly seen in Fig. 11.36B. As a result, the TRS of nonground carbide articles increases by up to 50%. Therefore, the fabrication of cemented carbide articles with such Co-enriched surface

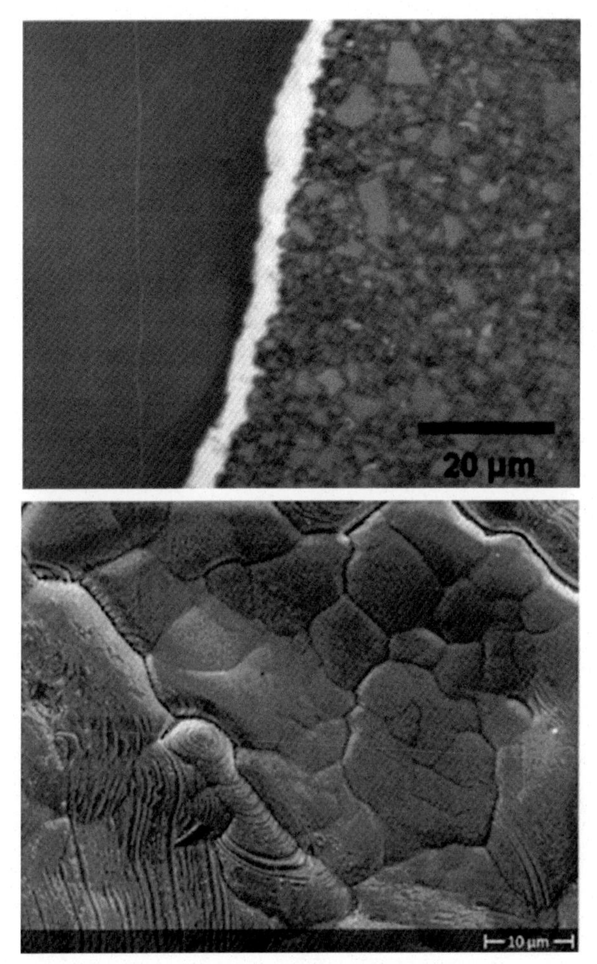

Figure 11.35 Microstructure and morphology of the Co-coated cemented carbides obtained by the new Co capping technology. *(Redrawn from Ref. [65].)*

layers comprising tough, ductile, and defect-free Co films has a similar effect on the TRS as mechanical polishing, which leads to the elimination of all surface defects.

The presence of the Co-enriched layers on the surface of carbide articles is found to remain during brazing and it leads to significant improvements in the wettability of such carbide articles by braze alloys and the relaxation of high stresses forming as a result of fast cooling the carbide articles after brazing. As a result, there is no problem with fabricating carbide tools by brazing such carbide articles of relatively brittle WC–6 wt.% Co grades usually containing little Co on the surface after conventional sintering.

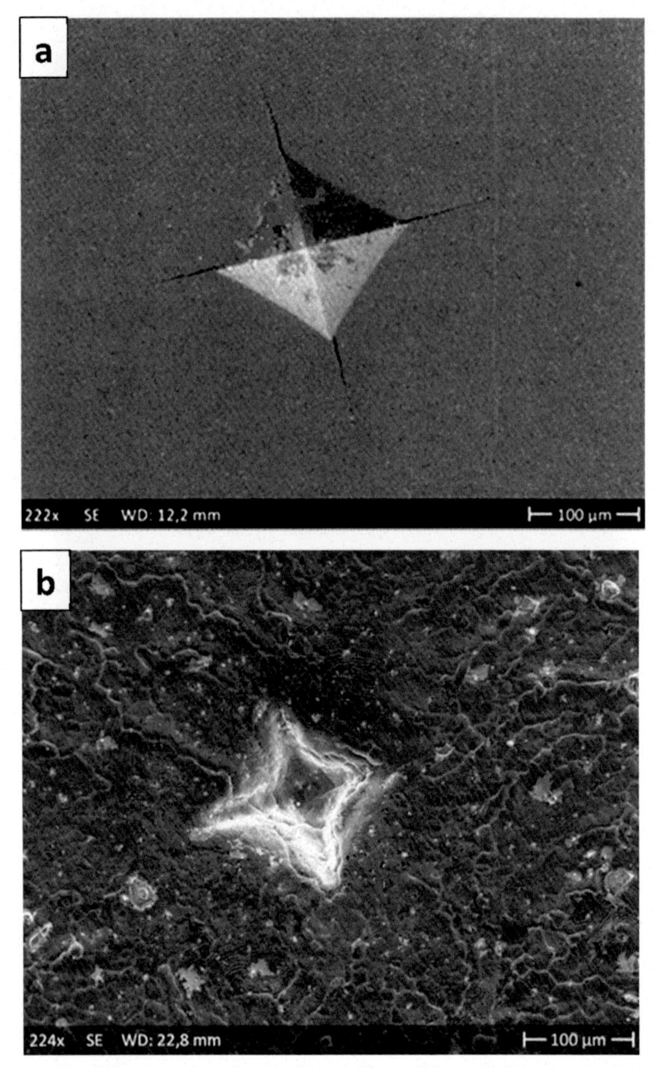

Figure 11.36 The Palmqvist cracks on the surface of the medium-grain WC–6 wt.%Co grade before (A) and after (B) performing the Co capping technology. *(Redrawn from Ref. [65].)*

References

[1] J. Garcia, V. Collado, C. Andreas, A. Blowqvist, B. Kaplan, et al., Cemented carbide microstructures: a review, Int. J. Refract. Met. Hard Mater. 80 (2019) 40–68.
[2] I. Konyashin, S. Hlawatschek, B. Ries, A. Mazilkin, Co drifts between cemented carbides having various WC grain sizes, Mat. Let. 167 (2016) 270–273.
[3] U. Fischer, E. Hartzell, J. Akerman, US Patent 4,743,515, 1988.

[4] B. Aronsson, T. Hartzell, J. Aakerman, Structure and properties of dual carbide for rock drilling, in: Proceedings of Adv. Hard Mater. Prod. Conf., MPR Publ. Serv. Ltd., Shrewsbury, UK, 1988, pp. 19/1–19/6.
[5] S. Rassbach, S. Moseley, W. Böhlke, Metallurgical fundamentals of macroscopic gradient hardmetals, in: L. Sigl, P. Rödhammer, H. Wildner (Eds.), Proc. 17th Int. Plansee Seminar, Plansee Group, Austria, 2009, pp. 48/1–48/13. V.2, Reutte.
[6] I. Konyashin, B. Ries, F. Lachmann, A.T. Fry, A novel sintering technique for fabrication of functionally gradient WC–Co cemented carbides, J. Mater. Sci. 47 (2012) 7072–7084.
[7] O. Eso, P. Fan, Z. Fang, A kinetic model for cobalt gradient formation during liquid phase sintering of functionally graded WC–Co, Int. J. Refractory Met. Hard Mater. 26 (2008) 91–97.
[8] O. Eso, P. Fan, Z. Fang, Kinetics of cobalt gradient formation during the liquid phase sintering of functionally graded WC–Co, Int. J. Refract. Met. Hard Mater. 25 (2007) 286–292.
[9] I. Konyashin, B. Ries, F. Lachmann, A.T. Fry, Gradient hardmetals: theory and practice, Int. J. Refract. Met. Hard Mater. 36 (2013) 10–21.
[10] L. Zhang, Y. Wang, X. Yu, et al., Crack propagation characteristics and toughness of functionally graded WC–Co cemented carbides, Int. J. Refract. Met. Hard Mater. 26 (2008) 295–300.
[11] J. Guo, F. Wang, et al., A Novel Approach for Manufacturing Functionally Graded Cemented Tungsten Carbide Composites, Advanced in Powder Metallurgy & Particulate Materials, Metal Powder Industries Federation, Princeton, NJ, USA, 2010, pp. 8/29–8/39.
[12] J. Guo, P. Fan, Z. Fang, A new method for making graded WC–Co by carburizing heat treatment of fully densified WC–Co, in: L. Sigl, P. Rödhammer, H. Wildner (Eds.), Proc. 17th Int. Plansee Seminar, vol. 2, Reutte, 2009, pp. 50/1–50/6.
[13] https://www.rocktechnology.sandvik/en/campaigns/powercarbide/, January 2021.
[14] Z. Fang, P. Fan, J. Guo, US Patent Application US2010/0101368A1, 2010.
[15] Y. Liu, H. Wang, J. Yang, et al., Formation mechanism of cobalt-gradient structure in WC–Co hard alloy, J. Mater. Sci. 39 (2004) 4397–4399.
[16] P. Fan, Z.Z. Fang, A review of liquid phase migration and methods for fabrication of functionally graded cemented tungsten carbide, Int. J. Refract. Met. Hard Mater. 36 (2013) 2–9.
[17] I. Konyashin, S. Hlawatschek, B. Ries, F. Lachmann, A. Sologubenko, T. Weirich, A new approach to fabrication of gradient WC–Co hardmetals, Int. J. Refract. Met. Hard Mater. 28 (2010) 228–237.
[18] I. Konyashin, S. Hlawatschek, B. Ries, A. Mazilkin Weirich, Novel industrial hardmetals for mining, construction and wear applications, Int. J. Refract. Met. Hard Mater. 71 (2018) 357–365.
[19] I. Konyashin, S. Hlawatschek, B. Ries, Engineered surfaces on cemented carbides obtained by tailored sintering techniques, Surf. Coat. Technol. 258 (2014) 300–309.
[20] B. Roebuck, M.G. Gee, R. Morrell, Hardmetals – microstructural design, testing and property maps, in: G. Kneringer, P. Rödhammer, H. Wildner (Eds.), Proceedings of the 15th International Plansee Seminar, vol. 4, Plansee Group, Reutte, 2001, pp. 245–266.
[21] I. Konyashin, A.A. Zaitsev, D. Sidorenko, E.A. Levashov, S.N. Konischev, M. Sorokin, S. Hlawatschek, B. Ries, A.A. Mazilkin, S. Lauterbach, H.-J. Kleebe, On the mechanism of obtaining functionally graded hardmetals, Mat. Let. 186 (2017) 142–145.

[22] I. Konyashin, A. Zaitsev, A. Meledin, J. Mayer, et al., Interfaces between model Co–W–C alloys with various carbon contents and tungsten carbide, Materials 11 (3) (2018) 404, https://doi.org/10.3390/ma11030404.
[23] M.V.G. Petisme, S.A.E. Johansson, G. Wahnström, A computational study of interfaces in WC–Co cemented carbides, Model. Simul. Mater. Sci. Eng. 23 (2015) 045001.
[24] H. Suzuki, K. Hayashi, Y. Taniguchi, The beta-free layer near the surface of vacuum-sintered tungsten carbide-beta-Co alloys containing nitrogen, Trans. Jpn. Inst. Met. 22 (11) (1981) 758–764.
[25] M. Schwarzkopf, H.E. Exner, H.F. Fischmeister, W. Schintlmeister, Kinetics of compositional modification of (W,Ti)C-WC-Co alloy surfaces, Mater. Sci. Eng. A 105/106 (1988) 225–231.
[26] K. Tsuda, A. Ikegaya, K. Isobe, M. Kitagawa, T. Nomura, Development of functionally graded sintered hard materials, Powder Met. 39 (4) (1996) 296–300.
[27] W. Lengauer, K. Dreyer, Functionally graded hardmetals, J. Alloys Compd. 338 (1–2) (2002) 194–212.
[28] J. Glühmann, M. Schneeweiß, H. van den Berg, D. Kassel, K. Rödiger, K. Dreyer, W. Lengauer, Functionally graded WC-Ti(C,N)-(Ta,Nb)C-Co hardmetals: metallurgy and performance, Int. J. Refract. Met. Hard Mater. 36 (2013) 38–45.
[29] W. Lengauer, K. Dreyer, Tailoring hardness and toughness gradients in functional gradient hardmetals (FGHMs), Int. J. Refract. Met. Hard Mater. 24 (2006) 155–161.
[30] R. Frykholm, M. Ekroth, B. Jansson, J. Ågren, H.-O. Andrén, A new labyrinth factor for modelling the effect of binder volume fraction on gradient sintering of cemented carbides, Acta Mater. 51 (2003) 1115–1121.
[31] S. Berger, R. Prat, R. Rosen, Nanocrystalline materials: a study of WC-based hard metals, Prog. Mater. Sci. 42 (1997) 311–320.
[32] G. Shao, X.-L. Duan, J. Xie, X. Yu, W. Zhang, R. Yuan, Sintering of nanocrystalline WC–Co composite powder, Rev. Adv. Mater. Sci. 5 (2003) 281–286.
[33] G. Goren-Muginstein, S. Berger, A. Rosen, Sintering study of nanocrystalline tungsten carbide powder, Nanostruct. Mater. 10 (1998) 795–804.
[34] Z. Fang, P. Mahesdhwari, X. Wang, H. Sohn, A. Griffo, R. Riley, An experimental study of the sintering of nanocrystalline WC–Co powders, Int. J. Refract. Met. Hard Mater. 23 (2005) 249–257.
[35] Z. Fang, X. Wang, T. Ryu, K.S. Hwang, H.Y. Sohn, Synthesis, sintering and mechanical properties of nanocrystalline cemented tungsten carbide – a review, Int. J. Refract. Met. Hard Mater. 27 (2009) 288–299.
[36] B. Caspers, T. Säuberlich, M. Reiss, Influence of manufacturing parameters on properties of nano tungsten carbide, in: Proceedings of the 2008 International Conference on Tungsten, Refractory and Hardmetals, vol. VII, 2008, pp. 50–58. Washington.
[37] A. Eiling, B. Caspers, G. Gille, B. Gries, Nanoskalige Hartmetalle mit Fe-basierten Bindern, in: H. Kolaska (Ed.), Pulvermetallurgie: Neue Anforderungen – Neue Produkte – Neue Verfahren, Fachverband Pulvermetallurgie, 2008, pp. 91–106.
[38] M. Brieseck, I. Hünsche, B. Caspers, B. Szesny, G. Gille, W. Lengauer, Optimised sintering and grain-growth inhibition of ultrafine and near-nano hardmetals, in: Proceedings of the International Conference on PM, vol. 1, EPMA, Copenhagen, 2009, pp. 239–246.
[39] I. Konyashin, B. Ries, F. Lachmann, Cemented Carbide and Process for Producing Same, WO 2011/058167, 2011.
[40] I. Konyashin, B. Ries, F. Lachmann, Near-nano WC–Co hardmetals: will they substitute conventional coarse-grained mining grades? Int. J. Refract. Met. Hard Mater. 28 (2010) 489–497.

[41] B.R. Konyashin, Microstructure, grain boundaries and properties of nanostructured hardmetals, Mater. Lett. 253 (2019) 128–130.
[42] A.J. Gant, I. Konyashin, B. Ries, A. McKiea, R.W.N. Nilen, J. Pickles, Wear mechanisms of diamond-containing hardmetals in comparison with diamond-based materials, Int. J. Refract. Met. Hard Mater. 71 (2018) 106–114.
[43] I. Konyashin, B. Ries, Y. Zhuk, A. Mazilkin, B. Straumal, Wear-resistance and hardness: are they directly related for nanostructured hard materials? Int. J. Refract. Met. Hard Mater. 49 (2015) 203–211.
[44] Y. Zhuk, Hardide: advanced nanostructured CVD coating, Int. J. Microstruct. Mater. Prop. 2 (1) (2007) 90–98.
[45] I. Konyashin, F. Lachmann, B. Ries, A.A. Mazilkin, B.B. Straumal, C. Kübel, L. Llanes, B. Baretzky, Strengthening zones in the Co matrix of WC–Co cemented carbides, Scripta Mater. 83 (2014) 17–20.
[46] I. Konyashin, F. Scfaefer, R. Cooper, B. Ries, J. Mayer, T. Weirich, Novel ultra-coarse hardmetal grades with reinforced binder for mining and construction, Int. J. Refract. Met. Hard Mater. 23 (2005) 225–232.
[47] I. Konyashin, B. Ries, S. Hlawatschek, A. Mazilkin, Novel industrial hardmetals for mining, construction and wear applications, Int. J. Refractory Met. Hard Mater. 71 (2018) 357–365.
[48] I. Konyashin, B. Ries, F. Lachmann, Cemented Carbide Material, WO2012/130851, 2012.
[49] L.-M. Berger, in: V. Sarin (Ed.), Coatings for Thermal Spray, Elsevier Science and Technology, 2014, pp. 471–506.
[50] H. Hinners, I. Konyashin, B. Ries, M. Petrzhik, E.A. Levashov, D. Parkc, T. Weirich, J. Mayec, A.A. Mazilkin, Novel hardmetals with nano-grain reinforced binder for hard-facings, Int. J. Refract. Met. Hard Mater. 67 (2017) 98–104.
[51] C. Bonjour, A. Actis-Data, Effect of ruthenium additions on the properties of WC–Co ultra micrograins, Powder Met. Tool Mater. 12 (2004) 543–549.
[52] T.L. Shing, S. Luyckx, I.T. Northrop, I. Wolff, The effect of ruthenium additions of the hardness, toughness and grain size of WC–Co, Int. J. Refract. Met. Hard Mater. 19 (2001) 41–44.
[53] https://pdf.directindustry.com/pdf/kennametal/kmt-stellram-expanded-milling-catalog/7354-609889-_2.html.
[54] www.multi-metals.com.
[55] S. Liu, Study on rare-earth doped cemented carbide in China, Int. J. Refract. Met. Hard Mater. 27 (2009) 528–534.
[56] I. Konyashin, S. Farag, B. Ries, B. Roebuck, WC–Co–Re cemented carbides: structure, properties and potential applications, Int. J. Refract. Met. Hard Mater. 78 (2019) 247–253.
[57] I. Konyashin, A. Schwedt, The influence of various Re:Co ratios on microstructures and properties of WC–Co–Re cemented carbides, Mat. Let. 249 (2019) 57–61.
[58] A. Nordgren, A. Ekmarker, S. Norgren, Rock Drill Button, US Patent Application US2018/0073108A1, 2018.
[59] I. Konyashin, M. Antonov, B. Ries, Wear behaviour and wear mechanisms of different hardmetal grades in comparison with polycrystalline diamond in a new impact-abrasion test, Int. J. Refractory Met. Hard Mater. 92 (2020) 105286.
[60] F. Bellin, A. Dourface, W. King, M. Thigpen, The current state of PDC technology, World Oil (November 2010) 67–71.
[61] I. Konyashin, B. Ries, S. Hlawatschek, H. Hinners, in: L. Sigl, H. Kestler, A. Pilz (Eds.), Novel Industrial Hardmetals for Mining, Construction and Wear Applications, Proc. 19th Int. Plansee Seminar, Reutee, Austria, 2017. HM6/1-HM6/12.
[62] I. Konyashin, B. Ries, F. Lachmann, Polycrystalline Diamond Composite Compact Elements and Methods of Making and Using Same, WO 2013/087773, 2013.

[63] I. Konyashin, W. Lengauer, in: G.S. Upadhyaya (Ed.), Sintering Mechanisms of Functionally Graded Cemented Carbides. Sintering Fundamentals II, Trans Tech Publications Ltd, Switzerland, 2016, pp. 116–198.
[64] I. Konyashin, S. Hlawatschek, B. Ries, F. Lachmann, M. Vukovic, Cobalt capping on WC–Co hardmetals. Part I: a mechanism explaining the presence or absence of cobalt layers on hardmetal articles during sintering, Int. J. Refract. Met. Hard Mater. 42 (2014) 142–150.
[65] I. Konyashin, S. Hlawatschek, B. Ries, F. Lachmann, M. Vukovic, Cobalt capping on WC–Co hard metals. Part II: a technology for fabrication of Co coated articles during sintering, Int. J. Refract. Met. Hard Mater. 42 (2014) 136–141.
[66] I. Konyashin, B. Ries, F. Lachmann, Cemented Carbide Article and Method for Making Same, WO2012/098102A1, 2011.
[67] I. Konyashin, Healing of surface defects in hard materials by thin coatings, Vacuum Sci. Technol. A 14 (2) (1996) 447–452.

CHAPTER 12

Applications of cemented carbides and mechanisms of their wear and degradation

12.1 Metal cutting

According to different estimates, nowadays the cost of machining amounts to more than 15% of the value of all manufactured products in industrialized countries. On this ground, metal cutting is an important part of modern manufacturing industries [1–3].

The majority of metal-cutting operations are performed by use of coated carbide indexable cutting inserts; such inserts with different wear-resistance coatings are shown in Fig. 12.1.

The indexable inserts' geometry is of prime importance because it directly affects the following [2,3]:

- Chip control. The tool geometry defines the direction of chip flow. This direction is important to control chip breakage and evacuation.
- Productivity of metal cutting. The cutting feed per revolution is considered to be the major resource to increase the machining productivity. This feed can be significantly increased by adjusting the tool cutting edge angle.
- Tool life. The geometry of the cutting inserts directly affects tool life, as it influences on the magnitude and direction of cutting forces, the sliding rate at the tool–chip interface, the distribution of the energy released during metal cutting, the temperature distribution on the cutting edge, etc.
- The direction and magnitude of the cutting forces. Four components of the cutting tool geometry, namely, the rake angle, the tool cutting edge angle, the tool minor cutting edge angle, and the inclination angle, define the magnitudes of the orthogonal components of the cutting force.

Cemented Carbides
ISBN 978-0-12-822820-3
https://doi.org/10.1016/B978-0-12-822820-3.00003-3

Figure 12.1 Appearance of indexable cutting inserts with different wear-resistant coatings.

- Surface roughness and quality of workpiece articles being machined. The influence of the cutting geometry on the machining residual stresses is easily realized if one recalls that this geometry defines to a great extent the state of the stresses in the deformation zone, i.e., around the tool.

The selection of the best carbide grade and insert geometry as well as the design of the tooling itself are sophisticated factors. A wide variety of new carbide grades and coatings available today make the choice of the proper carbide grade even more complicated for manufacturing engineers, as performance and tool lifetime relate to the machinability of workpiece materials, their hardness, desired productivity, and efficiency of the machining process as well as the needed surface quality and roughness. Usually, manufacturers of carbide indexable inserts offer a guide for the initial selection of the optimal carbide grade for each particular application range, as one shown in Table 12.1 [4].

According to Ref. [4], more that 85% of carbide indexable inserts are fabricated with CVD and PVD coatings, which are described in detail in Chapter 8. The major functions of the coatings are as follows:

- to provide increased surface hardness;
- to increase resistance to abrasive and adhesive wear (flank or crater wear);
- to reduce friction coefficients in order to ease chip sliding thus decreasing cutting forces and preventing adhesion of workpiece materials to the cutting insert surface;

- to reduce heat generated due to chip sliding thus decreasing cutting temperatures;
- to reduce the portion of the thermal energy that flows into the tool;
- to increase corrosion and oxidation resistance;
- to improve the surface quality of finished parts.

The wear process of carbide cutting inserts in metal cutting is reported in a number of publications (see, for example, Refs. [4–6]).

All the advantages of the coatings mentioned above can be achieved only when considering the peculiar wear behavior and wear patterns of tools when cutting most workpiece materials, in particular, cast iron and different steel grades, which is illustrated in Fig. 12.2. As it is shown in Fig. 12.2 the wear on the rake face of the tool (crater wear) and wear on the flank face (flank wear) are separated by a narrow strip of an unworn tool surface corresponding to the so-called "crater front distance" (Fig. 12.2); it is present on the rake face if the crater wear does not exceed a certain width. Figs. 12.3–12.5 show optical and HRSEM images of both the rake wear and flank wear of a carbide indexable insert with a coating comprising an alumina interlayer and a top golden-yellow TiN layer.

As one can see in Fig. 12.3A showing an initial stage of wear on the rake surface and in Fig. 12.4 showing the corresponding HRSEM image of the rake face, there is an unworn strip of the coating corresponding to the crater front distance, which is located between the crater wear and flank wear. Just the presence of such a strip ensures the high wear resistance of coated cutting inserts, as this unworn coating strip on the rake face allows easy chip sliding over the rake face thus decreasing cutting forces and preventing adhesion of workpiece materials to the insert rake face. In spite of the fact that the coating thickness is of the order of 10–30 μm, which is significantly lower than the wear on both the rake and flank surface, an angle at which the chip slides on the rake face is very small, so that, as one can see in Figs. 12.3 and 12.4, the contact zone between the sliding chip and the coating has a thickness of the order of several hundred microns. In such sliding wear conditions, tribofilms, which contain components of the wear-resistant coating and have a protective-lubricating effect, form between the sliding chip and the rake face in the region of the crater wear. As a result, the presence of the extremely hard and wear-resistance coating having lubricating properties determines the whole wear process, which is not significantly affected by the carbide substrate. A similar phenomenon occurs on the flank face, on which part of the coating at the edge remaining due to the presence of the unworn coating strip corresponding to the crater front

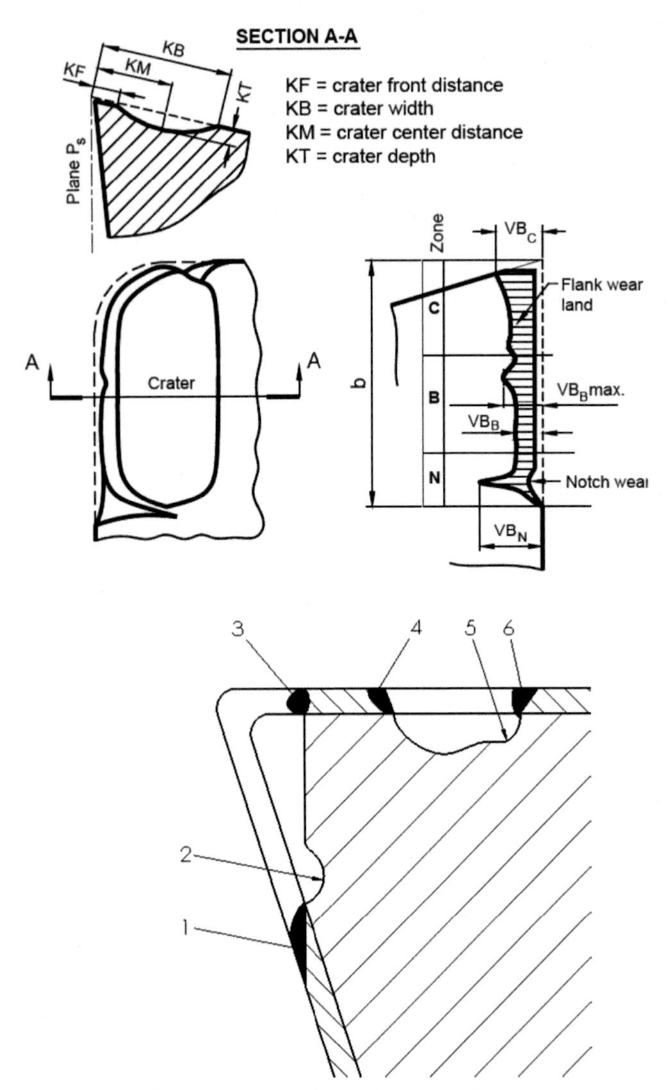

Figure 12.2 Above—different areas of the tool wear at the cutting edge according to the ISO standard ISO 3685:1993. Below—wear zones in the main section of a carbide insert with a wear-resistant coating after metal cutting according to Ref. [6]: 1—microstep formed by the coating at the edge of the wear area on the flank surface, 2—groove in the carbide substrate, 3—microledge formed by the coating on the rake surface protecting the flank surface and reducing its wear, 4—microledge protecting the rake surface and reducing the crater wear, 5—groove at the step of the far edge in the crater wear, and 6—microstep at the far edge of the crater wear. *(Redrawn from Ref. [2].)*

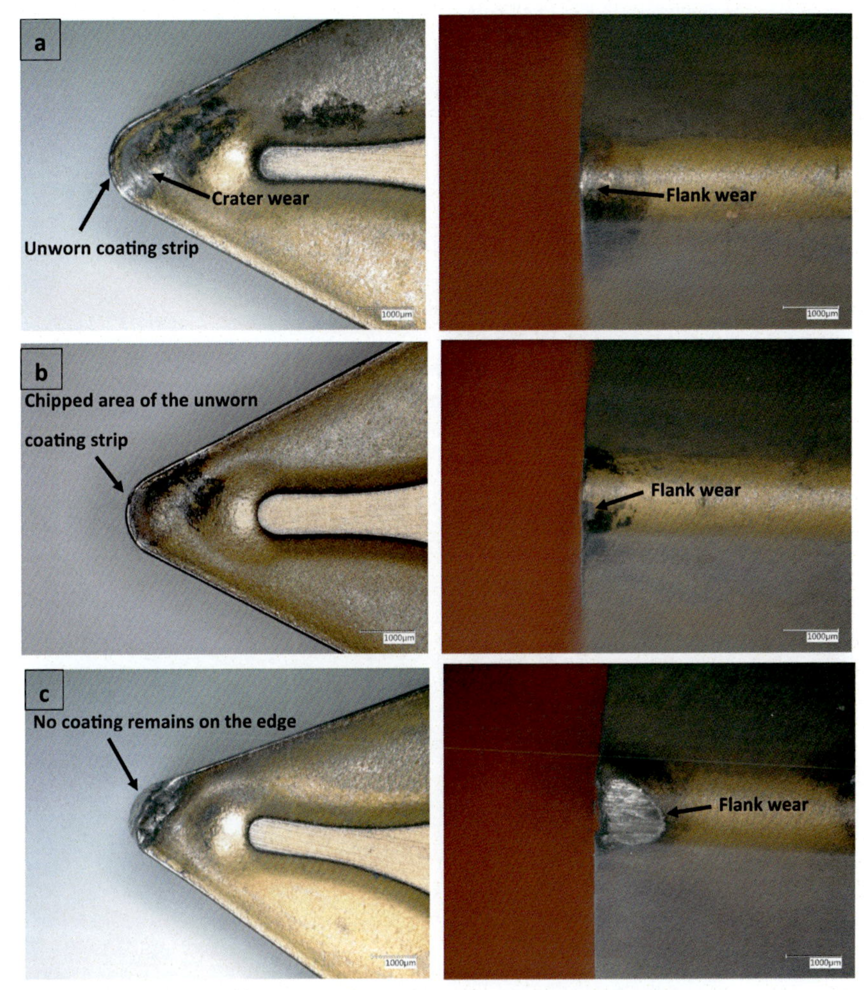

Figure 12.3 Different stages of the rake/crater wear (left) and flank wear (right) of a carbide indexable insert with a coating comprising an alumina interlayer and a top TiN layer: (A)—initial normal wear characterized by the presence of the unworn coating strip corresponding to the crater front distance; (B)—intermedium wear stage, at which chipping the coating strip starts; and (C)—final wear stage (catastrophic wear), at which no coating remains on the cutting edge leading to high rates of both rake wear and flank wear.

distance controls the whole wear process, thus dramatically reducing the flank wear. Usually, coated indexable cutting inserts operate until the end of the stage of initial wear, which is shown in Fig. 12.3A, at which the unworn coating strip between the crater wear and flank wear remains on

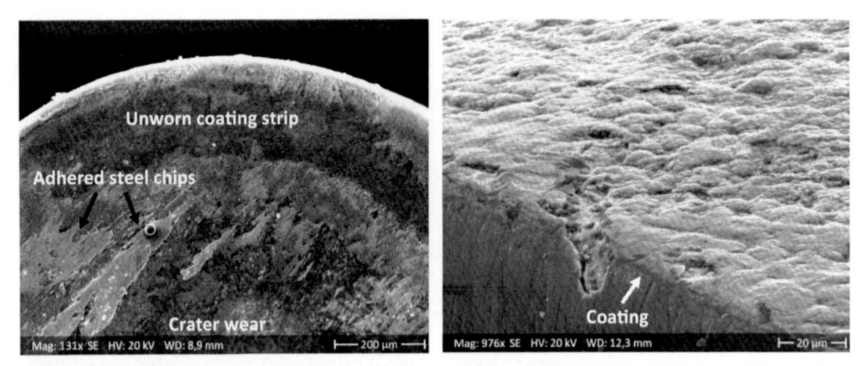

Figure 12.4 HRSEM images of the rake/crater wear (left) and flank wear (right) of the carbide indexable insert corresponding to the initial stage shown in Fig. 3.3A.

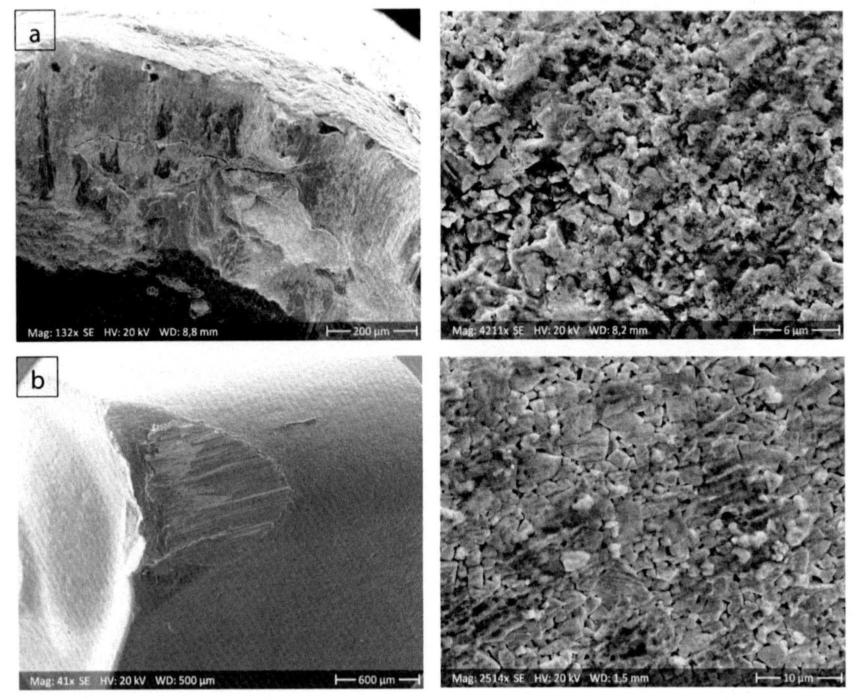

Figure 12.5 HRSEM images of (A) rake/crater wear at different magnifications, and (B) flank wear at different magnifications of the carbide indexable insert corresponding to the final wear stage (catastrophic wear) shown in Fig. 12.3C.

the rake face. Afterward, before this strip is worn out or damaged, the coated indexable inserts are replaced, which ensures the high surface quality and excellent dimension tolerance of parts being machined due to very low rates of flank wear.

According to the theory of adhesion-fatigue wear of cemented carbide cutting tools reported in Refs. [5,6], both the flank and rake surface of the coated cutting inserts are subjected to microfatigue phenomena related to cycling loads occurring as a result of continuous adhesion and removal of tiny chips of the workpiece material, shown on the rake surface in Fig. 12.4. The presence of the unworn strip of the wear-resistant coating mentioned above can significantly reduce such cycling loads. As it is shown in Fig. 12.2, as a result of its significantly increased hardness and wear resistance compared to whose of the carbide substrate, parts of the coating form microsteps and microledges at the edges of the wear area on the flank surface protecting the carbide substrate and dramatically reducing its wear; shallow grooves also form in the carbide substrate near the microsteps (Fig. 12.2). The grooves' formation is related to enhanced fatigue phenomena due to higher rates of the chips' adhesion to the cutting insert surface in the regions adjacent to the coating edges. Similar microledges and microsteps of the wear-resistant coating form on the rake surface near the edges of the crater wear resulting in a significantly reduced crater wear; shallow grooves similar to those on the flank surface also form on the rake surface. The depth of such grooves and rates of their formation can be reduced by use of carbide substrates comprising near-surface layers, which do not contain grains of cubic carbides, as such WC—Co layers are characterized by improved fracture toughness and fatigue resistance (see Chapter 11 for more details). Thus, the formation of tiny coatings' microledge and microsteps, which have a height of a few microns above the wear surfaces (Fig. 12.2 below) and can slightly so to say "remove" the workpiece from the bulk of the cutting insert, leads to a significant reduction of loads and fatigue impact occurring on the rake and face surfaces, thus protecting these surfaces and reducing their wear.

However, when indexable cutting inserts are not replaced when certain wear rates on the rake and flank faces are achieved, the unworn coating strip corresponding to the crater front distance becomes discontinuous as a result of wear or microchipping, which is illustrated in Fig. 12.3B. As one can see in Fig. 12.3B, in the area, where the coating strip on the rake face is chipped and damaged, the adjacent region of the flank wear significantly increases. Finally, as it can be seen in Figs. 12.3C and 12.5, when the whole unworn coating strip corresponding to the crater front distance is worn out and completely removed, the wear rates of both the rake surface and flank surface dramatically increase resulting in catastrophic wear. On this stage of the wear process, the coating does not play any role in the extremely rapidly progressing wear, so that distinct wear patterns of the cemented carbide substrate become visible on the worn rake and flank faces (Fig. 12.5). Thus, the presence of the unworn coating strip between the crater wear and flank wear is mandatory for the dramatic prolongation of tool lifetime due to the presence of the wear-resistant coatings. Such a strip does not form when machining some workpiece materials (for example, many nonferrous metals) resulting in a fact that conventional thin wear-resistant coatings become ineffective and are usually not employed in these cases.

As it can be seen in Fig. 12.5A, the wear pattern on the rake face is characterized by a combined extrusion/removal of the binder phase among carbide grains, so that they become unsupported leading to the detachment of separate carbide grains and their agglomerates as well as abrasive wear of both WC grains and grains of cubic carbides. These phenomena can be explained by the theory of the adhesion-fatigue wear of cemented carbide cutting tools mentioned above [5,6]. According to this theory, the wear process of uncoated cemented carbides in metal cutting occurs due to continuous adhesion and removal of microchips of the workpiece material leading to fatigue phenomena, which results in detachment of carbide grades, their fragments, and small agglomerates. In contrast to the rake wear, the flank wear shown in Fig. 12.5B is noticeably smoother and characterized by traces of selective binder removal and abrasion of carbide grains. As it was established in Ref. [5], the surface wear starts in the binder phase by a combined process of selective binder removal and microcutting followed by forming plastic strains in both the binder phase and carbide phase. Diffusion of the workpiece material into the cutting tool at high cutting temperatures is believed also to play an important role with respect to the tool wear, especially at high cutting speeds.

The images shown in Figs. 12.1—12.5 were obtained on worn cutting inserts after turning low-alloyed steels when using intensive cooling. As it can be seen in Fig. 12.6, generally, the wear behavior and wear mechanisms remain unchanged in dry turning of low-alloyed steel. The appearance of the flank wear surface and cutting edge in the case when the coating still remains on the rake surface indicates signs of mainly sliding wear.

Fig. 12.7 schematically illustrates different types of wear and damage of carbide cutting tools (which are partially shown above) including thermal cracking, mechanical fatigue cracking, chipping, buffing, etc., of the cutting edge.

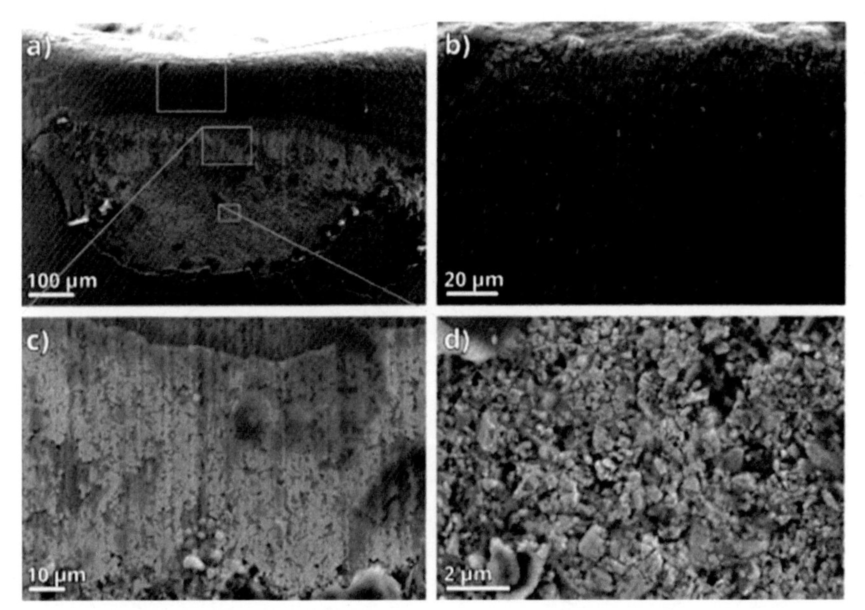

Figure 12.6 Flank wear after dry face turning of low alloyed SS2541 steel at cutting speed of 200 m/min, feed rate of 0.35 mm/rev and depth of 2 mm: (A) The flaked area of the flank wear surface, (B) the cutting edge where the coating still remains, showing signs of sliding wear, (C) part of the flaked area showing signs of sliding wear of the flank wear surface, (D) flaked region of the flank wear surface at a higher magnification. *(Redrawn from Ref. [1].)*

Generally, wear of cutting tools depends on the tool material and geometry, workpiece material properties, cutting parameters (cutting speed, feed rate, and depth of cut), cutting fluids, and machine-tool characteristics. Flank wear is most commonly used for wear monitoring. According to the standard ISO 3685:1993 for wear measurements, different areas of the cutting edge are divided into a number of major regions as shown in Fig. 12.2.

Thus, the major mechanisms determining tool wear in metal cutting are selective removal of the binder, abrasion, diffusion, detachment of WC grains and their fragments, and adhesion-fatigue phenomena. The wear is accelerated at higher speeds and the higher temperatures associated with them. The fundamentals of tool wear mechanisms are summarized in Refs. [7,8].

Usually, the cost of cutting tools corresponds to only nearly 3%—4% of the total production costs in machining, which is illustrated in Fig. 12.7;

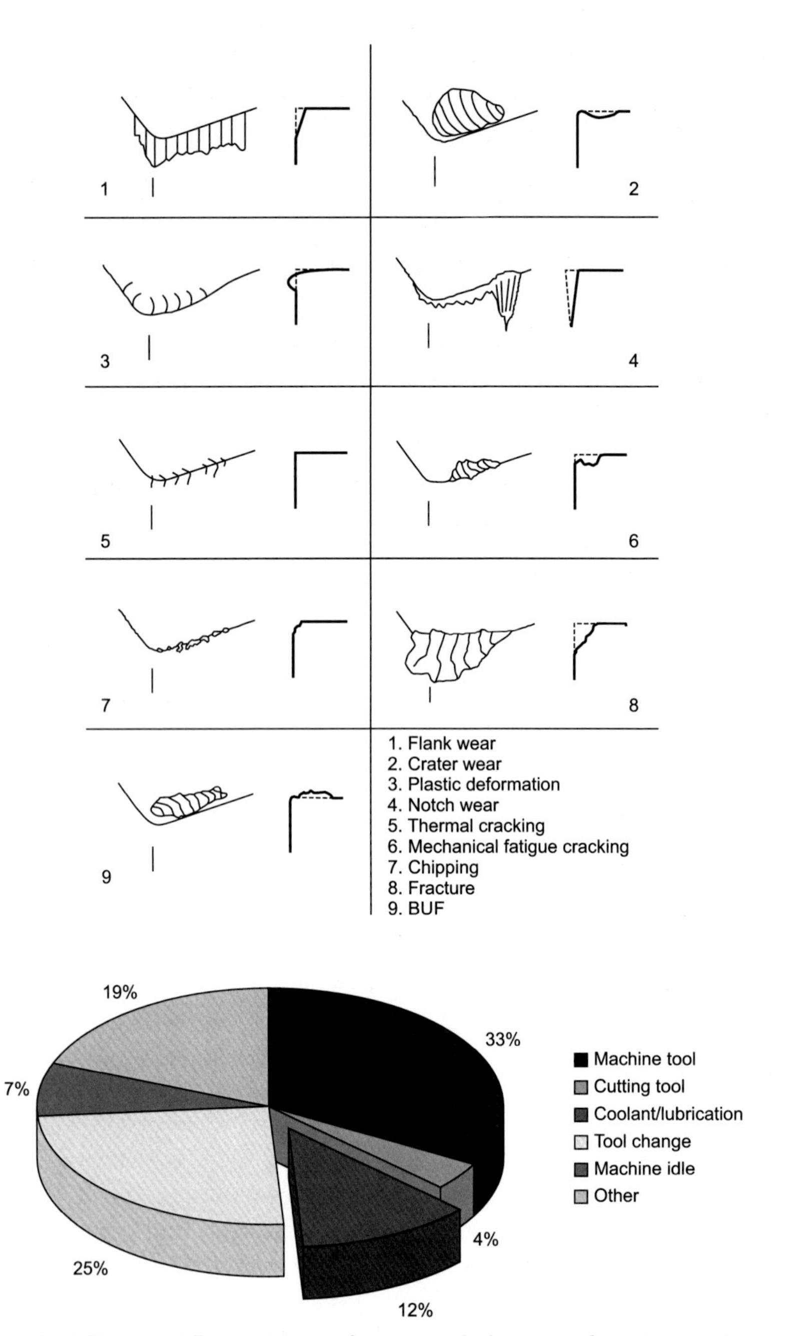

Figure 12.7 Above—different types of wear and damage of cutting tools, and below—general cost breakdown for metal-cutting operations. *(Redrawn from Ref. [2].)*

however, the proper choice of carbide grades, coatings, cutting parameters, machining tools, etc., can lead to significant savings (up to 40% or more) [2] (Table 12.1).

12.2 Mining and oil-and-gas drilling

Cemented carbides are widely employed for exploitation drilling, percussive drilling of deep holes in mines according to the general concept "drill-and-blast," geotechnical works, coal- and ore cutting by use of picks with carbide inserts, etc. Two following major fields of applications of cemented carbides in mining can be distinguished: (1) drilling holes and wells by use of drilling bits and (2) cutting stone, rock, ore, etc., by use of picks. The application conditions of cemented carbides for mining and construction are described in detail in Ref. [9].

Table 12.1 Rough guide for selection of carbide grades for metal cutting applications.

Cutting conditions	ISO code
Finishing steels, high cutting speeds, light cutting feeds, and favorable work conditions	P01
Finishing and light roughing of steels and castings with no coolant	P10
Medium roughing of steels, less favourable conditions. Moderate cutting feeds and speeds	P20
General purpose turning of steels and castings, medium roughing	P30
Heavy roughing of steels and castings, intermittent cutting, low cutting feeds and speeds	P40
Difficult conditions, heavy roughing/intermittent cutting, low cutting feeds and speeds	P50
Finishing stainless steels at high cutting speeds	M10
Finishing and medium roughing of alloy steels	M20
Light to heavy roughing of stainless steels and difficult to cut materials	M30
Roughing tough skinned materials at low cutting speeds	M40
Finishing plastics and cast iron	K01
Finishing brass and bronze at high cutting speeds and feeds	K10
Roughing cast irons, intermittent cutting, low speeds and high feeds	K20
Roughing and finishing cast irons and nonferrous materials. Favorable conditions	K30

12.2.1 Rock drilling

Drilling of holes and wells can be divided into the four following application ranges: (1) rotary drilling and overburden drilling by use of cutters or bits; (2) percussive drilling; (3) rotary-percussive drilling; and (4) rotary drilling by use of roller-cone bits.

Rotary drilling of the first type is typically used for making holes in rocks before blasting. Rocks of low to medium hardness and toughness, such as mineral salts, ores, coal, etc., are usually subjected to rotary drilling by use of cutters or bits. Typical drilling bits and cutters for rotary drilling and overburden drilling are shown in Fig. 12.8. The bits comprise carbide plates or inserts brazed to steel bodies, which rotate without percussion and cut the rock or coal. In this case, rock or coal is destroyed by mainly compression, crushing, and cutting. The rotation speed usually varies from nearly 100 rev/min in rock up to 1500 rev/min in coal. There is a certain value of force applied to the bit needed to overcome the rock toughness and get a certain rate of penetration into the rock. After obtaining certain wear of carbide plates, they must be reground to maintain the penetration rate. Bits for overburden drilling, which are mainly employed for geotechnical drilling, work similarly. In rotary drilling, cemented carbides are subjected to compression and bending stresses, as well as to abrasive wear and thermal shock. In some cases, e.g., in cutting mineral salts, water cooling is unacceptable, so that air cooling is used, which can lead to high contact temperatures on cutting edges.

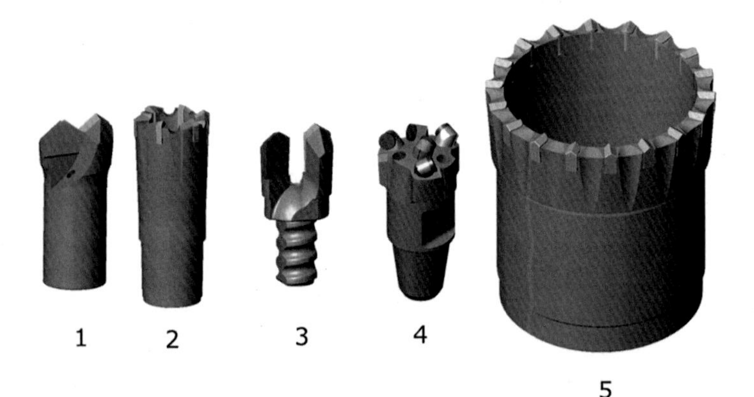

Figure 12.8 Typical bits for rock drilling: 1–3—bits for rotary drilling with carbide plates, 4—bit for rotary drilling with carbide/PCD (polycrystalline diamond) cutters, 5—bit for overburden drilling. *(Redrawn from Ref. [9].)*

Percussive drilling is employed for degradation of rocks with medium to high hardness and performed by creating high impact loads on the rock surface as a result of percussion. The bit strikes the rock and rotates at a certain speed to move the position of carbide insert or plate to another part of rock; the typical bit blow frequency is 2000—3000 bl./min and the rotation speed is 50—150 rev/min. Typical percussive drilling bits with carbide plates or inserts are shown in Fig. 12.9. Carbide plates are usually brazed; inserts are pressed into steel bodies at elevated temperatures, which allows formation of residual compression stresses between the steel body and insert after cooling down to room temperature. The major mechanism of rock degradation in percussion drilling is the formation of micro- and macrocracks in the rock under high-energy stroke and high compressive stress leading to rock crushing and chip formation. In some cases, e.g., in drilling iron ores, high contact temperatures on the carbide surface can occur, leading to severe thermal shock and thermal fatigue, which results in the formation of a network of thermal cracks.

Percussive-rotary drilling is a combination of rotary and percussive methods. A chisel bit is used, as in percussive drilling, but the bit is held against the rock surface under considerable load and tuned between impacts. The tool wedge angle is often nearly 90 degrees and its axis forms an oblique angle with the rock face to give an acceptable rake angle for the cutting action. The impact removes some material directly and forms cracks that facilitate chip formation in the cutting part of the cycle. The wear rate lies between the rates for rotary and percussive drilling, and percussive-rotary drilling is therefore useful for application to rocks that are too abrasive for rotary drilling. The bit remains for most of time under compression, and higher impact energies can be employed than in strait percussive drilling. Another type of percussive-rotary drilling is down-the-hole drilling, which is characterized by the presence of pneumatic or hydraulic shock workers directly near the drilling bit.

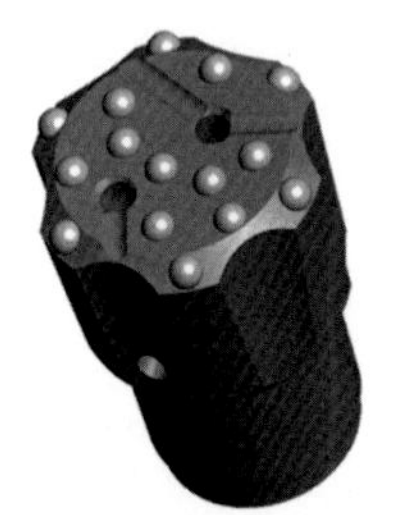

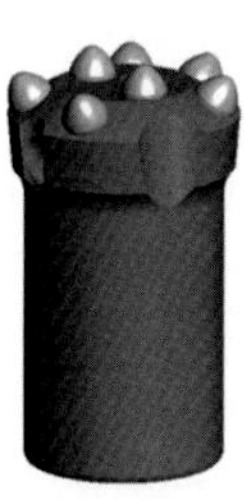

Figure 12.9 Typical bits for percussive drilling with either pressed carbide inserts or brazed carbide plates (in the middle). *(Redrawn from Ref. [9].)*

Rotary drilling performed by roller-cone bits is employed for drilling holes of large diameters, which are needed for oil-and-gas drilling or surface drilling of large holes in quarries by use of the general "drill-and-blast" method. Typical roller-cone bits for surface drilling in quarries are shown in Fig. 12.10. The roller cones rotate together with the drill rod and cut the rock without percussion under pressure provided by the rotating drill rod from the surface. Carbide inserts on the surface of the roller cones penetrate the rock under pressure and crush it. In this case, carbide inserts having a relatively long conical part are used, so that bending loads on them are high resulting in the need to employ carbide grades with the highest values of transverse rupture strength (TRS).

Wear mechanisms of cemented carbides in rock drilling were examined in detail and are presently well understood. The results on the carbide wear mechanisms in rock drilling are reported in Refs. [10–12]. The most important macromechanisms of wear of WC–Co cemented carbides in rock drilling are the following:

- impact spalling;
- impact fatigue spalling;
- sliding abrasion;
- thermal fatigue.

Impact and impact-fatigue wear are most important when hard rocks are drilled by percussive methods; abrasive wear is most important for drilling in softer, abrasive rocks; and thermal fatigue is most important for high-temperature generating rocks, e.g., iron ores. It was found that temperatures in a very thin near-surface layer of carbide inserts can reach or even exceed the melting point of quartz leading to the penetration of fine quartz particle into WC–Co and their intermixing with the Co binder of cemented carbides [11,12]. According refs. [11,12], the major micromechanisms of wear and degradation of cemented carbides in rock drilling, which are schematically shown in Fig. 12.11A–E, are the following:

- crushing WC grains and release of fragments;
- detachment of whole WC grains and their removal from the worn surface;
- crushing of binder/rock mixture formed as a result of rock penetration into the binder and release of fragments;
- tribochemical wear, scraping, and pounding off corroded or oxidized superficial layers on WC
- detachment of WC–Co chips comprising a number of WC grains.

Figure 12.10 Typical roller-cone bits for rotary drilling employed in mining and oil-and-gas drilling: before use (left) and after use (right). *(Redrawn from Ref. [9].)*

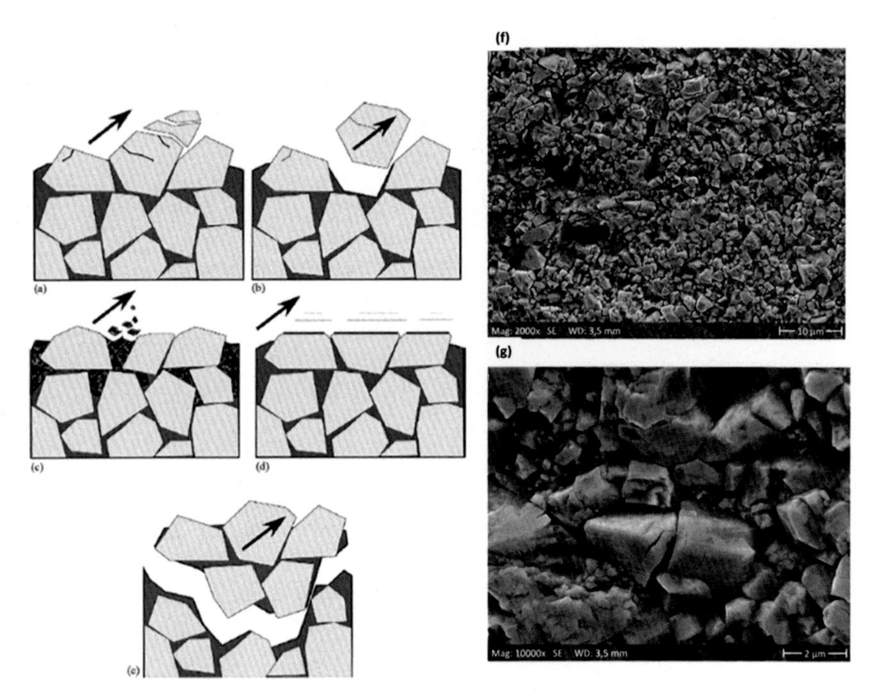

Figure 12.11 Schematic illustration of the most important mechanisms of material removal in cemented carbide for rock drilling: (A) crushing of WC grains and release of fragments; (B) detachment of whole or parts of WC grains losing their hold in the microstructure; (C) crushing of binder/rock mixture after the rock penetration and release of fragments; (D) tribological wear, scraping and pounding of corroded or oxidized superficial layers on WC; (E) detachment of composite-scale fragments; (F) extrusion and removal of the binder phase; and (G) abrasion of individual WC grains. *(A–E) Redrawn from Ref. [11]. (F and G) Redrawn from Ref. [9].)*

Besides these mechanisms, the two following wear mechanisms, which are illustrated in Fig. 12.11F and G, can also play an important role during rock drilling according to Ref. [9]:

- extrusion and removal of the binder phase;
- abrasion of individual WC grains.

Fig. 12.11F and D clearly show the binder removal, abrasion of individual WC grains, traces of detachment of whole WC grains or their agglomerates, and crushing and microchipping of WC grains as a result of percussive drilling of quartzite. It is also clearly seen that the binder is selectively worn out leaving unsupported WC grains on the carbide surface during percussive drilling.

Figs. 12.12 and 12.13 illustrate the appearance of a worn drilling bit and carbide insert after percussive drilling of quartzite and typical wear patterns of a conventional medium-coarse WC–6%Co grade usually employed for percussive drilling after a laboratory performance test on drilling quartzite [13]. The average Co content on the worn surfaces is roughly 2–3 wt.%; therefore, the binder is worn out from the carbide surface leaving unsupported WC grains. Their wear occurs by abrasion and microchipping

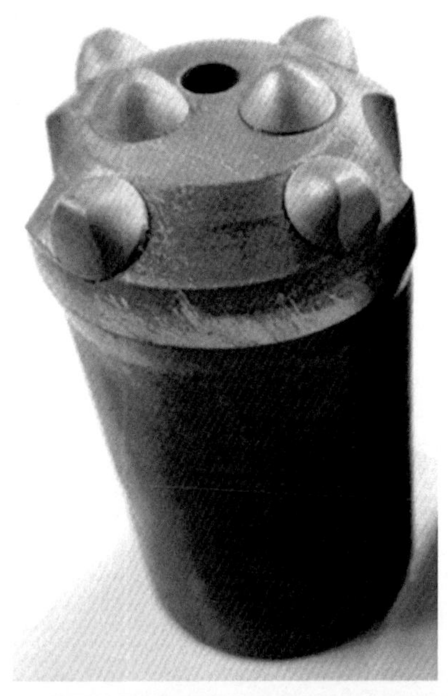

Figure 12.12 Appearance of a worn drilling bit and carbide insert (WC–6 wt.%Co) after percussive drilling of quartzite. *(With the permission of Element Six.)*

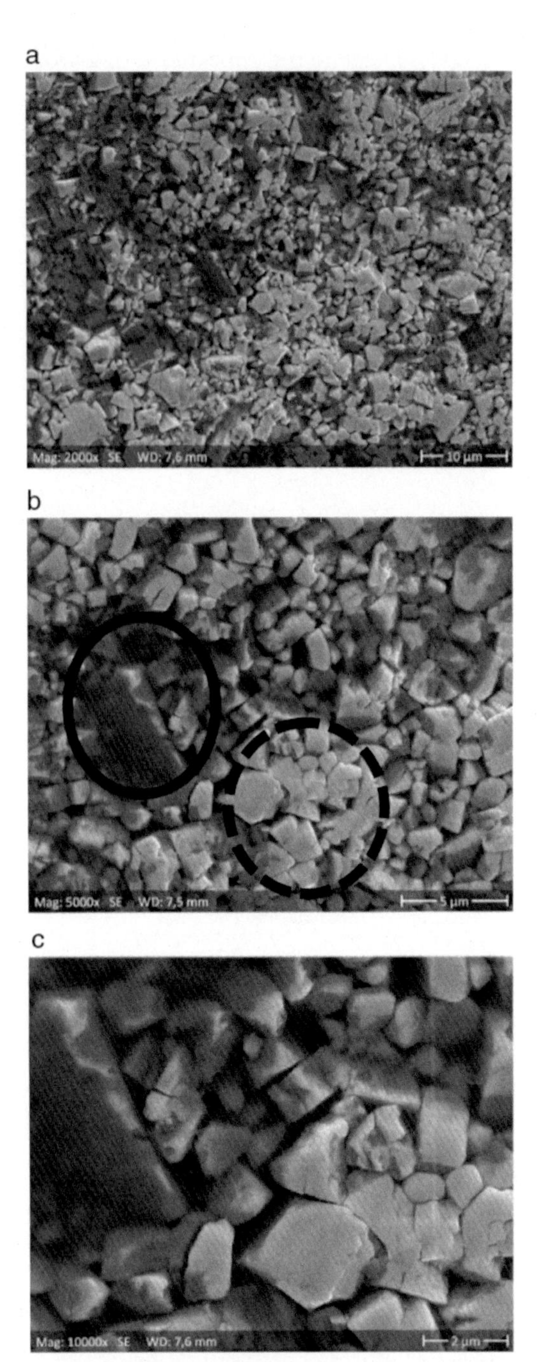

Figure 12.13 Worn surfaces of the medium-grain grade with 6% Co after percussive drilling of quartzite: (A)—overview at low magnification, (B)—worn surface at a higher magnification with the first region having a very low Co content indicated by a dotted circle and the second region having a high Co content indicated by a continuous circle, and (C)—the second region at a high magnification. *(Redrawn from Ref. [13].)*

leading to the fragmentation of the WC grains on the worn carbide surface and the detachment of small WC fragments. There are some regions on the worn surface, where the WC grains were in direct contact with the rock during drilling (the first region indicated by a dotted circle in Fig. 12.13B), and there are some regions lying slightly below the level of the first regions (the second region indicated by a continuous circle in Fig. 12.13B). The worn surface in the first region is characterized by a very low Co content (less than 1 wt.%) as a result of extrusion and removal of the Co binder. The Co content on the surface of the second region, which was not presumably in direct contact with the rock when drilling, is noticeably higher and equal to roughly 8–10 wt.% indicating that the binder was not predominately worn out from the carbide surface in this region. The surface of the second region presumably comprises traces of Co interlayers on the surface on underlying WC grains after the detachment of a large WC–Co fragment as a result of microfatigue. Thus, the major wear mechanisms of cemented carbide illustrated in Fig. 12.13 are the following: (a) removal of Co from the worn surface leaving unsupported WC grains, (b) abrasion and microchipping of the WC grains, (c) removal of small WC fragments forming as a result of WC microchipping, and (d) detachment of separate WC grains and large WC–Co fragments due to microfatigue phenomena on the carbide surface subjected to intensive abrasive wear and severe fatigue.

12.2.2 Rock- and coal cutting

Rock-, stone-, and coal cutting is performed by picks with cemented carbide inserts. The picks shown in Fig. 12.14 are inserted in large drums or heads, one of which is shown in Fig. 12.14. Such drums or heads rotate with rotation speed of nearly 2–6 m/s and press on the workpiece material to obtain the cutting feed of roughly 0.2–2 m/min. Picks for cutting coal and rock (as well as for trenching and soil stabilization belonging to the construction industry), which are shown in Fig. 12.14, are designed in such a way that they rotate either during or after each cut leading to continuous wear propagation in the whole volume of the carbide insert.

Cutting stone and rocks leads to the formation of high impact loads, intensive abrasion, and high cutting temperatures resulting in strong thermal shocks and extremely severe thermal fatigue. In coal-cutting itself, the abrasion wear and thermal shock are relatively insignificant; however, in many cases, layers of coal are located between layers of sandstone, which is

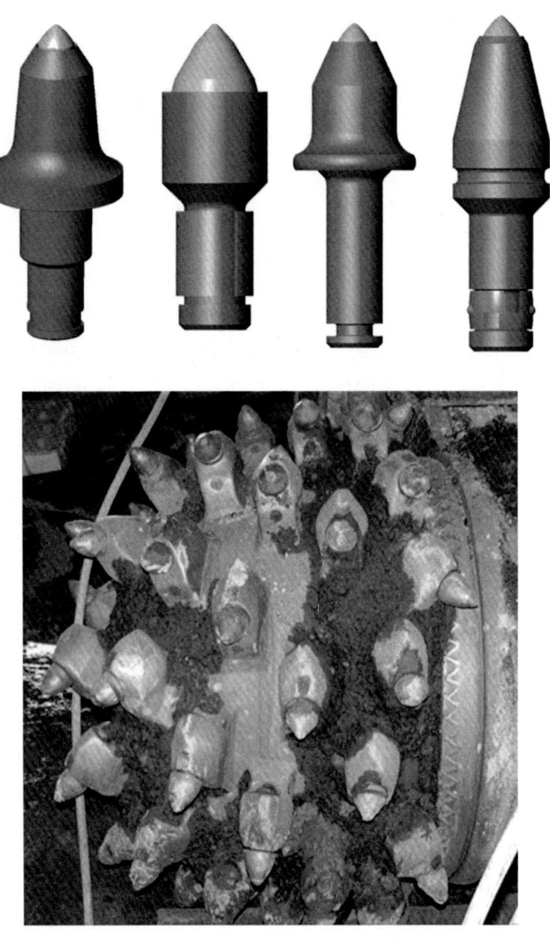

Figure 12.14 Picks for rock-, stone-, and coal-cutting (above) and a cutting head with such picks after use (below). *(Redrawn from Ref. [9].)*

an extremely heat-generating and abrasive rock. Sandstone cutting leads to formation of very high cutting temperatures visible by intensive sparking, which results in severe thermal fatigue. In many cases, water flushing cannot be employed for cooling carbide inserts of the picks and only water spraying for suppression of the dust formation is used, so that cutting temperatures on the carbide surface can be extremely high. As a result of this, the formation of network of thermal fatigue cracks, which are designated in the literature as "snake skin" or "reptile skin," is in many cases the major mechanism of carbide inserts' failures. Typical thermal cracks on the surface of carbide inserts after sandstone cutting are shown in Fig. 12.15. As it can

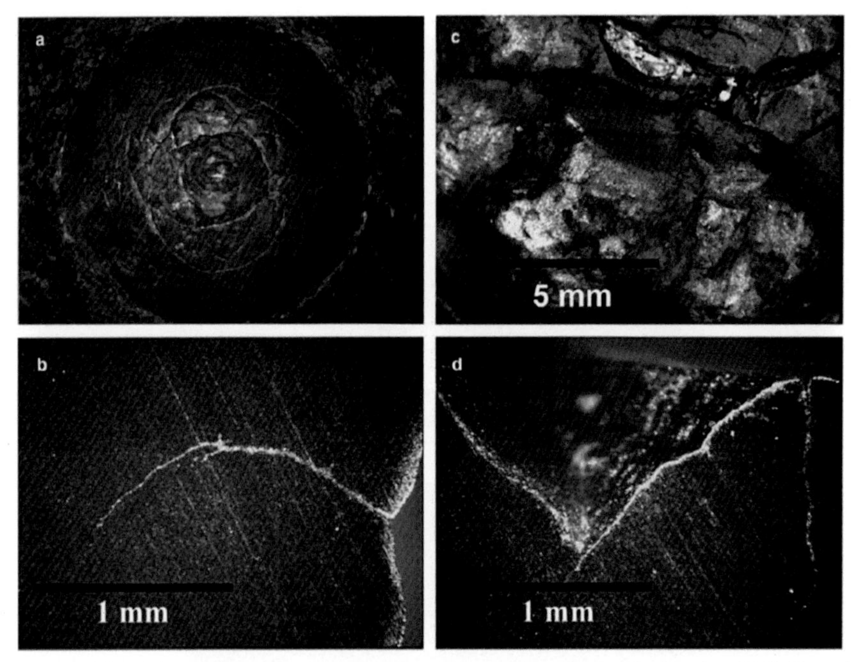

Figure 12.15 Thermal cracks on the surface of a coal-cutting pick ("snake skin") (A and C), and failures formed as a result of the propagation of the thermal cracks into cemented carbides (B and D). *(Redrawn from Ref. [14].)*

be seen in Fig. 12.15, the thermal cracks can easily propagate into the inserts under the influence of high mechanical impact loads and thermal shocks leading to chipping and damaging the inserts.

According to Refs. [13,14], another very important mechanism of wear of ultracoarse carbide grades in rock- and asphalt cutting is related to the presence of thick binder interlayer in their microstructure. In this case, the size of such interlayers becomes comparable with the size of fine abrasive particles forming in rock cutting. As a result, the binder is rapidly worn out leaving naked and unsupported WC grains on the carbide surface, which are shown in Fig. 12.16. It looks like that the worn surface of the coarse grade after granite cutting shown in Fig. 12.16A at a low magnification is relatively smooth and does not comprise grooves or deep cavities. Nevertheless, Fig. 12.16C showing the worn surfaces at a high magnification provides evidence for the predominant wear of the binder among the coarse WC grains. Such unsupported WC grains can be easily cracked, destroyed, and pulled out resulting in significant wear rates of the whole carbide insert.

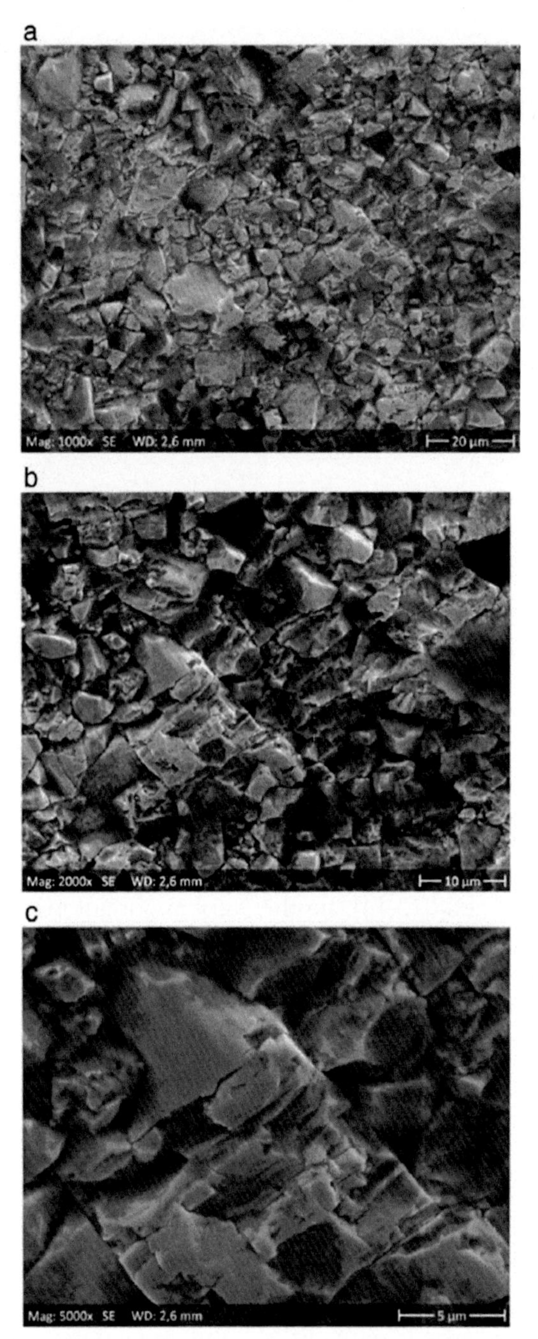

Figure 12.16 Worn surfaces of a coarse-grain carbide grade after granite cutting: (A) overview at low magnification, (B) worn surface at higher magnification, and (C) worn surface at high magnification. *(Redrawn from Ref. [13].)*

The employment of ultracoarse carbide grades with WC mean grain sizes of nearly 5 μm or greater, which are described in detail in Chapter 10, is the present state of the art in the fabrication of mining picks for rock cutting. On the one hand, one can dramatically improve the fracture toughness and resistance to the formation and propagation of thermal cracks by use of such carbide grades in rock cutting due to the presence of thick binder interlayers in their microstructure. On the other hand, the wear mechanism of these grades in rock- and stone cutting is characterized by some peculiarities, which are related to self-reinforcement of the binder phase in such thick binder interlayers by tiny WC particles forming as a result of microchipping the ultracoarse carbide grains present in the microstructure of the grades. Fig. 12.17 indicates that there are traces of microchipping of the ultracoarse WC grains on the worn surface. As a result of that, extremely fine WC fragments formed on the worn surface are embedded into the thick Co interlayers partially filling the gaps among the worn ultracoarse WC grains.

Also, the employment an approach of the binder hardening and reinforcement in special grades of cemented carbides for mining application can dramatically improve their performance and prolong tool lifetime. By the use of ultracoarse grades with nanograin reinforced binders (Master-Grade™) described in Chapter 11 in detail, the binder phase of cemented carbides for fabrication of mining picks is significantly hardened and strengthened resulting in dramatically prolonged life of tools for mining applications.

Figure 12.17 Worn surfaces of the coarse-grain grade after performing the ASTM B611 test. *(Redrawn from Ref. [15].)*

12.3 Construction

Picks with cemented carbide tips or inserts, which are shown in Figs. 12.18 and 12.19, are widely used for construction works for repairing roads and highways in order to remove worn surface layers of asphalt and concrete. The picks are inserted in large drums, shown in Fig. 12.19, which rotate with a rotation speed of nearly 1000–2000 rev./min, feed of 10–20 m/min, and depth of nearly 5–25 cm; the picks are cooled by flashing water. The picks are designed in such a way that they rotate during each cut leading to continuous wear propagation in the whole volume of the carbide inserts. The picks for asphalt milling shown in Fig. 12.19 comprise carbide tips or "caps" brazed to a steel pick body and the picks for concrete milling comprise carbide inserts brazed into the steel pick body.

Carbide inserts and tips are subjected to intensive thermal and mechanical fatigue as well as severe abrasive wear during asphalt- and concrete milling. In some cases, the formation of network of thermal fatigue cracks, which are similar to the "snakeskin" shown in Fig. 12.15, occurs. The formation of such thermal cracks shown in Fig. 12.20 is in many cases the major mechanism of degradation of WC–Co cemented carbides in operations of asphalt- and concrete milling. The thermal cracks can easily propagate into the cemented carbide insert under the influence of high impact loads and thermal shocks leading to microchipping the carbide inserts or tips, which can result in their fast wear and premature failures. As in

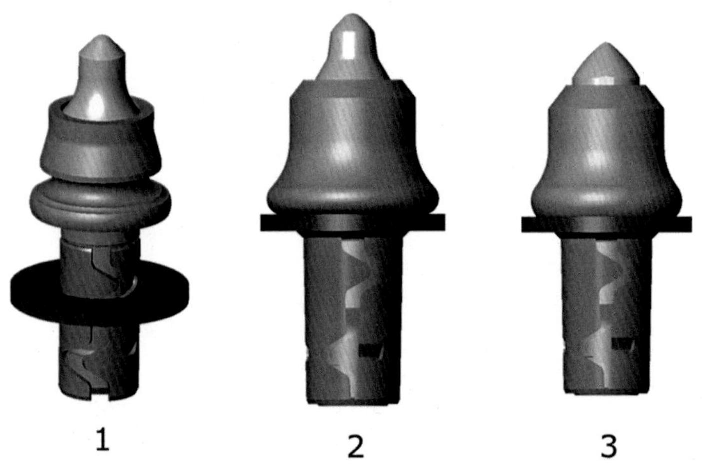

Figure 12.18 Typical road-planing picks for construction: 1 and 2—picks for asphaltmilling, 3—pick for concrete milling. *(Redrawn from Ref. [9].)*

Figure 12.19 Typical drum for road-planing (above) and picks for asphalt milling (below): left—new pick and right—worn pick. *(Redrawn from Ref. [16].)*

the case of mining picks mentioned above, almost exceptionally ultracoarse carbide grades are employed for road-planing picks, so that the major mechanism of wear and degradation of cemented carbides in asphalt cutting is related to the presence of thick binder interlayer in the microstructure of such ultracoarse grades. In this case, the size of such soft binder interlayers becomes comparable with the size of fine abrasive particles forming due to concrete- and asphalt milling. As a result, the binder is rapidly worn out leaving naked and unsupported WC grains on the carbide surface, which are shown in Fig. 12.21A. Such unsupported WC grains can be easily

Figure 12.20 Thermal cracks formed as a result of asphalt milling. *(Redrawn from Ref. [16].)*

microchipped and pulled out during asphalt- or concrete milling resulting in rapid wear of the whole cemented carbide part. As in the case of rock cutting, tiny fragments of large WC grains can be embedded into the thick Co interlayer partially protecting them from abrasive wear and degradation, which is clearly seen in Fig. 12.21B.

By use of the ultracoarse grades with nanograin reinforced binders (MasterGrade^{TM}) described in Chapter 11, the binder phase of cemented carbides for road-planing can be significantly hardened and reinforced resulting in dramatically prolonged life of tools for road-planing.

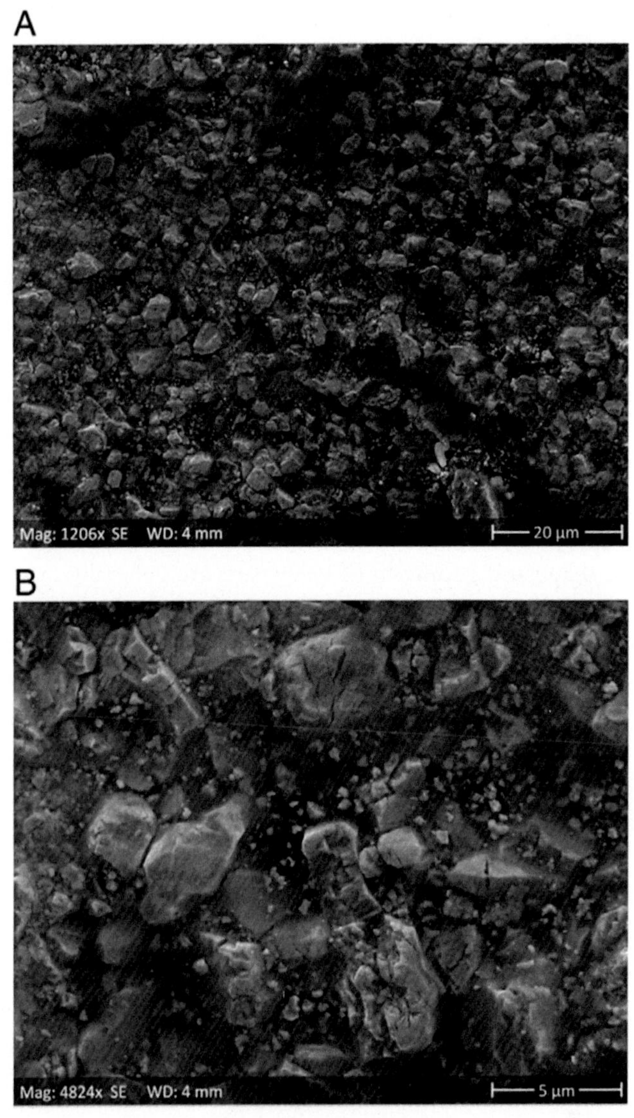

Figure 12.21 The surface of a worn carbide tip after asphalt milling at different magnifications. *(Redrawn from Ref. [17].)*

Tools for other construction works (soil stabilization, trenching, etc.) are similar to those shown in Fig. 12.14 and their wear mechanisms lie in between of those occurring in mining and road-planing; the major wear mechanism of cemented carbides in each concrete application depends on the operation conditions and workpiece materials.

12.4 Wear parts

Fig. 12.22 shows some wear parts made of cemented carbides employed in various machines, tools, and aggregates, part of which is subjected to intensive abrasion wear and/or erosion. Numerous carbide wear parts include; for example, valves of pumps operating in abrasive media, nozzles and blades of sand-blasting equipment, nozzles of paint-spraying equipment, molds for pressing plastics, ceramics, bricks and tiles, etc. In many cases, e.g., in molds for pressing, the fracture toughness of cemented carbides does not play an important role, so that carbide grades with high hardness and relatively low fracture toughness can be employed. In some applications, where wear parts are subjected to impact loads, medium-grain grades with medium hardness but increased fracture toughness are used. Wear parts made of cemented carbides vary from very small articles, such as balls for ball-point pens, to large and heavy parts, such as punches, dies, or large valves for pumps for the oil-and-gas industry with weights of up to 100 kg. The shape and geometry of wear parts are generally limited by the necessity of producing them from WC−Co powders. Nevertheless, the limitations can be overcome by use of fine diamond grinding, polishing, wire erosion, as well as by diffusion bonding of sintered and ground carbides articles to produce large hollow parts with complicated channels.

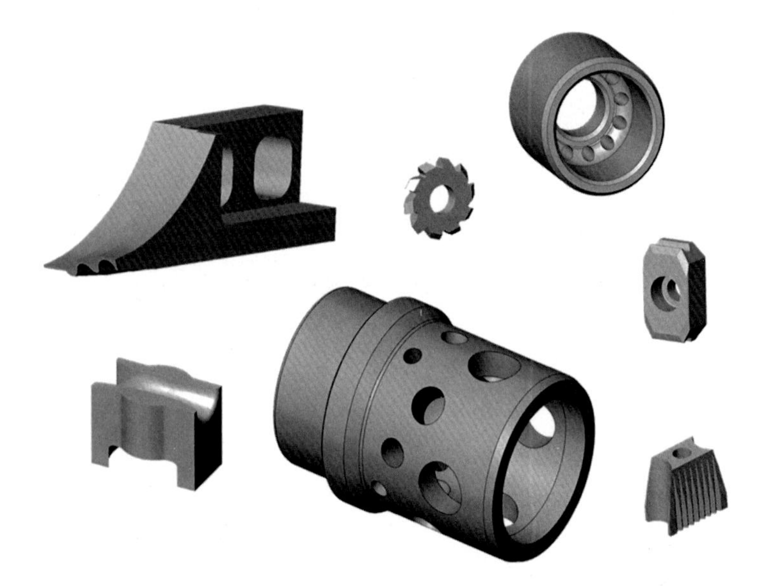

Figure 12.22 Typical cemented carbide wear parts. *(Redrawn from Ref. [9].)*

Generally, the pure abrasive wear of WC–Co materials in applications, in which wear parts are not subjected to impact loads or thermal shocks, is roughly indirectly proportional to the material hardness, which is illustrated in Fig. 12.23. Nevertheless, even in such applications of wear parts, wear mechanisms can significantly vary depending on the environment, loads, friction forces, etc. The different wear mechanisms comprising abrasive wear, adhesive wear, fretting, cavitation erosion, particle erosion, detachment of WC grains, and their fragments as a result of fatigue, rubbing erosion, etc., are illustrated in Fig. 12.24.

In spite of the fact that the hardness of cemented carbides usually plays a decisive role with respect to wear rates of different carbide grades employed for fabricating wear parts, some peculiarities of the microstructure, composition, and fine structure of WC–Co materials can also have an influence on wear rates and performance of wear parts. To establish this influence, three experimental grades having nearly the same hardness (12.8 ± 0.1 GPa) but different combinations of the WC mean grain size (150 nm, 0.8, and 4.6 μm) and Co content (correspondingly 24, 18, and 3 wt.%) were examined in the two major ASTM wear tests—the gentle G65 test and more severe and "aggressive" B611 test [15] (the ASTM wear tests are described in Chapter 7).

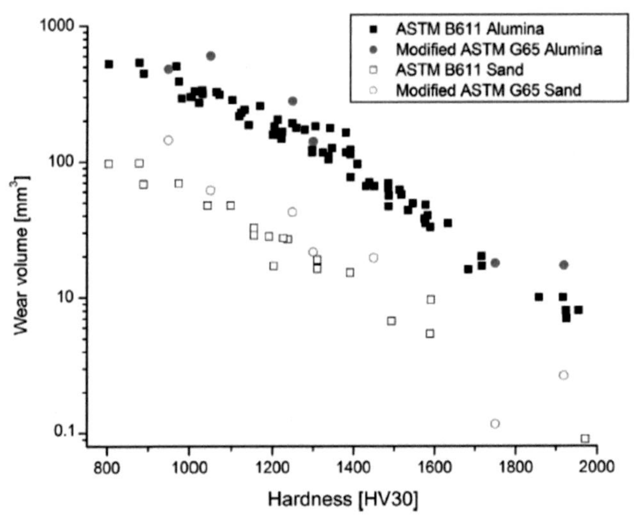

Figure 12.23 Relationship between abrasion volume loss and hardness for different WC–Co grades. *(Redrawn from Ref. [18].)*

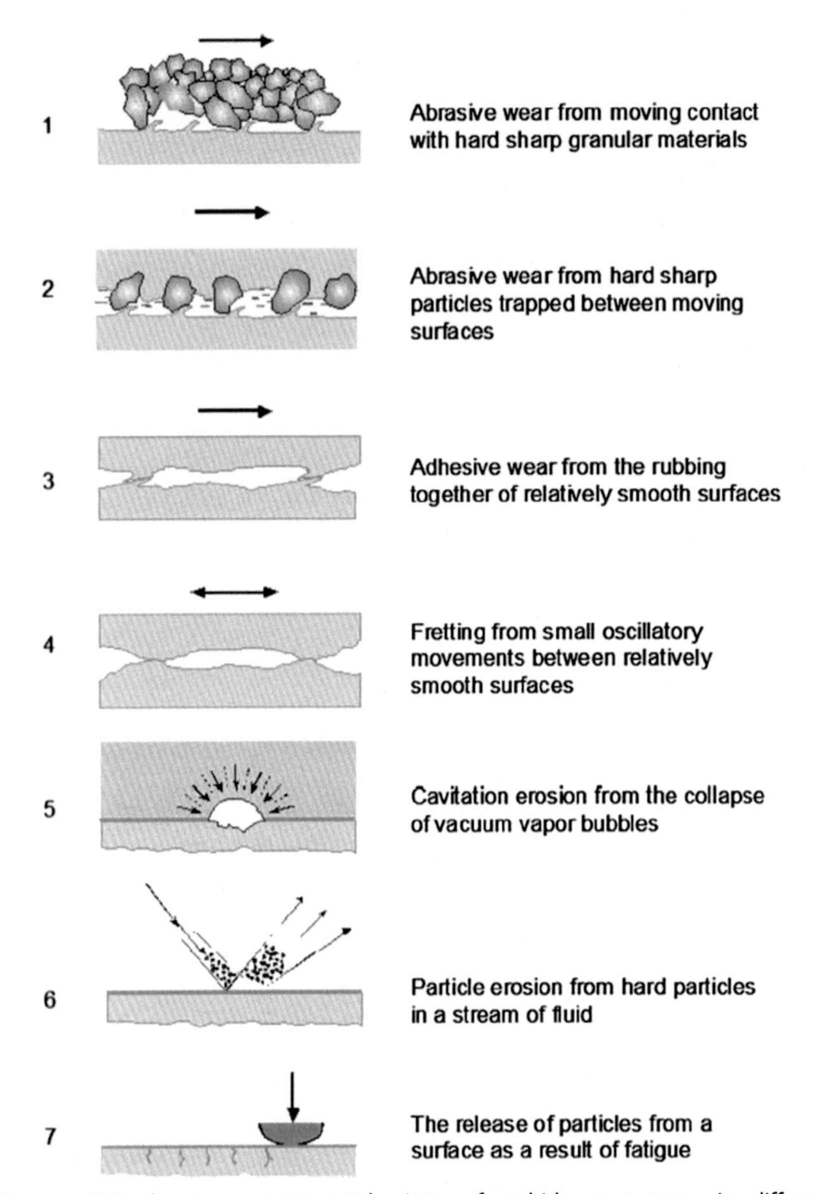

Figure 12.24 The major wear mechanisms of carbide wear parts in different applications. *(Redrawn from Ref. [18].)*

Fig. 12.25 shows dependences of wear rates of the cemented carbides with the various combinations of WC mean grain sizes and Co content as a function of (A) WC mean grain size, (B) Co content, and (C) fracture toughness in the gentle ASTM G65 test. As one can see in Fig. 12.25, the

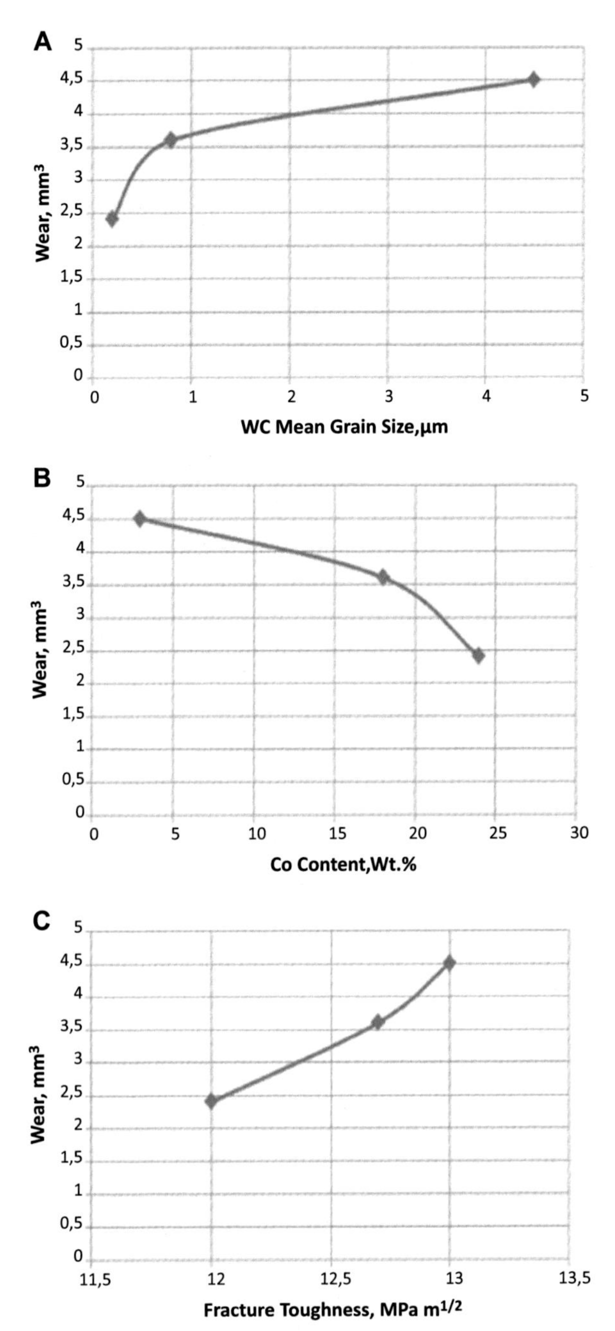

Figure 12.25 Wear of the cemented carbides with the various combinations of WC mean grain sizes and Co content as a function of (A) WC mean grain size, (B) Co content, and (C) fracture toughness in the ASTM G65 test. *(Redrawn from Ref. [15].)*

wear resistance in this case increases with decreasing the WC mean grain size and is not affected by the fracture toughness. As a result, in spite of the highest binder content and lowest fracture toughness, the finest grade characterized by extremely thin Co interlayers has the greatest wear resistance. This is related to the fact that the worn surface of this grade is smooth and not characterized by the predominant binder wear (Fig. 12.26C), as the thickness of Co interlayers in this case is smaller than the size of the finest abrasive silica particles present when conducting the ASTM G65 test. In contrast, the coarsest grade comprising thick Co interlayers has a poorest wear resistance as a result of the binder removal from the wear surface by fine abrasive particles forming in the ASTM G65 test (Fig. 12.26A). The wear pattern of the carbide grade with the WC grain size of 0.8 μm is characterized by an insignificant rate of the selective binder wear, which ensures its high wear resistance in the ASTM G65 test (Fig. 12.26B).

Fig. 12.27 shows dependences of wear rates of the cemented carbides with the various combinations of WC mean grain sizes and Co content as a function of (A) WC mean grain size, (B) Co content, and (C) fracture toughness in the relatively aggressive and severe ASTM B611 test. As it can be seen in Fig. 12.27, the wear resistance in this case increases with increasing the WC mean grain size and is affected by the fracture toughness of the carbide grades examined. As a result, the ultracoarse grade characterized by the highest fracture toughness and lowest Co content has the greatest wear resistance. In this case, as it can be seen in Fig. 12.28A, the wear pattern of the ultracoarse grade having the highest fracture toughness is relatively smooth and does comprise traces of macrofailures in form of, for example, deep grooves. Fig. 12.28A indicates that there are traces of microchipping of ultracoarse WC grains on the worn surface. As a result of that, extremely fine WC fragments formed on the worn surface partially fill the gaps among the worn coarse WC grains resulting in the effect of the binder self-reinforcement by such fine WC fragments described in Section 12.2, which plays an important role.

In contrast to that, the worn surface of the submicron grade with the WC mean grain size of 0.8 μm is rougher than that of the ultracoarse grade and comprises traces of microchipping, groves, and cavities (Fig. 12.28B). They form as a result of microchipping and detachment of relatively large WC—Co fragments due to severe microfatigue phenomena occurring in the ASTM B611 test, as the fracture toughness of the submicron grade is noticeably lower than that of the ultracoarse grade. The worn surface of the finest grade shown in Fig. 12.28C is very rough in comparison with that of

Figure 12.26 Worn surfaces after performing the ASTM G65 test: (A)—grade with 3% Co and WC mean grain size of 4.5 μm, (B)—grade with 18% Co and WC mean grain size of 0.8 μm, and (C)—grade with 24% Co and WC mean grain size of 150 nm.

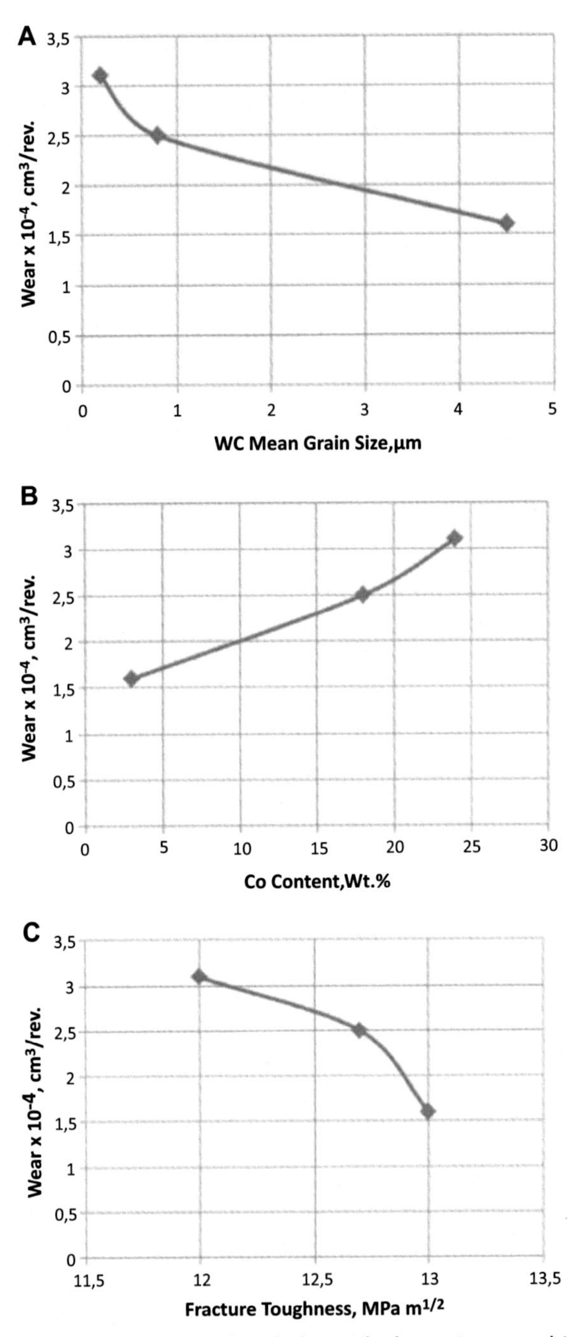

Figure 12.27 Wear of the cemented carbides with the various combinations of WC mean grain sizes and Co content as a function of (A) WC mean grain size, (B) Co content and, (C) fracture toughness in the ASTM B611 test. *(Redrawn from Ref. [15].)*

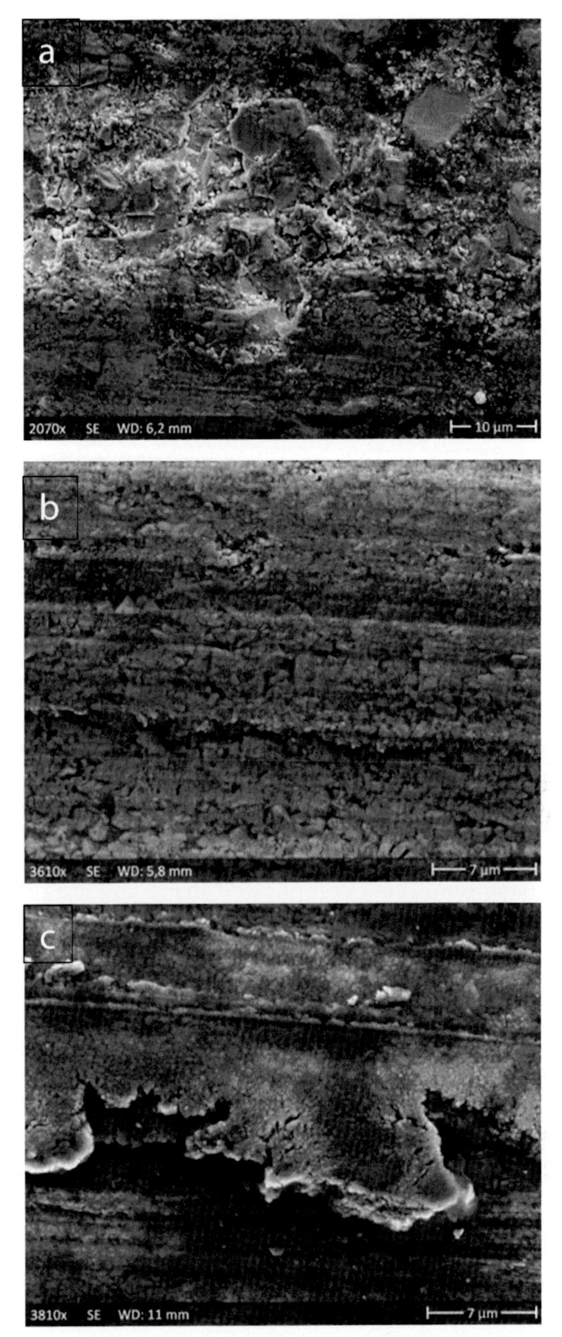

Figure 12.28 Worn surfaces after performing the ASTM B611 test: (A)—grade with 3% Co and WC mean grain size of 4.5 μm, (B)—grade with 18% Co and WC mean grain size of 0.8 μm, and (C)—grade with 24% Co and WC mean grain size of 150 nm. *(Redrawn from Ref. [15].)*

both the ultracoarse grade and the submicron grade and comprises deep groves. It can be clearly seen in Fig. 12.28C that there are numerous traces of chipping and microcracking on the worn surface of this grade providing evidence that the microfatigue phenomena play a very important role in this case. They lead to microfractures and microcracks on the worn surface resulting in the detachment of huge WC—Co fragments when performing the ASTM B611 test. This is related to the reduced resistance of the finest grade with the WC mean grain size of 150 nm to microfatigue phenomena as a result of its relatively low fracture toughness, as this grade must contain large amounts of grain growth inhibitors negatively affecting its toughness (see Chapter 9).

Thus, each concrete application of carbide wear parts is characterized by a certain combination of pure abrasion and cyclic loads of various intensity leading to fatigue phenomena affecting the wear process, so that a carbide grade with the optimal combination of WC mean grain size and binder content must be chosen for each wear application.

12.5 Metal forming

Nowadays, tools for cold metal forming (stamping) consisting of steel parts and carbide inserts have a complex geometry; cemented carbide grades are usually chosen empirically based on the level of impact loads occurring during metal forming. The carbide grades must have a high resistance to high shear forces, impact loads, severe fatigue, and abrasive wear. The carbide articles have typically a complicated shape and are characterized by very rigid dimensional tolerance during operation (in the range of a few microns). To secure low tolerance deviations during operation, cemented carbide tools with a high abrasive wear resistance are preferable. Up-to-date stamping tools in combination with high-tech machines operating at several hundreds strokes per minute allow highly efficient producing steel parts with complex geometry ensuring a great number of such parts to be produced per minute of operation. Stamping tools place carbide parts under extreme stress and in many cases under the influence of abrasive media. For special applications, corrosion-resistant cemented carbide grades are required, so that carbide grades with Ni or Ni—Co binders frequently alloyed with chromium, molybdenum, and other grain growth inhibitors are employed in many cases. For example, the corrosion-resistant grades designated as "CF grades" fabricated by Ceratizit [19] exhibit excellent corrosion resistance combined with high mechanical and physical properties required in the tool and die industry.

The high abrasion resistance in combination with high impact resistance needed in some operations of metal forming is achieved due to using submicron, fine-grain, and coarse-grain carbide grades with different Co contents. For example, Ceratizit [19] produces three submicron grades (CTS12, CTS20, and CTS30) with Co contents varying from 6% to 15%, Vickers hardness varying from 1790 to 1320 Vickers units, and fracture toughness varying from 8.2 to 11.9 MPa $m^{1/2}$. Additionally, Ceratizit fabricates a range of fine-grain grades (CTF11 to CTF54) with binder contents varying from 5.6% to 27%, Vickers hardness varying from 1730 to 940 Vickers units, and fracture toughness varying from 9.2 to 22 MPa $m^{1/2}$. Ceratizit also produces three medium-grain grades (CTM17, CTM30, and CTM40) with Co contents varying from 8.5% to 20%, and a range of coarse-grain grades (CTE20 to CTE60) with Co contents varying from 10% to 30%, Vickers hardness varying from 1120 to 680 Vickers units, and fracture toughness varying from 18 to 27 MPa $m^{1/2}$. The need for such a great number of carbide grades for metal forming tools is caused by their very different operation conditions including impact loads varying from medium to extremely high. For each concrete application, the optimal carbide grade is usually chosen empirically or semiempirically. Fig. 12.29 shows a drawing of a typical metal forming tool for cold stamping of screws comprising two inserts of medium-coarse carbide grades with different Co contents. Stamping tools with carbide inserts are widely employed in the automotive and aircraft industries, construction, furniture industry, electric, and electronic industry, etc.

12.6 Rolling and drawing of wires and bars

Cold and hot rolling and drawing of wires and bars wire or bar drawing are important applications of cemented carbide dies and mill rolls. The unique combination of high wear-resistance, compressive strength, fracture toughness, and thermal fatigue resistance allows the employment of cemented carbides in such applications of cold and hot rolling and drawing.

Conventional low-alloyed steel grades, high-strength steels, and stainless steels can be drawn into semifinished wires or bares as initial articles for the

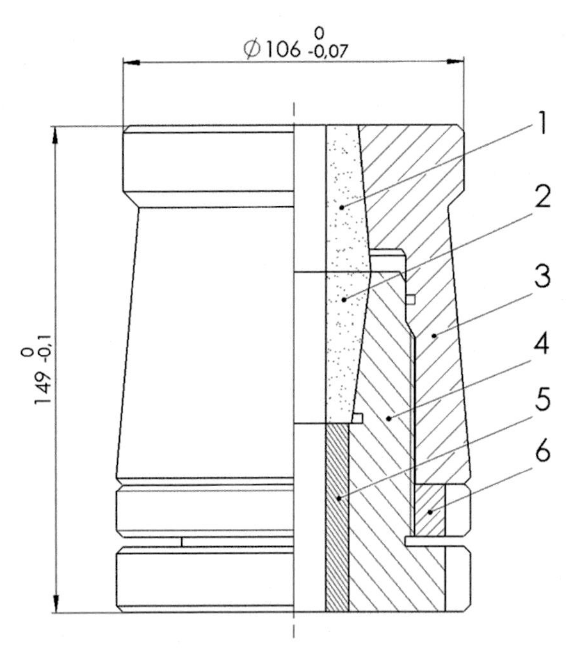

Figure 12.29 Typical stamping tool for stamping screws comprising two inserts of medium-coarse carbide grades with different Co contents: 1—insert of WC—20% Co, 2—insert of WC—15% Co, and 3—6—steel components.

fabrication of different products, for example, tubes for pipelines, or rails for railways. Round wires, bares for steel reinforced concrete, flat bars, square bares, hexagonal bars, and steel bars having special shapes are available on the market in a vast number of grades of construction steels, low- or high-alloyed steels, and stainless steels as well as nonferrous metals. The steel market is huge and amounted for 169.2 million tons in March 2021 according to the estimate of the World Steel Association [20].

Steel mills around the globe employ different tailored technologies and equipment like conventional cam-type drawing machines or chain track drawing machines to produce high-quality steel products. For wires having diameters in the range of 4.5—26 mm, high speed and low-cost single stand equipment can be employed. High speeds of up to 120 m/s combined with relatively low tolerances of ±0.1 mm are employed in up- and downstream automation systems like a loop laying system shown in Fig. 12.30 combined with a coil handling system ensure a high productivity level.

Bars having diameters of 6—130 mm can be produced by use of the so-called "free size rolling systems." Precision milling by use of machines

Figure 12.30 Loop laying heads and loop cooling system employed in the coil production. *(With the permission of Element Six.)*

comprising three rolls combined with gas damping in the stand allows producing bars with diameters of up to 130 mm. Such bars are available in the following shapes and dimensions:

Round, 2–60 mm in diameter
Square, 7–45 mm in size
Hexagonal, 8–50 mm in size
Tubes, 16–76 mm in diameter.

Thermomechanical rolling of wires and bars is characterized by a combination of mechanical rolling and heat treatment at elevated temperatures followed by cooling to achieve well-structured high-strength materials. Hence, the high diversity of the product portfolio needs the production systems characterized by high flexibility on the one hand or must be tailored with respect their specific requirements on the other hand. By employing combined uncoiler systems to feed the steel bars automatically into the machine and torsion free bar bundle systems, high efficiency and safety of the modern steel rolling are achieved.

The total worldwide market share for cemented carbide rolls is estimated to lie in the range of approximately 370 million USD per annum and is steadily growing due to the strong worldwide construction boom. When breaking down the world mill roll market, a significant number of carbide mill rolls for hot milling are produced nowadays; this number is much greater than that of mill rolls for cold rolling and drawing systems. Cold rolling is often used for small wires varying from 20 mm down to approximately 0.4 mm in diameter. To produce wires varying from 0.4 to 2 mm in diameter, two-roller stands with tandem rolls twisted by 90 degrees are usually employed. Modern drawing systems can produce fine wires having diameters of as low as 0.05 mm or even smaller. For fabricating the wires varying from 2 to 20 mm in diameter, stands with three rolls are preferably used; the three-roll stand with a gas damping system is shown in Fig. 12.31. The modern three-roll stands are equipped with advanced inline measurement systems that guarantee the high product quality.

The introduction of the nontwist blocks in the mid-1960s paved the way for the employment of cemented carbide milling rolls for hot rolling of wires and rods [21]. Hot forging allows faster forging at significantly higher

Figure 12.31 Three-roll stand with a gas damping system. *(With the permission of Element Six.)*

forging rates. The process of hot rolling leads to large changes in cross-sectional sizes and shapes of the articles being rolled. This process hence involves "squeezing" the rolled material, as the milling rolls apply the necessary forming loads. Such forming loads are nearly directly proportional to the flow stress of the material being rolled. Depending on the shape and size of the articles being rolled relatively complex stress states can be expected in the roll material, on the pass grove surface and close to it [21].

Mill rolls work under conditions of high impact loads, abrasive and sliding wear, rolling and mounting stresses, severe thermal fatigue, and corrosion, which occur simultaneously leading to damages of the pass groove in its near-surface layer and finally resulting in roll failures. These factors are nowadays well identified and deeply understood, so that the development of new grades and improvement of rolling conditions are performed on a regular basis.

During operation of mill rolls at elevated temperatures and high loads, numerous thermal cracks form mainly on the mill roll surface or close to it as it is illustrated in Fig. 12.32. Impact loads are intensified in some special cases, for example, as a result of too small pass groove diameters or poorly

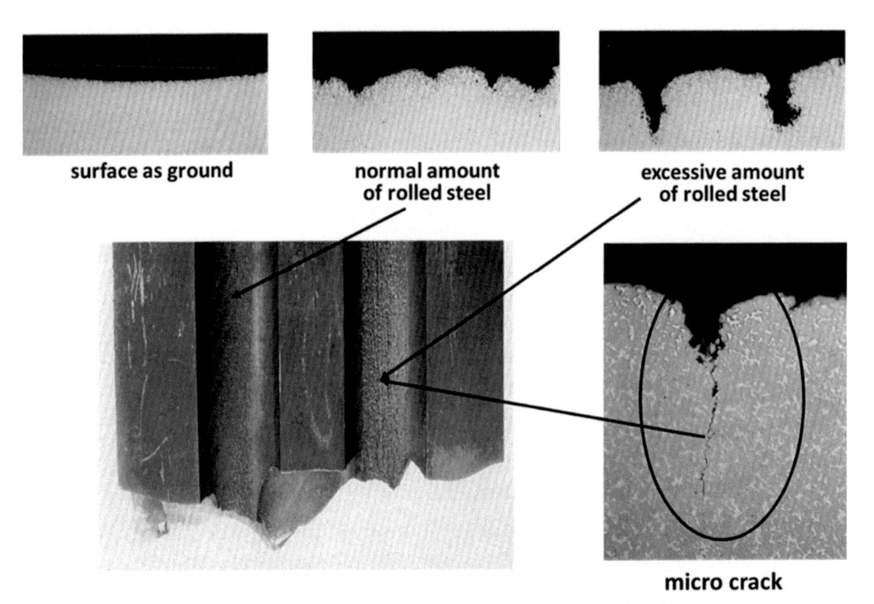

Figure 12.32 Typical wear and damage patterns of carbide mill roll grooves occurring as a result of formation of thermal cracks and their propagation into the bulk of the roll. *(With the permission of SAAR-HARTMETALL und WERKZEUGE GmbH.)*

aligned guides. Sudden changes in size, shape, and temperature of rods being rolled due to the presence of cold front parts or tail end parts, cobles, or improper handling can cause premature failures of mill rolls.

Insufficient cooling leads to the fast growth of thermal cracks through the material due to the limited thermal shock resistance of cemented carbides, so that the cracks become large and visible, which might result in breakages of the rings in the worst case. During the hot rolling process, the steel meets the roll surface at a temperature of about 1100°C. When the roll rotates properly, the temperature of a section of the rod being rolled is reduced, as the rod enters the zone subjected to cooling. During the period when there is a contact between the roll and the rod, the significant heat transfer to the roll takes place followed by cooling the roll surface, which results in severe thermal fatigue. High billet temperatures result in obtaining enormous residual stresses in the roll after its cooling down to room temperature. Low rolling speeds lead to the enhanced heat penetration into deep regions of the roll. This effect can cause the formation of deep cracks propagating into the mill roll body. High material reduction rates and hard rod materials generate additional heat, which is detrimental with respect to the roll performance. Large rolls have a large contact area and therefore a longer contact time with the rods resulting in a deeper heat penetration. Inadequate cooling is one of the most important factors determining roll life. The presence of pitting and corrosion on the roll surface depends mainly on the cooling water system and should not been underestimated. Cooling is typically achieved by spraying the pass groove surface with water under high pressure. The pressure should be sufficiently high to permit the water to penetrate to the roll surface through the vapor barrier, but not too high to bounce off the roll surface without removing significant heat. The optimum pressure of water is in the range of 4–6 bar measured at the point of discharge from the water header. The cooling water containing impurities can chemically react with the roll material causing corrosion, which might lead to premature roughening of the pass groove surface. In addition to surface roughening, pits can also form on the roll surface as a result of corrosion. The surface defects caused by corrosion can act as crack initiation sites and corrosion can therefore cause premature breakages of the rolls. In addition, the surface defects can also accelerate the thermal cracking process [21].

Beside the rolling stresses, mill rolls can also be subjected to mounting stresses. Transversal cracks initiated by roll mounting stresses can cause a noticeable overload as a result of unproper maintenance, poor conditions of the roll ring bore, mismatching between roll bore surfaces, and mandrels surfaces, which lead to appearing local stress concentration sites and hoop stresses present in the ring. Hoop stress levels may exceed the value of the tensile strength of cemented carbide, resulting in the formation of radial cracks. Too high mounting pressures are an additional factor that can increase a risk of the crack formation on the bore side when the roll bearings and gearboxes operate in such a way that the excessive heat generation occurs. Modern mill roll shaft clamping systems are designed to prevent hoop stresses due to the presence of axial hydraulic pressure.

Fig. 12.33 illustrates different types of degradation of mill rolls during their life cycle and outlines phenomena causing the mill rolls' degradation.

The selection of the best cemented carbide grade for a concrete operation of rolling is based on the rolling parameters, rolled or drawn material properties, and unique local operation conditions of each steel mill. It is recommended that the selection of the cemented carbide grades for each stand should include testing carbide grades with different binder contents, characterized by high fracture toughness and a reduced risk of premature breakages. After gathering some experience in terms of wear, crack initiations, and roll breakages, cemented carbide grades with lower or higher binder contents should be tested. Using cemented carbide grades with lower binder contents results in a higher hardness and improved wear resistance but reduced fracture toughness and thermal fatigue resistance. The focus regarding the mill roll performance is the wear of the passing groove in relation to the service life expectation.

Medium-coarse cemented carbide grades with various cobalt contents are usually employed for cold rolling applications. When drawing wires of small diameters, it is possible to apply submicron cemented carbide grades with binder contents in the range of 10–15 wt.%.

For hot rolling applications, coarser carbide grades must be employed. The microstructure of coarser carbide grades comprises significantly less WC/binder grain boundaries, which results in an improved thermal conductivity and increased thermal fatigue resistance. The continuously occurring processes of expansion and contraction caused by numerous heating and cooling cycles coupled with the high temperature gradient can cause the formation of high stresses in the surface layer of the roll near the pass groove resulting in the thermal crack formation. The thermal cracking resistance of materials is directly proportional to its fracture toughness and thermal conductivity and inversely proportional to its coefficient of thermal expansion and elastic modulus [21]. Improved thermal properties of the

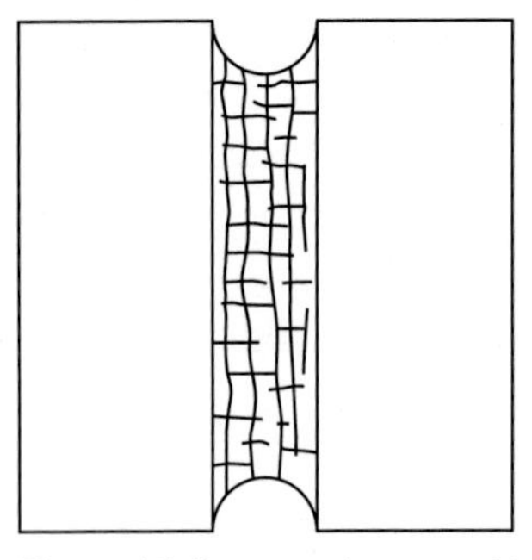

Thermal fatigue cracks caused by repeated heating and cooling of the caliber surface during hot rolling

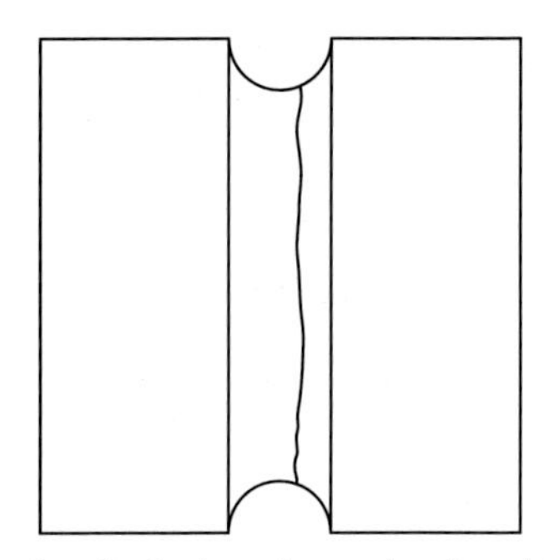

longitudinal crack can developed from late reworking of thermal fatigue. Often not seen until the roll is reground

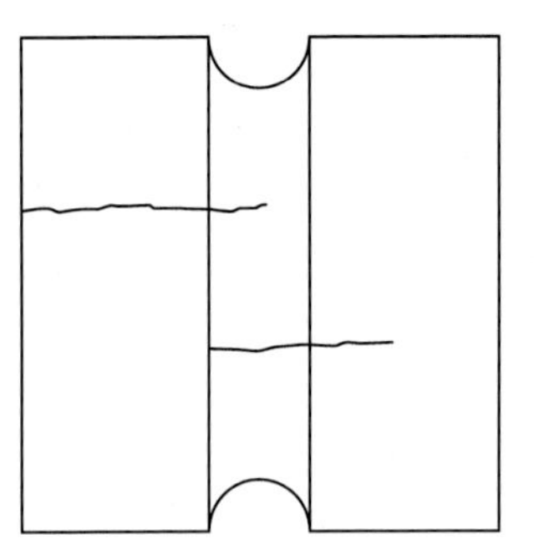

Cracks across the roll either starting from the edge, the bore or the pass caused by excessive tensile stress

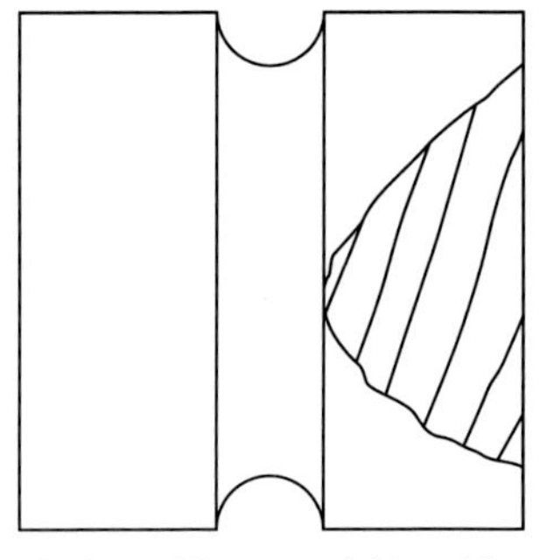

Broken of fragment initiated from the edge of the pass groove due to insufficient side clamping forces

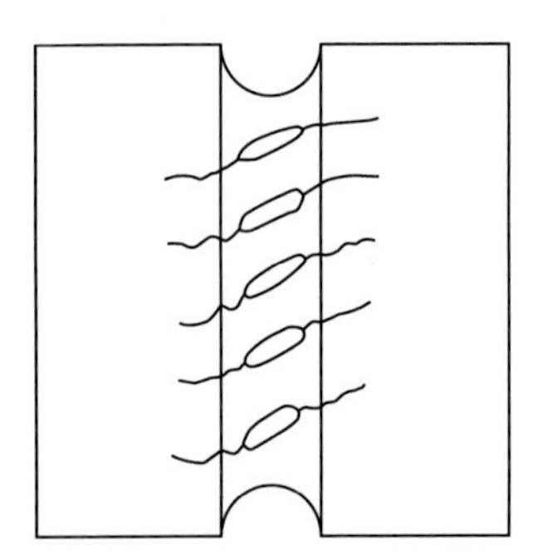

Cracks initiated from the sharp corners of the nicks caused by fatigue due to stress concentration

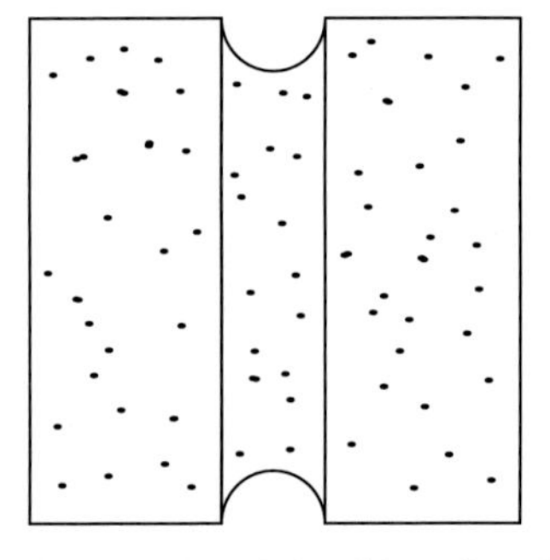

Pits over the whole of the roll surface caused by corrosion due to poor process water quality

Figure 12.33 Schematic diagrams illustrating different types of degradation of mill rolls during their life cycle and phenomena causing the mill roll degradation. *(With the permission of Element Six.)*

cemented carbide in combination with proper cooling conditions significantly reduce the intensity of thermal fatigue.

Also, the use of Co/Ni or Co/Ni/Cr binder systems leads to an improved corrosion resistance of cemented carbides for the fabrication of mill rolls (Fig. 12.34). The high corrosion resistance of such binder systems ensures the high resistance of the binder phase with respect to the corrosive impact by the cooling water containing in many cases traces of aggressive chemical compounds.

Any catastrophic failure of mill rolls due to the crack formation and propagation generally leads to costly unscheduled stand stoppages. Timely inspections or inline control systems and proper redressing of the mill roll surface to removed traces of thermal and corrosive surface degradation are mandatory. Refurbishing of the worn surfaces is possible by grinding, hard turning, or electro-discharge processing. Turning can be applied for cemented carbide grades with binder contents of 25 wt.% or greater. Regrinding and eroding are mainly used for the finer and harder cemented carbide grades employed in the fabrication of wire drawing rolls. Timely reworking is strongly recommended to prevent the fast crack propagation and catastrophic failures of the entire mill roll. Cracks initiated through unproper machining must be taken into consideration for the overall process optimization. Turning, grinding, and EDM cutting with well-adjusted machining parameters are necessary to achieve the best performance of the mill rolls. High energy input and heat generation during EDM processing at inadequate parameters result in a risk of surface cracking. Any crack left after redressing will propagate very rapidly when the roll is repeatedly inserted into the mill stand, resulting in premature

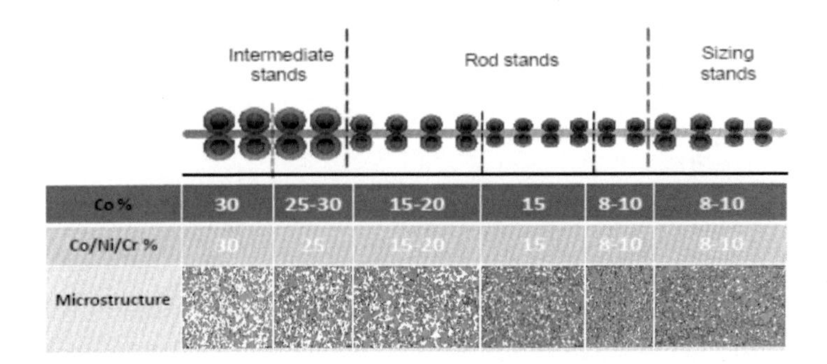

Figure 12.34 Overview of recommended cemented carbide grades for different hot rolling stands and their microstructures. *(With the permission of Element Six.)*

catastrophic failures of the roll. Redressing depths varying from 0.5 to 2 mm are usually employed to remove pass groove surface cracks. The best tools for regrinding the rolls are diamond wheels or c-BN and polycrystalline diamond (PCD)-based cutting inserts.

Numerous shapes of mill rolls with diameters varying from 50 mm up to 600 mm and thicknesses varying from 10 mm up to 200 mm, which are flat or comprise several pass grooves, are produced on a large scale; they are customized for different products and production systems (Fig. 12.35). The fabrication of mill rolls consumes a large quantity of cemented carbides, so that an economic production requires access to low-cost raw materials in combination with the optimized production technology. Near-net-shape pressing and advanced final machining technologies like high speed grinding are key factors needed for the efficient fabrication of cemented carbide rolls.

12.7 High-pressure high-temperature components

The need for synthetic diamond in different grain sizes and PCD continues to grow worldwide. Synthetic diamond grits and PCD can be manufactured with the same specialized high-pressure high-temperature (HPHT) equipment types. Increasing requirements for oil and gas drilling in the energy sector are continuously replacing cemented carbide drilling tools by

Figure 12.35 Typical mill rolls and guiding rolls. *(With the permission of Element Six.)*

significantly longer lasting PCD drilling tools. Turning and drilling of modern composite materials in the aerospace sector is also a growing market for superhard materials. Complex cemented carbide/PCD interface designs, including diamond gradients and optimized stress management, enable the use of PCD with hardness over 40 GPa in harsh applications, in which fracture toughness greater than 15 MPa $m^{1/2}$ is necessary for a safe use. PCD has made it into such demanding areas as road milling and mining. Latest developments like shaped PCD, 3D PCD, and ultrahigh pressure PCD increase the area of application ever further.

In 2019, the worldwide production of synthetic diamonds amounted to at least 14.6 billion carats [22]; this amount of synthetic diamonds is fabricated by HPHT processes, in which carbide tools and their performance are crucial. Firstly, cemented carbide tools were successfully used in commercial applications of HPHT processes in the 1950s and became one of the main enablers to synthesize artificial diamond. The outstanding combination of high compressive strength, high temperature creep resistance, hot hardness, TRS, fracture toughness, and wear resistance of WC−Co cemented carbides allows their use at extreme conditions. Specially developed cemented carbide tools can withstand the necessary temperatures for high-pressure sintering processes in the range of 1300°C−2500°C and the pressures of 5.5−10 GPa cost-effectively. A frequently used example of how to better understand the pressure−temperature conditions is of an upside-down Eiffel Tower (10,100 t) with all its weight concentrated on a soft drink can, while heating it up to 1500°C at the same time. Under these conditions, a relatively short synthesis time of about 5−30 min can be employed depending on the desired diamond grit grain size.

Most natural diamond grows extremely slowly at depths of 160−400 m below the Earth surface, where the upper mantle pressure reaches up to 13 GPa (equivalent of 130,000 atm). The growth of natural diamonds in the Earth takes millions of years, and they are being brought to the surface typically in rock formations known as kimberlites.

The interest in synthetic diamond started in the early 19th century, and for nearly 150 years many scientists spend sometimes nearly their whole professional life trying to convert graphite into diamond without much success. Trials included crystallization out of a carbon rich liquid, carbon rich oil, blasted steel tubes, high applied current, quenching of graphite oversaturated liquid iron, some form of high pressures, or high temperatures. Over time, scientists learned more about the formation conditions of natural diamond and started using significantly increased pressures and high temperatures. Various designs of HPHT equipment were developed, but for a long time, the materials used were not able to simultaneously support the minimum pressures and temperatures required for the diamond nucleation and growth.

In the 1950s, cemented carbides were first employed and became a major enabler to successfully synthesize artificial diamonds. Modern technology drastically speeds up the transformation process from graphite into diamond. The diagram shown in Fig. 12.36 illustrates temperature and pressure regions where diamond and graphite are stable forms of carbon. It also shows the temperature—pressure region, in which diamond synthesis with aid of catalysts consisting of Fe group metals is employed in industry.

Depending on the employed press system, a few minutes reaction time can be enough for synthesizing industrial diamonds. Nowadays, the major types of equipment employed in the diamond industry are the belt, cubic, and toroidal systems.

The successful and economic synthesis of diamond depends largely upon the life of the HPHT components of the manufacturing equipment; in particular, the pressure generating tools comprise cemented carbide anvils and dies [23].

The Belt uniaxial high-pressure device consists of a large cemented carbide die surrounded by a system of multiple interface-fit high strength alloy rings. The steel rings provide radial support to the die by applying

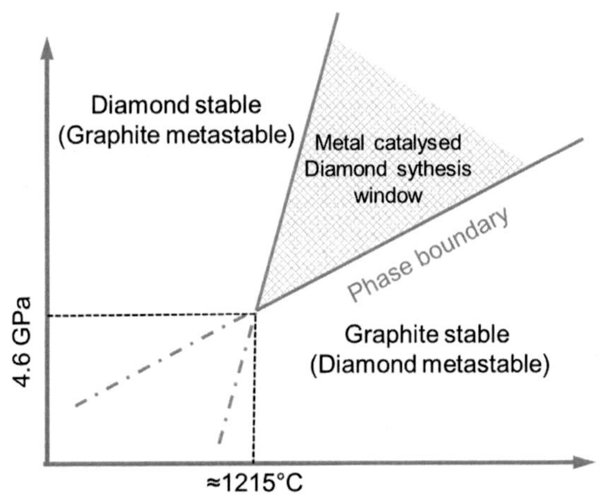

Figure 12.36 Typical temperature—pressure diagram for obtaining synthetic diamonds. *(With the permission of Element Six.)*

uniform compressive stress on the cemented carbide die, necessary to withstanding the high tensile forces occurring during the synthesis process. The loads are generated by elevated hydraulic forces, in the range of thousands of tones, applied to the top and the button anvil. The resulting capsule pressure is typically in the range of 4—10 GPa, and it depends on the capsule size and the response from the pressurized surfaces shown in Fig. 12.37. The contraction within the synthesis volume during the graphite → diamond transformation is the most critical phase for the die, due to the short and fast volume drop; hence, the resulting load change could result in the crack formation. The temperature inside the capsule can reach 1600°C, and the resulting deformation, phase changes, and the capsule components' compressibility also affects dynamically the forces applied toward the cemented carbide anvils and die.

The optimum cemented carbide grade selection for the belt anvils and dies is complex and strongly dependent on the press system. For the belt dies, WC-Co cemented carbide grades with a WC mean grain size of 2—3 μm and binder contents of 8%—14% are typically used. This results in a hardness range from 1200 to 1400 Vickers units [23]. Depending on the size of the designated reaction volume, the belt dies can be as heavy as

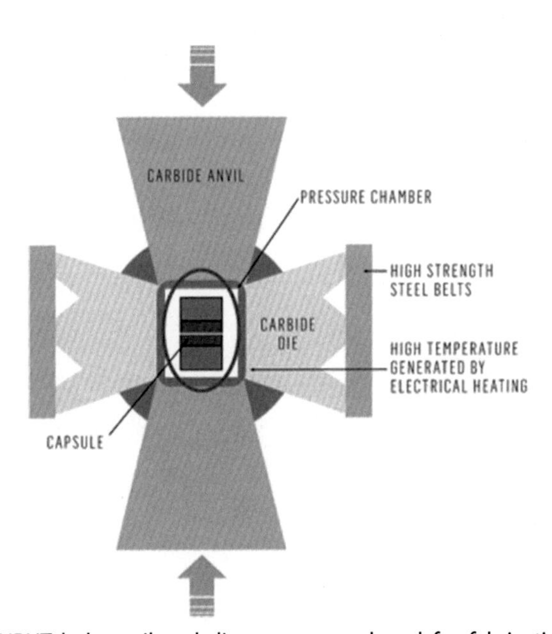

Figure 12.37 HPHT belt anvil and die system employed for fabrication of synthetic diamond products. *(With the permission of Element Six.)*

800 kg. Belt anvils are usually harder and typical consist of carbide grades with a WC mean grain size lying in the range of 1.0—1.5 μm and binder contents of 6 wt.%—10wt.% having hardness values of 1500—1800 Vickers units.

Cubic anvil systems consist of six geometrically identical cemented carbide anvils. The cemented carbide anvils are press fitted into steel fixtures connected to six hydraulic pistons applying equal loads on a cubic-shaped capsule. The main benefit of the cubic system is the constant pressure load on all the six anvils, resulting in compressive stresses in the anvils at maximum loads when the diamond product is synthesized/sintered. The cubic anvils can be harder than the belt anvils and typical consist of carbide grades with a WC mean grain size in the range of 0.8—1.0 μm and binder contents varying between 5 wt.% and 9 wt.%; they are characterized by hardness values varying between 1600 and 1800 Vickers units.

A unique combination of different properties of cemented carbides like high hot hardness, compressive strength, and creep resistance can be further improved by applying additives to the cobalt binder like molybdenum or rhenium.

Cemented carbide anvils and dies, shown in Fig. 12.38, are fabricated by use of similar production methods as those described in Chapter 6. Due to the tool size and weight, dedicated massive production machines for green machining and grinding combined with adequate handling systems are

Figure 12.38 Typical cemented carbide components for HPHT applications like belt dies and belt anvils of different sizes as well as cubic anvils. *(With the permission of Element Six.)*

required. To prevent any critical material contamination, clean room conditions are standard production environment for the cemented carbide tools for HPHT applications. Hot isostatic pressing as an additional quality improvement process is often applied.

The HPHT process creates two different issues with respect to performance and lifetime of cemented carbide articles. Firstly, extreme conditions of compressive and tensile stresses, creep, and thermal mechanical fatigue pose a risk of failure of the carbide articles. Secondly, instabilities can manifest themselves in unequal loading or spontaneous releases of capsule pressure (gasket failure).

The bore diameter of the Belt Dies is exposed to high synthesis pressures and elevated temperatures, which can lead to formation of vertical microcracks and plastic zones within the HPHT components. These microcracks can grow from synthesis run to synthesis run and can finally cause a vertical crack within the bore propagating through the plastic deformation zone. Once the crack has propagated about 2/3 of the outer diameter of the die, the plane of a vertical crack is turned through 90 degrees caused by tensile stresses in the horizontal plane as a result of the transverse moment of the steel rings. The horizontal, so-called "pancake crack," shown in Fig.12.39, propagates from the turning position outward close to the outer diameter of the die and inward up to the plastic deformation zone close to the inside diameter. A crack like this can be stable for an extensive number of runs.

Usually the die is replaced when the vertical cracks in the die bore reach a critical size which could compromise the product quality, or the stability of the synthesis cycle is endangered, which is illustrated in Fig. 12.39.

Roughly 90% of the dies start forming vertical cracks, which propagate in the above described sequence.

Occasionally, a breakthrough of the pancake crack across the plastic deformation zone or into the gasket area can be observed. Such a propagation needs to be treated with great caution, as the die stands the risk of collapsing thereby endangering the safety of operators, carbide anvils, and presses (Figs. 12.40 and 12.41).

In the case of the Belt press anvils, the region below the planar surface gets plastically deformed. Tensile stresses occur at the transition area between plastic and elastic deformation zones. After a certain number of synthesis cycles, the tensile stresses can cause microcracks on the surface of the anvils, which later propagate into the bulk of the anvils. Finally, the

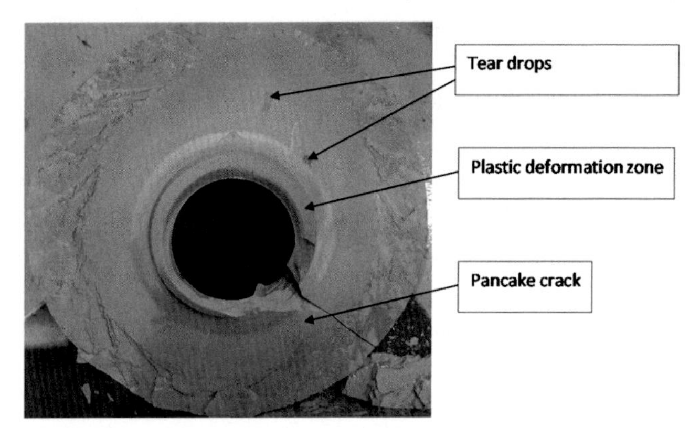

Figure 12.39 Typical postmortem fracture surface of a belt die showing the plastic deformation zone close to the bore, a vertical crack from the bore to the outside diameter and final horizontal "pancake crack" with the darker tear drops. *(With the permission of Element Six.)*

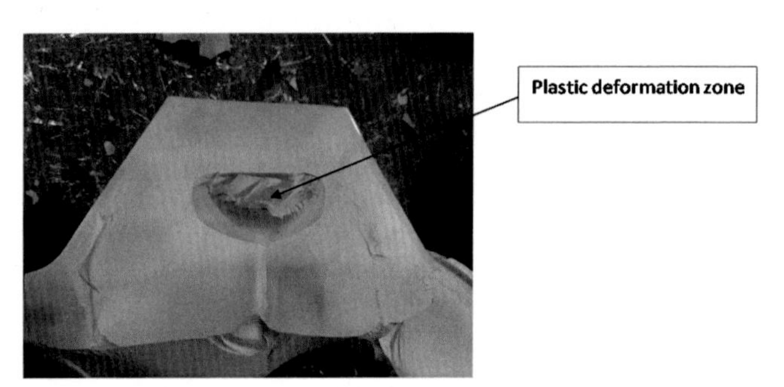

Figure 12.40 Typical postmortem fracture surface of a belt anvil showing the plastic deformation zone below the nose of the anvil. The darker area can be designated as the crack start area. The crack then propagates to the surface of the anvil and finally across the nose of the anvil. *(With permission of Element Six.)*

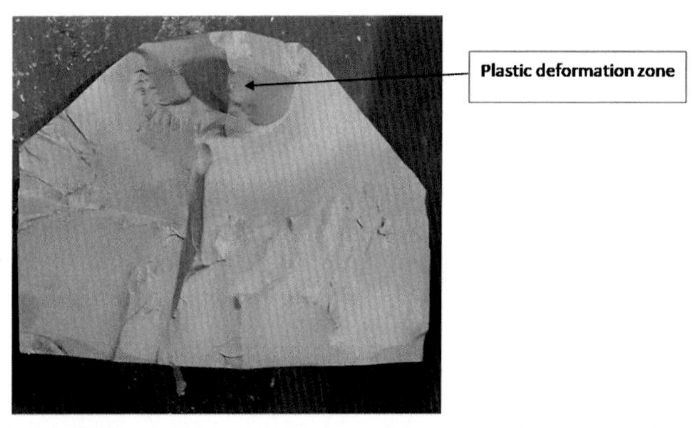

Figure 12.41 Typical postmortem fracture surface of a cubic anvil showing the plastic deformation zone at the nose of the anvil. *(With the permission of Element Six.)*

cracks burst through the surface around the plastic deformation zone. This is usually the time to replace the anvils to prevent any dangerous events.

Depending on the process, the failure mechanism can also be dictated by the process stability itself. Pressure losses at high loads are very likely to show clear adverse effects on life of anvils.

The carbide anvils used in the Cubic presses usually suffer from very similar breakage patterns as those occurring in the Belt anvils. As a consequence of deviatoric stresses within the anvil at final loads, a plastic zone forms underneath the nose of the anvil. At the interface between plastic and elastic zones, tensile stresses form, which can eventually lead to the crack formation. Largest difference to the crack formation in the Belt anvils is that cracks in the Cubic anvils usually originate at the surface of one of the anvils' faces. Once cracks are visible at the surface, the anvil is declared as a cracked anvil; its repeated use can cause a collapse of the following run.

It is obvious, that the Finite Element Analysis (FEA) models can be very useful to optimize the design and cemented carbide grade needed to prolong lifetime of HPHT components (see Section 14.3).

References

[1] L. Toller-Nordström, J. Östby, S. Norgren, Towards understanding plastic deformation in hardmetal turning inserts with different binders, Int. J. Refract. Metals Hard Mater. 94 (January 2021) 105309, https://doi.org/10.1016/j.ijrmhm.2020.105309.
[2] L. Prakash, Fundamentals and General Applications of Hardmetals, Comprehensive Hard Materials, Elsevier Science and Technology, 2014, pp. 29—89. Editor-in-Chief V.Sarin.
[3] P.J. Davim (Ed.), Machining fundamentals and recent advances, Springer, 2008, ISBN 978-1-84800-212-8, https://doi.org/10.1007/978-1-84800-213-5 e-ISBN: 978-1-84800-213-5.
[4] V.P. Astakhov, J.P. Danim, Tool (geometry and material) and tool wear, in: P.J. Davim (Ed.), Machining Fundamentals and Recent Advances, Springer, 2008, pp. 29—57.
[5] D.M. Gurevich, Adhesion-fatigue wear of cemented carbide cutting tools, Sov. Eng. Res. 6 (5) (1986) 35—38.
[6] D.M. Gurevich, Wear of coated carbide inserts in turning, Vestnik Mashinostroeniya 59 (6) (1979) 45—47.
[7] M.C. Shaw, Metal Cutting Principles, Oxford Science, Oxford, UK, 1984.
[8] E.M. Trent, P.K. Wright, Metal Cutting, Butterworth-Heinemann, Boston, USA, 2000.
[9] I. Konyashin, Cemented Carbides for Mining, Construction and Wear Parts, Comprehensive Hard Materials, Editor-in-Chief V.Sarin, Elsevier Science and Technology, 2014, 425-251.
[10] J. Larsen-Basse, Wear of hard-metals in rock drilling: a survey of the literature, Powder Metall. 16 (1973).

[11] U. Beste, S. Jacobson, A new view of the detioration and wear of WC/Co rock drill carbide buttons, Wear 264 (2008) 1129–1141.
[12] U. Beste, S. Jacobson, Rock penetration into cemented carbide drill buttons during rock drilling, Wear 264 (2008) 1142–1151.
[13] I. Konyashin, B. Ries, Wear damage of cemented carbides with different combinations of WC mean grain size and Co content. Part II: laboratory performance tests on rock cutting and drilling, Int. J. Refract. Metals Hard Mater. 45 (2014) 230–239.
[14] I. Konyashin, F. Scfaefer, R. Cooper, B. Ries, J. Mayer, T. Weirich, Novel ultra-coarse hardmetal grades with reinforced binder for mining and construction, Int. J. Refract. Metals Hard Mater. 23 (2005) 225–232.
[15] I. Konyashin, B. Ries, Wear damage of cemented carbides with different combinations of WC mean grain size and Co content. Part I: ASTM wear tests, Int. J. Refract. Metals Hard Mater. 46 (2014) 12–19.
[16] I. Konyashin, B. Ries, S. Hlawatschek, A. Mazilkin, Novel industrial hardmetals for mining, construction and wear applications, Int. J. Refract. Metals Hard Mater. 71 (2018) 357–365.
[17] I. Konyashin, B. Ries, Y. Zhuk, A. Mazilkin, B. Straumal, Wear-resistance and hardness: are they directly related for nanostructured hard materials? Int. J. Refract. Metals Hard Mater. 49 (2015) 203–211.
[18] M. Gee, Wear of Hardmetals, Comprehensive Hard Materials, Elsevier Science and Technology, 2014, pp. 363–383. Editor-in-Chief V.Sarin.
[19] Internet site www.ceratizit.com. Dated April 2021.
[20] Internet site www.worldsteel.org. Dated April 2021.
[21] P.K. Mirchandani, E.L. Klaphake, T.M. Gallagher, SinterMet. Strategies for Obtaining Optimum Performance from Cemented Carbide and Bar Mill Rolls, Inc. Kittanning, Pennsylvania U.S.A.
[22] Internet site https://pubs.usgs.gov/periodicals/mcs2020/mcs2020.pdf. Dated April 2021.
[23] K. Wada, Tungsten carbide based hard metals used for high pressure experiments, Rev. High. Pres Sci. Technol. 28 (1) (2018) 9–16.

CHAPTER 13

Recycling cemented carbides

In the last decades, recycling of used cemented carbide articles back to their original powdered form has become more and more popular; as a result, the recycling is now a reliable and extremely important source of secondary raw materials for the cemented carbide industry. The recovery of secondary raw materials has not only restructured the pricing dynamics of a highly competitive cemented carbide market but also paved the way for the eco-friendly material production with a better carbon footprint.

The costs of the cemented carbide scrap are dependent on its availability and quality. Quality is mainly driven by the scrap homogeneity and purity.

Despite its economic benefits, incorporating recycled secondary raw materials into the production stream can lead to certain challenges. As the chemical composition of each batch of the recycled carbide powder is highly dependent on that of the original carbide scrap, workarounds must be found to counteract harmful chemical impurities or smooth out batch-to-batch fluctuations with respect to the chemical composition. Thus, for a long time, it was unthinkable to integrate such recycled powders into the production stream without sacrificing the cemented carbide quality and consistency.

As a result of recent developments in the cemented carbide industry, better sorting techniques and technological processes were elaborated to sort out different scrap batches according to their binder contents and WC mean grain sizes. Thus, the quality of the secondary raw materials started increasing, which resulted in the large-scale employment of secondary raw materials in the cemented carbide industry (Fig. 13.1).

Collecting enough tungsten-containing scrap components to sustain the industry's need for raw materials is a complicated task; however, it is the only way of closing the tungsten life cycle loop and ensuring the sustainability of the tungsten resources. Even despite enormous efforts, the International Tungsten Industry Association bulletin [1] reports that as much as approximately 70% of the worldwide tungsten-containing scraps are still wasted by dissipation, dilution, or improper scrapping. Nowadays,

Cemented Carbides
ISBN 978-0-12-822820-3
https://doi.org/10.1016/B978-0-12-822820-3.00013-6

Figure 13.1 X-ray fluorescence assisted sorting the cemented carbide scrap. *(With the permission of Tikomet OY.)*

the share of secondary raw materials utilized in the production of cemented carbides in Europe is about 40%–50% and these numbers rapidly grow.

While it looks like that there is still ample room for improvement, it must be emphasized that the efficiency of the recycling process itself is highly dependent on the state of the scrap material. High-performance cutting tools, for example, are often coated with a variety of coatings containing Ti(CN), Al_2O_3, polycrystalline diamond, etc., which would be considered as impurities if they are found in cemented carbides. As many companies adopt customized coating arrangements, it can be challenging to design standardized surface removal procedures, especially if the utilized materials are characterized by a wide range of characteristics. Similarly, carbide/steel tools for mining and road-planing must be debrazed and chemically cleaned followed by removal of steel parts.

Contrary to the abovementioned, much scrap created during the production process of cemented carbides is very well suited for recycling. Powder scrap from the powder preparation, sintered scrap due to unsatisfactory material quality, or grinding sludge from postprocessing of carbide articles are examples of carbide scrap types that can be easily recovered back to high-quality raw materials by chemical and mechanical treatment. Fig. 13.2 shows scraps of different types after sorting and screening.

Fig. 13.3 shows an overview of the most important technologies employed for recycling tungsten-containing scraps, particularly cemented carbide scrap. Estimation of major advantages and disadvantages of the common recycling technologies and typical sources of tungsten-containing scrap are given in Table 13.1. The advantages of each method are indicated by green color (light gray in print) and their disadvantages are indicated by red color (dark gray in print).

Figure 13.2 Cemented carbide scrap of different types after sorting and screening. *(With the permission of Wolfram Bergbau und Huetten AG.)*

13.1 Full chemical recycling

Full chemical recycling is a universal indirect recycling technique for all kinds of tungsten-containing scraps [2,3] (Fig. 13.4). The process can handle a wide range of cemented carbide scraps regardless of their quality. Grinding slurries, powder scraps, or any low-quality scraps can be mixed and treated simultaneously, regardless of any impurities contained in them, such as grain growth inhibitors or traces of braze alloys, coatings, glues, or oils. Consequently, quality requirements regarding the scrap sorting process are minimal if not obsolete.

As the full chemical recycling process yields high purity Ammonium Paratungstate (APT), it can then be postprocessed and refined identically to tungsten oxide and tungsten metal powders produced from ores. As a result, full chemical recycling allows customizations of the raw material by manipulating grain sizes and grain shapes according to the product requirements.

The quality of raw materials obtained via full chemical recycling is therefore identical to that of virgin materials. However, due to the high

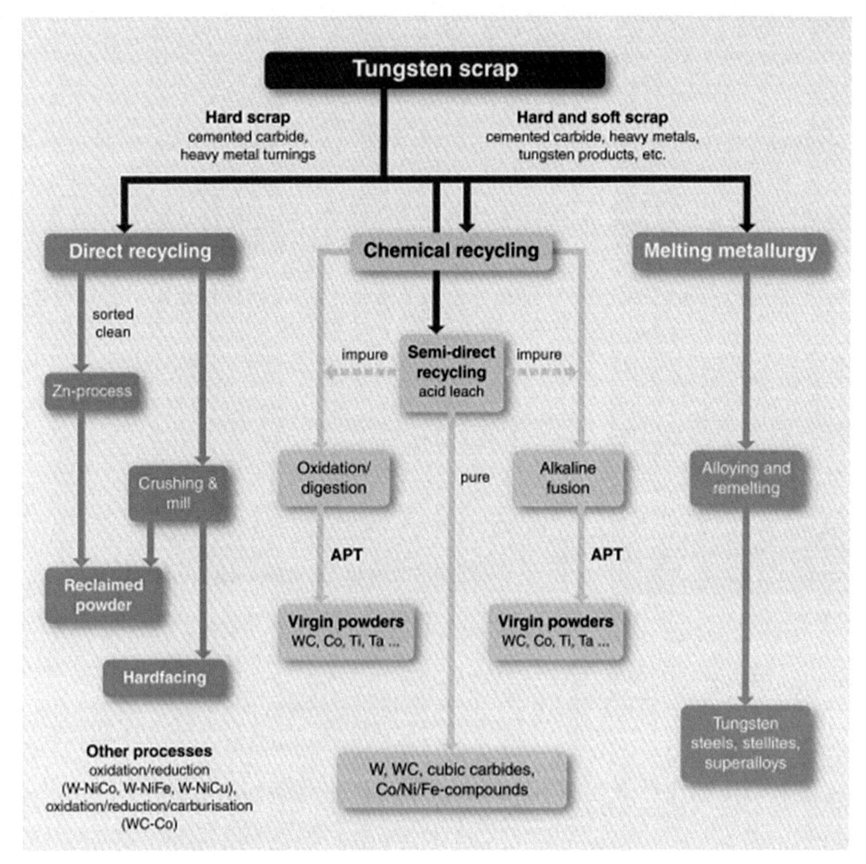

Figure 13.3 The most important technologies for recycling tungsten-containing scraps (the melting metallurgy method belongs to recycling tungsten metal scraps). *(Redrawn from Ref. [3].)*

process temperatures and a number of chemical processing steps, the production costs of full chemical recycling are relatively high. All other components present in the scrap (cobalt, nickel, chromium, vanadium, titanium, tantalum, niobium, etc.) are recovered in separate processing lines to improve the process economics.

Powder scrap after roasting the scarp is treated by the use of the alkaline fusion process. When recycling slurries, liquid must be first separated from the grinding slurry; in a second step, the wet grinding mud must be dried before roasting. After completing the alkaline fusion process, the solidified melt is crushed and dissolved in water.

Table 13.1 Major features, advantages, and disadvantages of the modern recycling techniques.

Source/method/feature	Full chemical recycling	Zn recycling	Leaching	Cold stream process
Powder	✕		✕	
Hard scrap	✕	✕	✕	✕
Grinding slurry	↗		✕	
Direct recovery		✕	✕	✕
Indirect recovery	✕			
Size of scrap	←	↑	↑	→
Binder content	↙	↗	↖	→
Processing time	→	↑	↑	←
Energy consumption	↗	↗	←	←
Yield	↖	↙	↖	↖
Reliable process	←	←	↑	↑
Quality of powder	↙	←	↑	→
Price of product	↑	←	↙	↙
No additional process for metal binder or additive recovery required	↘	↙	↘	↙

During a further process step, APT is produced followed by its calcination step resulting in obtaining tungsten oxide. The tungsten oxide is reduced to tungsten metal followed by its carburization to produce tungsten carbide according to conventional technologies described in Chapter 6. The full chemical recycling process is illustrated in Fig. 13.4.

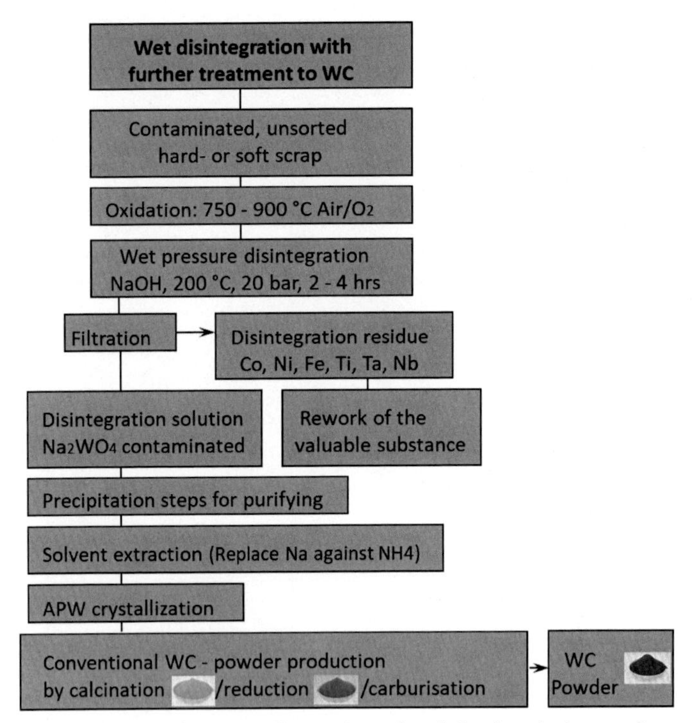

Figure 13.4 Schematic diagram illustrating the full chemical recycling process. Oxidation of cemented carbide secondary raw materials occurs at 750–900°C and followed by wet pressure–assisted disintegration in NaOH and filtration. Purifying the aqueous sodium tungsten solution is performed in the same aggregates as those employed for primary raw materials and results in obtaining high-purity APT employed for the fabrication of tungsten oxide, tungsten metal, and tungsten carbide powders.

13.2 Leaching carbide binders

Selective leaching of carbide binders in used cemented carbide articles by acids is mainly applied for carbide scrap that is a source of WC for its further reuse in the fabrication of the same or similar cemented carbide products [1]. This technique is widely employed for recycling of used high-pressure high-temperature components, like anvils and dies for diamond and PCD synthesis at ultrahigh pressures. Several mineral acids are used industrially; in most cases hydrochloric acid, but sometimes nitric acid, sulfuric acid, or phosphoric acid are employed. When using high concentrated mineral acids for the leaching process, the caution in terms of work and environmental safety is mandatory.

Well-sorted worn or damaged cemented carbide articles are used for the leaching process. There is a major requirement in terms of scrap sorting to ensure obtaining high-quality recovered WC powders. It is essential to sort the scrap in such a way that all the used carbide articles being recycled have nearly the same WC mean grain size. Chemical leaching can be employed for recycling powder scraps and carbide articles containing different grain growth inhibitors. Surface contaminants, like braze alloys, glue, oil, etc., do not have a significant influence on the leaching process. Carbide articles with high binder contents and coarse grain sizes are the preferred secondary material sources. Faster leaching is achieved as a result of employing carbide articles with large binder mean free paths. By dissolution of the metal binder, the integration of the cemented carbide structure is significantly reduced, and a follow-up milling process can easily break up the remaining parts of the original carbide articles being recycled. The binder-containing solutions remaining as a result of leaching are usually sold to companies dealing with the fabrication of Co and Ni powders (Fig. 13.5).

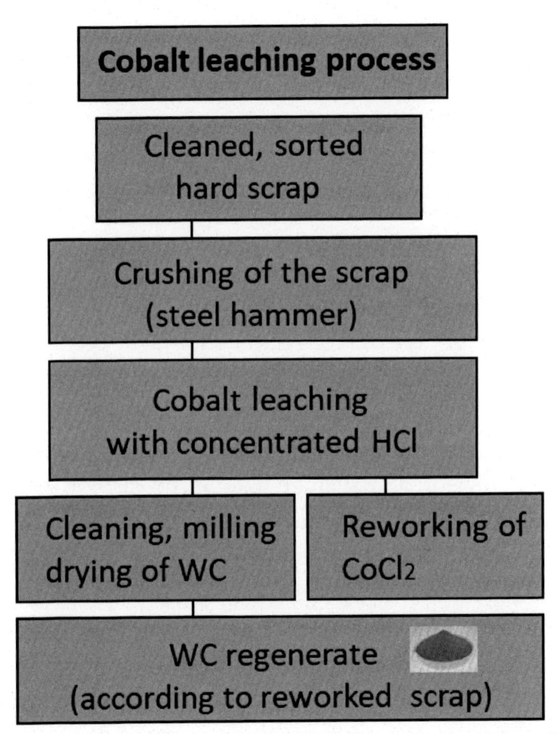

Figure 13.5 Schematic diagram illustrating the leaching technology. The sorted used cemented carbide articles are used for leaching the cobal binder in concentrated HCl. Cleaning, milling, and drying of the product obtained in such a way result in obtaining a pure secondary WC powder.

Figure 13.6 Typical pressure digestion autoclave. *(With permission of Wolfram Bergbau Huetten AG.)*

Fig. 13.6 shows a typical pressure digestion autoclave employed in the leaching technological process. After filling the sorted cemented carbide scrap into such a large autoclave, recirculated acid solutions dissolve the binder phase of the carbide scrap, whereas WC grains remain undissolved. The reaction time depends on the type and concentration of the mineral acids, process parameters (temperature and pressure) as well as on the WC grain size, binder content, and average sizes of the carbide scrap pieces being recycled. Typical reaction times at boiling temperatures of the mineral acids lie between several hours and several days. Once the leaching process is completed, the acid solution is pumped out of the tank and the remaining pieces containing WC agglomerates of different sizes are subjected to water-cleaning, ball-milling, and drying. The WC powder obtained in such a way can be used for the fabrication of carbide articles with nearly the same WC mean grain size as that of the carbide scrap. When using well-sorted scraps, the recycled WC powders can be of medium to high quality. The carbon footprint of the leaching process is by far better in comparison to the full chemical recycling technique.

13.3 Zinc reclaim process

The zinc reclaim process is an 80-year-old technique dedicated to recycling fully sintered cemented carbide components that have reached the end of their life cycle. In contrast to other recycling techniques, the Zn process merely breaks down solid cemented carbides into WC–Co powders without altering their average composition [1].

As it is shown in Fig. 13.7, the process can be divided into four steps. First, carbide scrap parts are sorted according to similar applications groups and similar grain sizes. As zinc is completely miscible with cobalt, nickel, or iron, zinc is then melted as a result of heating above the Zn melting point and its melt reacts with the carbide parts in a vacuum furnace. The zinc–Co/Ni/Fe eutectics melts at 870–900°C and undergoes a notable volume expansion causing the disintegration of the carbide parts, which

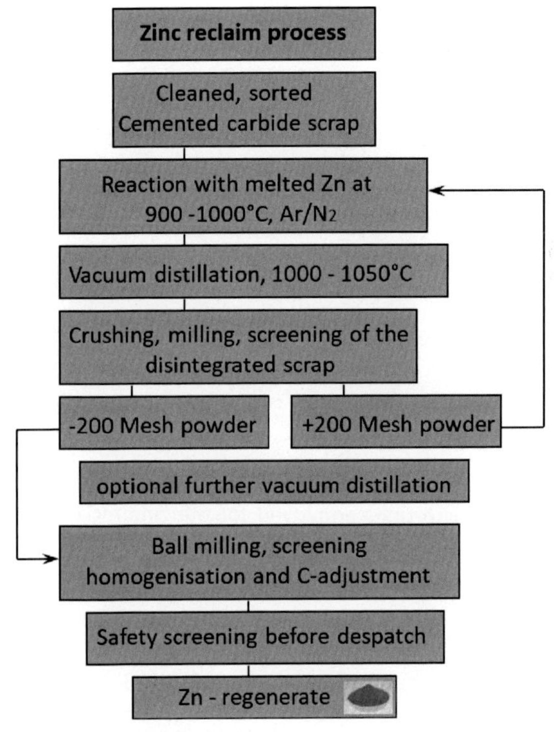

Figure 13.7 Sorted cemented carbide secondary raw materials are subjected to melted zinc, which allows obtaining a porous sponge as a result of vacuum distillation of zinc vapors. By employing the vacuum distillation, the melted zinc can be almost completely removed followed by a final milling step, which allows releasing the pure WC–Co secondary powder.

results in the formation of a porous sponge. Finally, zinc is evaporated from the sponge by decreasing the pressure down to 2–10 mbar and increasing the temperature up to 1030°C. If the process is performed properly, only about 20 ppm zinc remains in the recycled carbide powder. Typical units employed for the Zn reclaim process are shown in Fig. 13.8. The porous carbide sponge obtained as a result of performing the zinc reclaim process, which is shown in Fig. 13.9, is subsequently crushed and milled to the desired grain sizes.

The process of recycling cemented carbides becomes substantially more cost effective at lower binder contents or smaller sizes of the carbide parts being recycled, as the infiltration of the cemented carbide binder by liquid Zn at 900°C is merely a function of time. Following the same logics, the Zn recycling technique allows processing large cemented carbide parts without the need for their crushing or fragmentation simply by processing for longer times at elevated temperatures. All remaining coarse agglomerates that are not disintegrated during the Zn reclaim process can be afterward recirculated and reprocessed. The recycling processes based on

Figure 13.8 Typical modern units employed for the zinc reclaim process. *(With the permission of Tikomet OY.)*

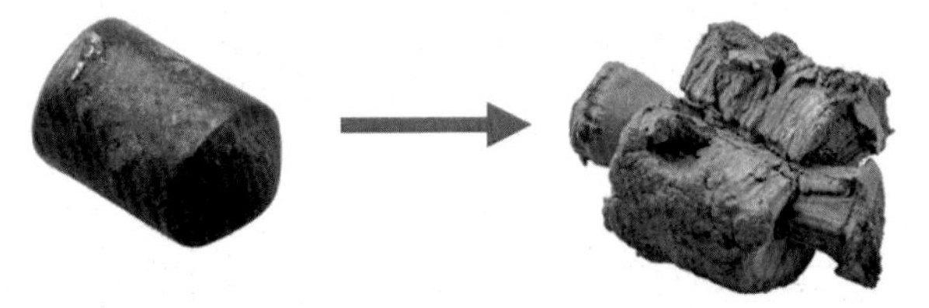

Figure 13.9 Illustration indicating how the zinc reclaim process operates: left – an initial carbide scrap article (insert) and right – the carbide porous sponge after completing the zinc reclaim process. *(With the permission of Tikomet OY.)*

crushing big carbide parts become more and more uncommon, as they are energy and time-consuming; they are steadily substituted by the Zn reclaim process.

As nothing is added to or subtracted from the carbide scrap being recycled except for zinc, it is assumed that the main reason for compositional fluctuations and grain size variations of the zinc reclaim processes is related to sorting and cleaning the initial carbide scrap. Any unwanted impurities (braze materials, glue, oil, dust, etc.) must be removed prior to the scrap processing.

Thus, in contrast to the chemical leaching process, achievable qualities of the recycled WC–Co powders are directly related to the perfectness of the sorting process with respect to obtaining similar grain sizes and compositions of different scrap batches. A well-equipped metallurgical laboratory is therefore crucial for obtaining reliably high quality of the recycled WC–Co powders.

Analyzing the quality of the recycled powders is more important than that of "virgin" powders, so various studies regarding the fluctuations of carbon content and WC grain sizes in the recycled powders are needed. Variations of carbon or cobalt contents in the recycled powders can be addressed simply by adjusting the production recipe when calculating the composition of production batches, while impurities or additions of grain growth inhibitors, such as Cr_3C_2, TaC, or TiC, cannot be eliminated or reduced by the use of the existing technology.

In summary, the relative simplicity of the Zn reclaim process requiring only a vacuum furnace, a ball mill, and a sieving machine in combination with the low process temperatures and high yields of >95% makes the zinc reclaim process a highly efficient recycling technique with one of the lowest carbon footprint in comparison with the other recycling technologies. However, even though the overall quality of the zinc-reclaimed powders is continuously improving, considerable quality and compositions' variations of the recycled powders can take place.

13.4 Cold stream process

Like the Zn reclaim process, the cold stream process is a direct recycling method for converting solid tungsten carbide articles at the end of their service life back into the powdered form, which is shown in Fig. 13.10.

The process was developed and patented in 1965 [4]; it is based on a process of accelerating carbide scrap parts in a high-pressure air flow which

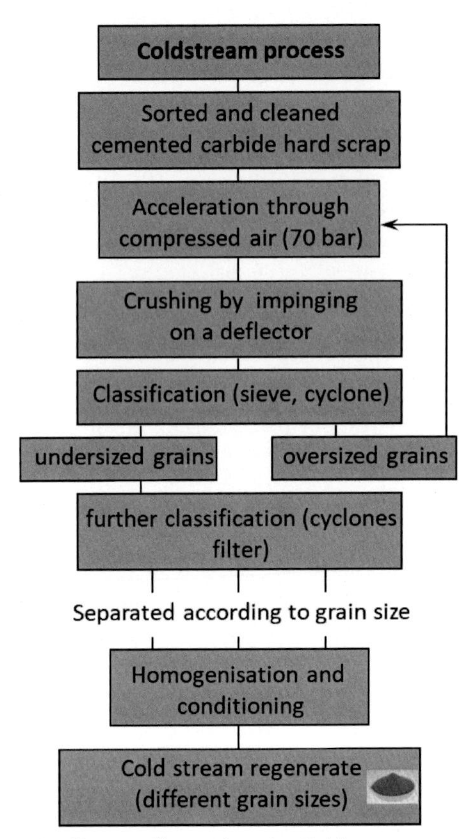

Figure 13.10 Schematic diagram illustrating the Cold stream process, which includes crushing the sorted and cleaned cemented carbide scrap followed by the classification and homogenization of the carbide powder obtained in such a way.

are then collided against a stiff target. Naturally, individual scrap parts must be light enough to be transported by the air stream and, thus, the cold stream process must comprise several crushing and screening steps. To ensure the proper fragmentation of the scrap particles during collision, a venturi nozzle accelerates the particles up to the speed that is greater than the speed of sound by a factor of two; it also protects the particles from oxidization due to its cooling effect. The fragmented powder is then screened, and the oversized fraction is recirculated back into the cold stream system. Fig. 13.11 illustrates the design of the cold stream process apparatus and the morphology of the carbide particles obtained by this method.

The cold stream process does not require chemical or thermal treatment of the scrap material. Thus, the environmental footprint of this method is

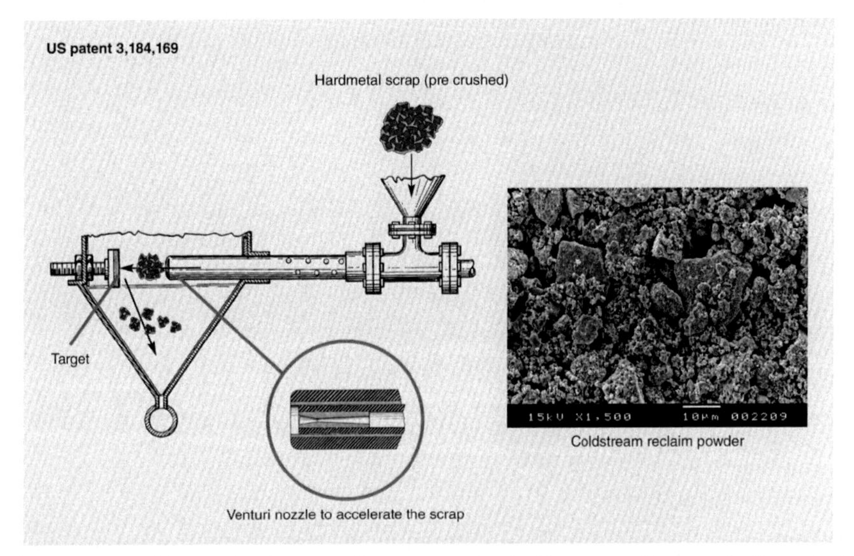

Figure 13.11 Schematic diagram illustrating the Cold stream Process (left): sorted, cleaned, and precrushed carbide scrap is fed into a high-pressure and high-velocity stream of air and accelerated against a target with high energy to cause its disintegration and pulverization; and SEM image of the pulverized powder (right). *(Redrawn from Ref. [3].)*

even lower than that of the Zn reclaim process. Nevertheless, mechanical pulverization techniques such as the cold stream technology can only produce a minor share of the worldwide recycled cemented carbide raw materials. This is largely due to high machine wear rates caused by particle erosion; the powder obtained by this method is contaminated by iron, even when using the cemented carbide collision target. Nevertheless, the recycled material can be employed in applications not demanding fine and high-quality WC−Co powders, for example, for producing hard-facings.

References

[1] W.-D. Schubert, B. Zeiler, in: B. Zeiler (Ed.), Recycling of Tungsten, the Technology-History, State of the Art and Peculiarities, ITIA Newsletter, 2019.
[2] B. Zeiler, W.-D. Schubert, in: B. Zeiler (Ed.), Recycling of Tungsten, Current Share, Economic Limitations and Future Potential, ITIA Newsletter, 2018.
[3] B. Zeiler, A. Bartl, W.-D. Schubert, Recycling of tungsten: current share, economic limitations, technologies and future potential, Int. J. Refract. Metals Hard Mater. 98 (2001) 105546.
[4] L.A. Adams, I.L. Friedman, L.S. Friedman, K.C. Zagielski, Apparatus for Pneumatically Pulverizing Material, 1965. US patent 3,184,169.

CHAPTER 14

Modeling cemented carbides

14.1 Modeling the structure of cemented carbides on the micro-, nano- and atomic level by ab initio calculations

As it is shown in Chapter 9, submicron, ultrafine, and near-nano cemented carbides are fabricated by the use of additions of various grain growth inhibitors, mainly carbides of transition metals (VC, Cr_3C_2, Mo_2C, TaC, etc.), which can be generally designated as "MC." The structure of such cemented carbides was investigated by transmission electron microscopy (TEM), high-resolution transmission electron microscopy (HRTEM), and atom probe tomography (APT) on the nano- and atomic level and reported in numerous publications indicating their segregation at WC/Co grain boundaries in form of mixed cubic carbides (M,W)C (see Chapter 9). Nevertheless, the exact mechanism behind the WC grain growth inhibition during liquid phase sintering is still not fully understood, so it is needed to be clarified by modeling and simulations. Also, some phenomena of wetting tungsten carbide by liquid Co-based binders, in particular by those with various carbon contents, significantly affecting the cemented carbides' structure on micro- and nano level, should be explained by ab initio calculations of WC/Co interfaces.

Modeling cemented carbides on the micro-, nano-, and atomic level by the use of first-principles calculations in the framework of density functional theory (DFT) was performed in Refs. [1–6].

The tendency of transition metals of groups IV(B) to VI(B) being added to WC–Co materials to segregate at WC/Co interfaces was determined by their effect on the WC/Co interface energy. The binder phase was approximated to consist of only fcc Co and all phases were assumed to be ideally stoichiometric. The structure of WC/Co interfaces was modeled in such a way that they would either comprise or not comprise cubic carbide films of atomic thickness N. Values of the interface energy in cases of (1) the presence of the film (Υ_{film}) and (2) the absence of the film ($\Upsilon_{WC/Co}$) were compared. The film formation is favored when $\Delta\Upsilon = \Upsilon_{film} - \Upsilon_{WC/Co}$

Cemented Carbides
ISBN 978-0-12-822820-3
https://doi.org/10.1016/B978-0-12-822820-3.00011-2

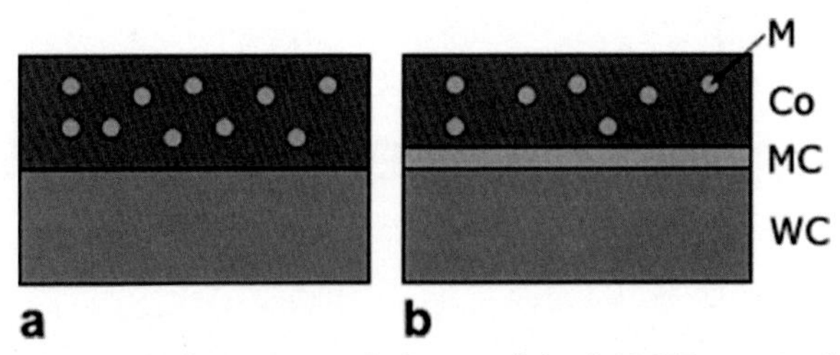

Figure 14.1 Schematic illustration of the model of WC/Co grain boundaries in cemented carbides containing grain growth inhibitors (MC), metal atoms of which are dissolved in the liquid binder phase: (A) – an interfacial MC film does not form at the WC/Co grain boundary and (B) – an interfacial MC film forms at the grain boundary. *(Redrawn from Ref. [4].)*

is negative. The film is assumed to form coherently with the WC crystal lattice, and an elastic energy per metal carbide layer is therefore needed to strain the film into coherence [1–4].

Generally, the simplified model of WC/Co grain boundaries in cemented carbides containing grain growth inhibitors (MCs), atoms of which are dissolved in the liquid binder phase (Fig. 14.1), is used for the first-principle calculations [1–4].

The results of Ref. [1] predict that for the basal (0001) crystal plane of the prismatic WC single crystals almost all the examined metal carbides (TiC, VC, Cr_3C_2, ZrC, NbC, Mo_2C, HfC, and TaC) might form thin interfacial films, as $\Upsilon_{MC/Co} < \Upsilon_{WC/Co}$ for the majority of the cubic carbides (except for ZrC and HfC). The predicted trend for the vanadium carbide to form (VC)–based films on the prismatic WC crystal planes should be considerably smaller than that on the basal planes, and the formation of the TiC, NbC, and TaC films on the prismatic crystal planes should be energetically unfavorable. It is predicted that under realistic doping conditions, such ultrathin films may exist at WC/Co interfaces at high temperatures of liquid-phase sintering thus suppressing the WC grain growth. It must be noted that the predicted significant difference in the tendency to form thin films on the basal and prismatic WC crystal planes (especially in the case of TiC, NbC, and TaC) should lead to the formation of platelet-like WC grains in the doped cemented carbides; however, such platelet-like grains are not observed in the microstructure of the cemented carbides containing such grain growth inhibitors.

The formation of VC thin films at WC/Co interfaces in the VC-doped cemented carbides was modeled in Refs. [2–4] since VC is known to be

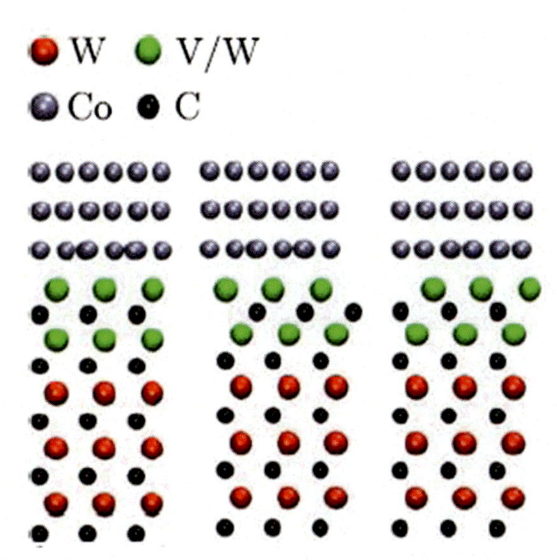

Figure 14.2 The different structural models viewed from the side along the $[2\bar{1}\bar{1}0]_{WC}$ direction. The lower parts show the hexagonal WC layers (*red and black dots*), the middle parts are the layers of mixed (W,V)C carbide (*green and black dots*), and the upper parts are the Co binder phase (*blue dots*). *(Redrawn from Ref. [4].)*

the strongest and mostly widespread grain growth inhibitor. The number of V layers and the possibility of forming mixed (W,V)C carbides were theoretically predicted. The most stable configuration of the interfacial VC films is predicted to consist of only two V layers when modeling the films' configuration in a large temperature range, which is illustrated in Fig. 14.2.

To assess the effect of composition, calculations of the possibility of forming mixed ($V_{0.75}$, $W_{0.25}$) C interfacial films were performed in Ref. [2]. According to the calculations' results, such mixed carbides are expected to form at elevated temperatures and be stabilized by the interfacial effects. Compared with the original interface in the absence of interfacial thin films, there is an energy gain associated with the creation of the two subinterfaces involved in the thin film formation. The interface is predicted to contain a thin V-rich film for a wide range of temperatures and chemical potentials of V corresponding to VC additions in amounts below the $(V,W)C_x$ solubility limit [3].

It should be noted that, as it can be seen in Fig. 14.3A and B, the microstructures of the WC−10wt.% Co samples containing either 1.2 wt.% NbC or 1.2 wt.% VC do not contain platelet-like WC grains, the presence of which can be expected when taking into account the results of the

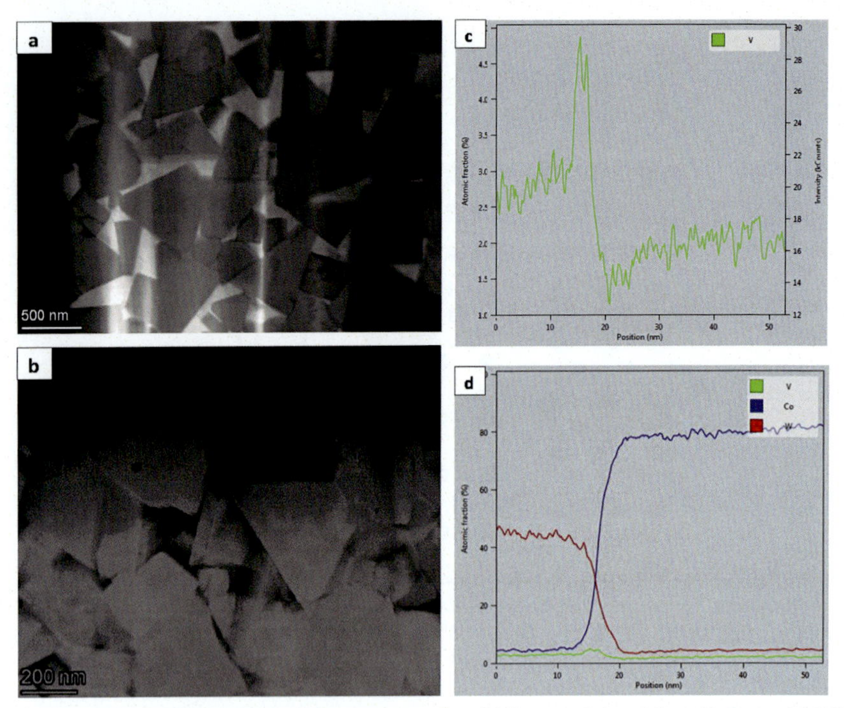

Figure 14.3 (A) STEM image of the cemented carbide containing 10 wt.% Co and 1.2% NbC, (B) STEM image of the cemented carbide containing 10 wt.% Co and 1.2% VC, and (C, D) curves indicating the distribution of V, Co, and W at the WC/Co interface of the cemented carbide containing 10 wt.% Co and 1.2 wt.% VC. *(With the permission of Element Six.)*

ab initio calculations. The thickness of the V-containing interfacial film at the WC/Co grain boundary is found to be nearly 2–3 nm (Fig. 14.3C), which is significantly greater than the thickness of the theoretically predicted interfacial VC films comprising two layers of vanadium atoms.

Based on first-principles simulations, it is also possible to calculate an equilibrium interfacial phase diagram or the so-called "complexion diagram" for the phase boundaries in doped cemented carbides. In this case, in addition to DFT, the employment of Monte Carlo simulations based on the DFT results for obtaining configurational free energies and thermodynamic modeling for describing properties of adjoining bulk phases are needed. An example of such a complexion diagram for cemented carbides containing VC is given in Ref. [4].

One can also conduct first-principles calculations of interfacial and surface energies in WC–Co cemented carbides not containing grain growth inhibitors [5]. In this case, the DFT calculations must be combined with a number of computational methods including quasiharmonic approximation, temperature, and thermodynamic integration, as well as calculations of liquid surface tension and work of adhesion of phase boundaries. In Refs. [5,6], it was predicted that the wettability of WC by liquid Co should be complete in W-rich conditions corresponding to low carbon contents and incomplete in C-rich conditions at high carbon contents. Outcomes of these first-principles calculations on the wettability of WC by liquid Co well correspond to experimental results on wetting tungsten carbide by liquid Co-based binders with various carbon contents [7] (see Chapter 9). The theoretical predictions also indicated the presence of a carbide skeleton in WC—Co materials, so that there should be no continuous Co interlayers at WC/WC grain boundaries independently on the carbon content in cemented carbides [5,6]. Outcomes of these theoretical predictions do not well agree with the experimental results obtained in Refs. [7,8]. As one can see in Fig. 14.4, the WC/WC grain boundary of the WC—Co cemented carbide with a medium—low carbon content comprises a continuous Co-based film of nearly 3 nm in thickness, which is significantly greater than the theoretically predicted value.

Thus, presently, ab initio calculations allow obtaining qualitative information about microstructural features of cemented carbides; however, the general principles of such calculations should be further developed on the basis of their experimental validations to obtain quantitative information on the composition and structure of cemented carbides, particularly, of the WC/WC and WC/Co grain boundaries.

14.2 Modeling mechanical properties of and degradation processes in cemented carbides

Modeling different properties of cemented carbides is usually based on the finite element method (FEM). The FEM is a numerical method for modeling different properties, including mechanical properties, which comprises finding a solution of differential equations or functional minimization. A domain of interest is represented by a combination of finite elements (FEs). Approximation functions in the FEs are determined in terms of nodal values of a physical field which must be modeled. A physical task is transferred into a discretized FE task with unknown nodal values. For a linear task, a system of linear algebraic equations must be solved. Values inside the FEs can be expressed in terms of nodal values [9].

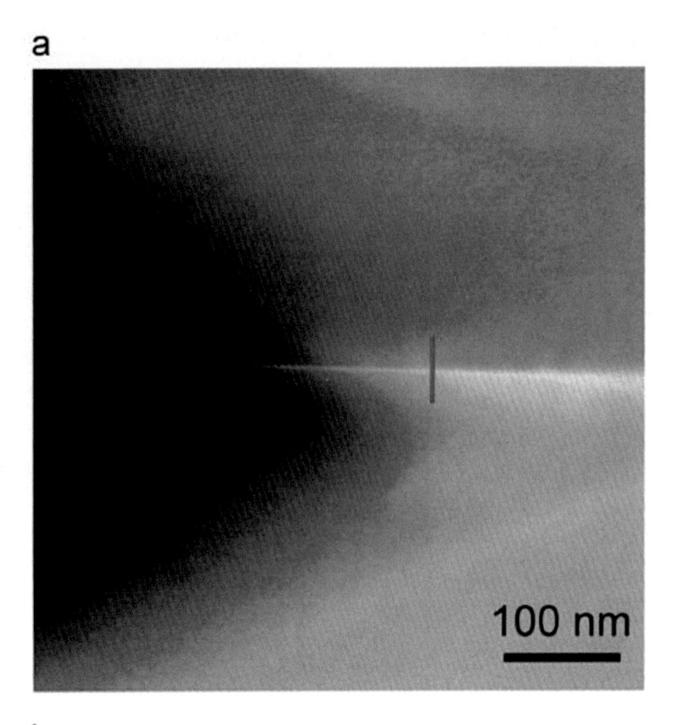

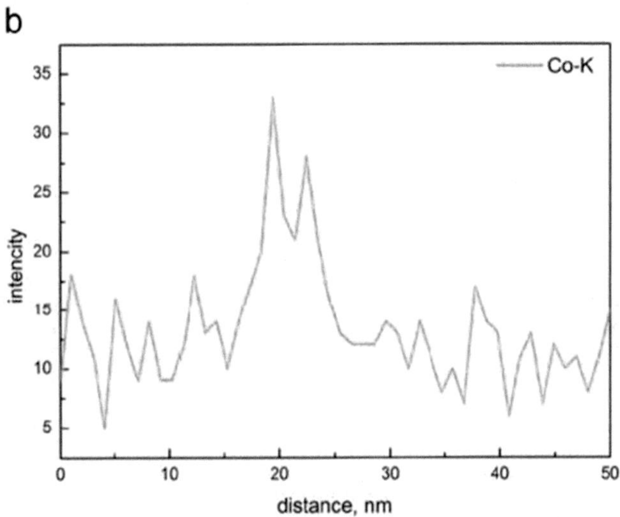

Figure 14.4 (A) STEM HAADF image of the WC—Co cemented carbide with a medium—low carbon content showing the contract area between two WC grains (bright) and the Co-based binder (dark, removed as a result of preparation of the TEM sample); the WC/WC grain boundary comprises a continuous Co-based film (*white*) of nearly 3 nm in thickness, and (B) curve obtained by EDX indicating the Co distribution across the WC/WC grain boundary in the area shown by the *red line* (gray line in print) in (A). *(Redrawn from Ref. [8].)*

Mechanical properties of cemented carbides are determined by their microstructural characteristics, composition, and topology. Microstructure engineering can be used as an approach to the optimization of the cemented carbides' performance in different applications.

Several features of cemented carbides measured on the microlevel determine their macroscopic material properties such as strength, fracture toughness, etc. These features are, e.g., second-order residual stresses, strains in the metallic binder, and carbide grains' contiguity, which are dependent on the binder content [10]. Numerous research projects dealing with cemented carbide structural modeling were performed over the last decades.

FE models established for different applications can be employed for defining the macroscopic mechanical properties of cemented carbides, such as the elastic modulus, yield point, and plastic behavior. Some FE models are elaborated to simulate microstructural features of cemented carbides. Simulations can be considered as a powerful tool allowing time saving when developing new carbide grades and tools in combination with the simultaneous validation. Prolongation of lifetime of cemented carbide components needs deep understanding of their failure mechanisms. Advanced computer systems and modeling software are the key enablers for simulations of the cemented carbides' microstructure, properties, and performance. A basic workflow for modeling the microstructure and degradation modes of cemented carbides is shown in Fig. 14.5.

There are a number of published works of cemented carbides' structural imaging by use of thin lamellae obtained by focused ion beam (FIB), which are examined in high-resolution scanning electron microscopes (HRSEMs) known in the literature (see, e.g., Ref. [11]). A typical FIB lamella indicating a 2D plane view of the WC−6 wt.%Co cemented carbide is shown in Fig. 14.6. The EBSD grain size analysis of such lamellae by their layer-by-layer cutting with the aid of FIB allows obtaining and analyzing a 3D structure of cemented carbides, which is shown in Fig. 14.7, ensuring their microstructural modeling. Simulations of interfaces and grain boundaries for the characterization and stochastic model generation combine the evaluation of WC/Co grain boundaries and interfaces [11].

Abrasion is thought to be one of the major degradation mechanisms of cemented carbides (beside erosion, corrosion, thermal fatigue, etc.). However, despite recent published works, the dependence of abrasive wear on microstructural features of cemented carbides is not completely understood. In this respect, in situ scratch testing and abrasion simulation of

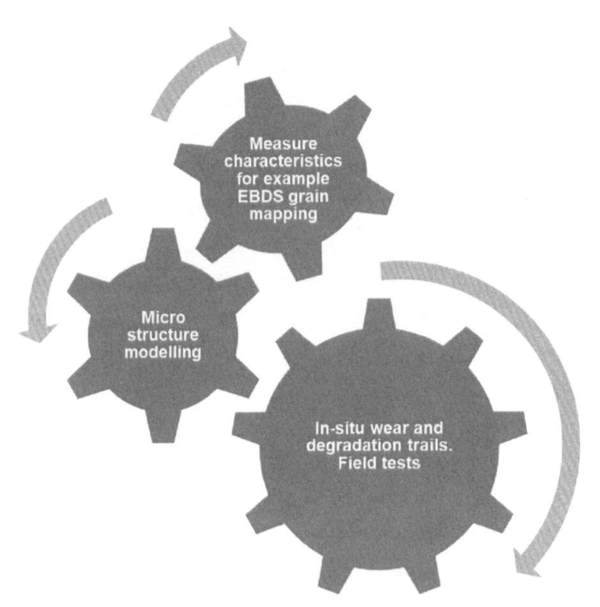

Figure 14.5 Basic workflow for successful modeling of the microstructure and degradation modes.

Figure 14.6 Typical cemented carbide FIB lamella (subjected to Ar-ion thinning) indicating a 2D plane view of the WC–6 wt.%Co cemented carbide. *(Redrawn from Ref. [12].)*

cemented carbides in a SEM provide a powerful tool to understanding the mechanisms of abrasive wear [14,15]. The scratch testing studies were performed by the use of a spherical diamond intender and the results obtained provide a very interesting view on the surface degradation. Through the employment of miniature apparatus, real-time observations of

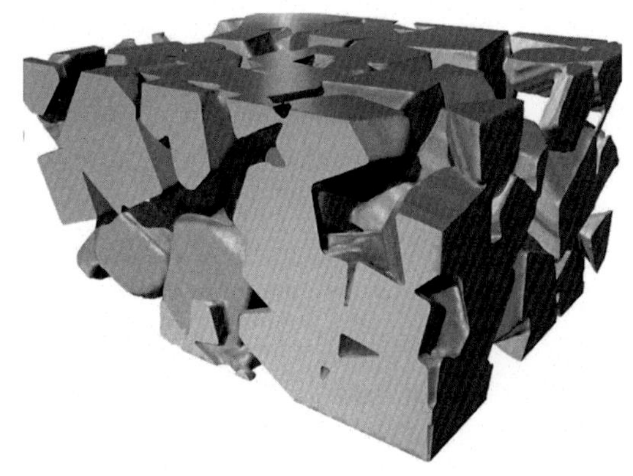

Figure 14.7 3D reconstruction of the cemented carbide microstructure. *(Redrawn from Ref. [13].)*

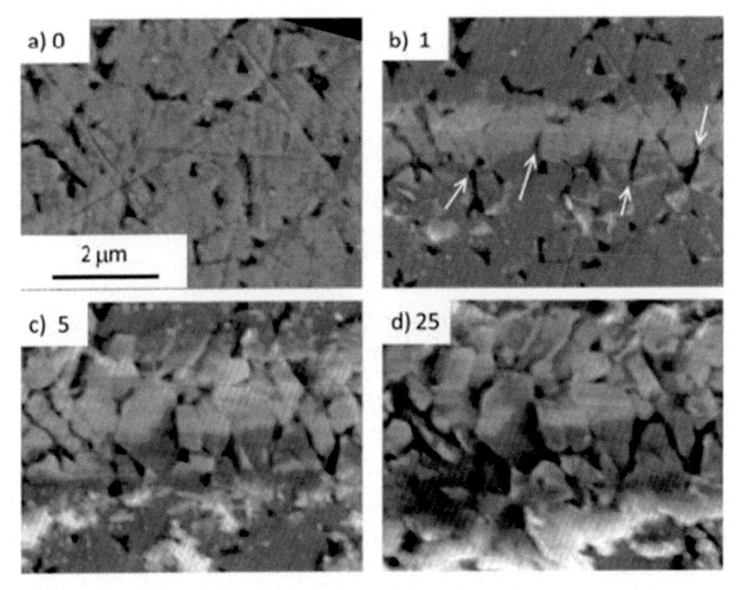

Figure 14.8 Typical groves on the cemented carbide surface obtained as a result of scratching the carbide surface carried out under an applied load of 50 mN. Numbers refer to complete cycles (out and back). *(Redrawn from Ref. [14].)*

degradation effects like surface or subsurface WC grain microcracking, pull out, debris generation and reembedment, grain and binder plasticity, and surface oxidation open a new perspective with respect to understanding phenomena of cemented carbides' abrasion. Pictures of typical grooves obtained as a result of the *in situ* scratch tests are shown in Fig. 14.8 [14,15].

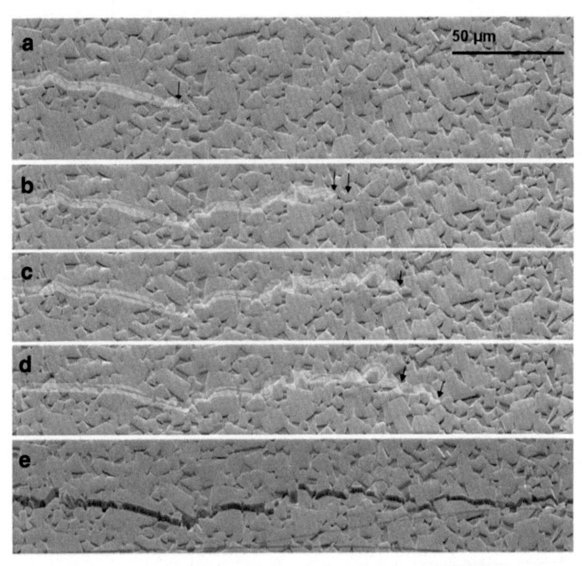

Figure 14.9 Subcritical crack growth shown in secondary electron images taken at (A) Stable position prior to application of full load (high resolution image); (B) first growth step after applying load to the *left arrow*; further growth to the *second arrow* observed after further 3.2 s; (C) second growth step after further loading for 7.2 s; (D) "Final" position of the crack after 300 s of loading when no further crack growth was observed; and (E) crack after its leaving overnight for the EBSD examination. The crack path is highlighted for clarity. *(Redrawn from Ref. [8].)*

Another interesting approach to examining the controlled crack growth obtained by an ultrastiff test rig and observed *in situ* in a SEM allows one to examine the crack initiation and propagation in cemented carbides under the impact of static loads. Double beam cantilever fracture specimens were places between an anvil and wedges made of alumina ceramics, which are subjected to a load of 4.5 kN. Cracks were observed to propagate in jumps of between 20 and 70 μm, with each jump occurring in less than 200 ms (Fig. 14.9). High-resolution SEM and EBSD techniques are used to characterize the microstructure, orientation of the WC grains on the crack path, and degree of plastic deformation in both the binder phase and carbide phase [16].

HRSEM in situ observations ensure understanding the crack propagation paths through different consistent parts of cemented carbides. Consequently, advanced modeling approaches allow one to predict the crack propagation mode and crack paths through cemented carbide samples with a very good agreement to the real experimental results [17].

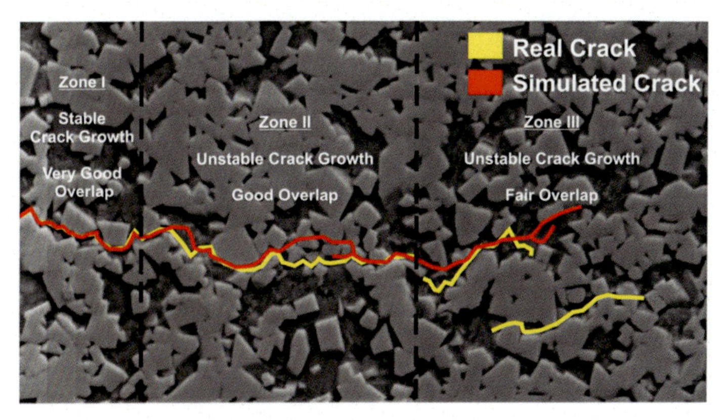

Figure 14.10 Comparison between the real crack paths (*yellow curves*) and simulated crack pattern (*red curve*). *(Redrawn from Ref. [17].)*

One needs precise information on the material properties to model and predict the crack propagation mode, which is also necessary for the follow-up model verification. Fig. 14.10 illustrates the real experimental fatigue crack path in comparison to the simulated fatigue crack path. In this respect, the FEM simulations based on the approach of continuum mechanical damage were implemented in practice. Separate simulations of damage and brittle failures under conditions of severe fatigue are usually employed for the WC phase and Co phase.

The crack patterns observed can be characterized in terms of the microstructure, WC grains' shape, and orientation. The binder phase strain accumulation leads to transgranular crack propagation paths, whereas the plastic deformation ahead of the crack tip is clearly seen in the regions of the binder phase [17].

In contrast to the FE modeling, empirical approaches are usually more expensive and time consuming. Nevertheless, besides all the up-to-date modeling options, there is a need to prove the models by their experimental validation.

14.3 Modeling the stress distribution in cemented carbide articles

As it is mentioned above, the Finite Element Analysis (FEA) and FEM are numerical methods for simulations of different material characteristics, which can be used for modeling various properties of carbide articles. They are useful for modeling articles of complicated geometries, load distributions, or inhomogeneous properties, when it is difficult to obtain experimental results. The FEM is originated from the need to solve complex elasticity and structural analysis tasks in civil and aeronautical engineering. With the development of computer technologies, FEA and FEM began to be employed in numerous applications, including machine building, aerospace, automobile, ship, electronic industries, etc.

Common FEA applications are as follows:

- Mechanical/Aerospace/civil/automotive engineering
- Structural stress analysis
  - Static/Dynamic
  - Linear/nonlinear
- Fluid flow
- Heat transfer
- Electric magnetic fields
- Soil mechanics
- Acoustics
- Biomechanics

The most important advantages of FEA are the reduction of modeling times and costs, optimization of production costs, lowering materials costs, early indication of weaknesses, quality improvements, design optimization, etc. [18].

Transferring the performance of real tools into a simplified, idealized physical model and mathematical model employs discretized element types shown in Fig. 14.11.

The principal stresses and the so-called "von Mises stresses" are among the most commonly used in the field of mechanical behavior simulations and can be generally described as follows.

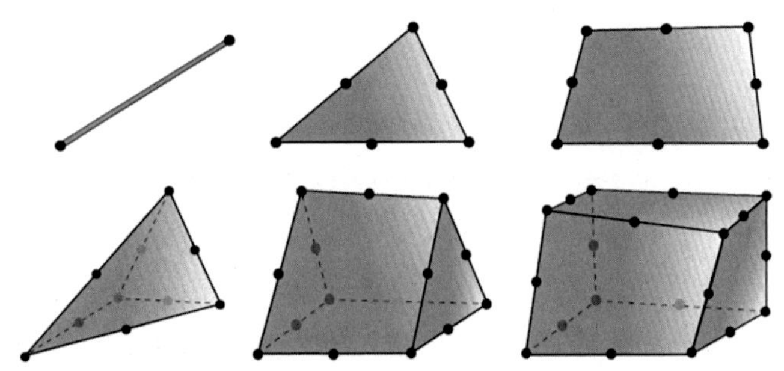

Figure 14.11 Common element types used for the finite element analysis. *(With the permission of Element Six.)*

14.3.1 Principal stresses

At every point of a stressed body, there are at least three planes called principal planes, with normal vectors, principal directions, where the corresponding stress vector is perpendicular to the plane, i.e., parallel to the normal vector, and where there are no normal shear stresses. The three stresses normal to these principal planes are called principal stresses.

14.3.2 von Mises stresses

The von Mises stress (von Mises criterion) uses the principal stresses to compute an equivalent tensile stress in the material for a simple comparison with the tension allowable for the material. Physically, it predicts when the stress reaches a critical value and there is enough strain energy in the material to obtain its yield. This value is designated as the "maximum distortion strain energy criterion."

Using the von Mises stress is typically employed to establish the pass/fail check.

The following examples taken from the carbide manufacture illustrate the potential of the FEA.

14.3.3 Modeling of compaction

Designing, manufacturing, and commissioning a new carbide article often determine its delivery time. To speed up this process, computer simulations of the die compaction and sintering are frequently employed. Nowadays, FE simulations of the compaction process in combination with the appropriate material (carbide grade) allow quantitative predictions of the density distribution in the green body. The shape of the carbide article after sintering can be predicted by the use of appropriate simulations of the material behavior during sintering. The sinter distortions caused by inhomogeneous density distributions in the green body can be minimized by optimizing the pressing kinematics and compensated by appropriately shaped punch surfaces.

One of the most important production routes employed for fabricating cemented carbides and other materials (metals or ceramics) is uniaxial die compaction followed by sintering. Except for the very simple geometries, it is not possible to achieve a homogeneous green density distribution by the die compaction. The green density is relatively inhomogeneous depending on the part geometry, tool design, and friction between the powder and die walls. As a result, the part undergoes shape distortions or even cracking during sintering. Sinter distortions of carbide articles can also be caused by gravity effects or friction in the region of their contact with the supporting sinter trays during sintering.

The commissioning of a new pressing tool needed to fabricate a new carbide article is a very time-consuming process incurring high costs of the press downtime. During compaction, the tooling must endure high stresses and intensive wear. Thus, tool fracture occurring during production is another important factor; the fracture probability is related to the complexity of the tooling and must be avoided. To reduce the development time and costs when producing a new carbide article, the whole production process can be investigated and optimized by computer simulations. FE simulations of the compaction and sinter processes in combination with the appropriate material optimization allow quantitative predictions of tool loadings, green density distributions, and sinter distortions for articles having a complex geometry (Fig. 14.12). The different colored areas in Fig. 14.12 indicate the green part density inhomogeneity leading to distortions during consequent sintering, so that the visible distortion is highest in the red-colored region. The prediction of crack formation and its avoidance is even a more complicated task. This example should give an idea how to avoid the undesired sintering distortions, which develop as a result of inhomogeneous density distributions in the green article

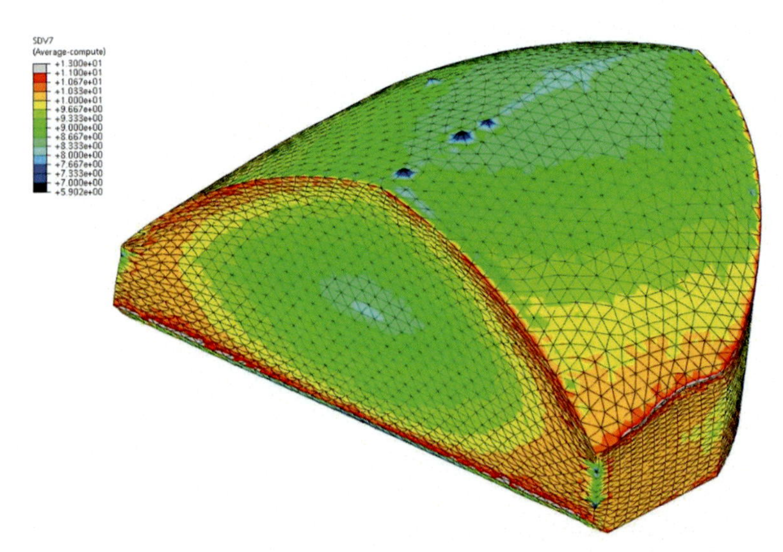

Figure 14.12 Distribution of the green part density in a typical carbide insert causing distortions during its consequent sintering. *(With the permission of Element Six.)*

due to compensating them with appropriately shaped punches. These shapes can be obtained by numerical simulations of the compaction and sintering processes. The method is usually applied to producing indexable cutting inserts; as a result, the tool optimization can be performed [19].

14.3.4 Furnace gas flow model

The heat treatment of green carbide parts is one of the major process steps in the cemented carbide industry. The FEA is an ideal method to achieve the optimization of furnace design and sintering regimes. The attempt to simulate the gas and heat distribution in a furnace is quite complicated; nevertheless, as it is shown in Fig. 14.13, the plan view of the gas flow around the parts on a sinter tray during sintering in a controlled gas atmosphere can be obtained by the FEA simulations. The pink and blue colored regions around the parts show the lowest gas flow velocity. When varying the gap between the trays, the following results are obtained: at the gap of 2.5 mm, most of the gas flows around the sintered parts, while increasing the gap up to 10 mm or more leads to the fact that the most of the gas flows over the parts. The results obtained provide evidence that the gas convection is mainly affected by the spacing between the trays and only slightly affected by the dimensions of the furnace itself.

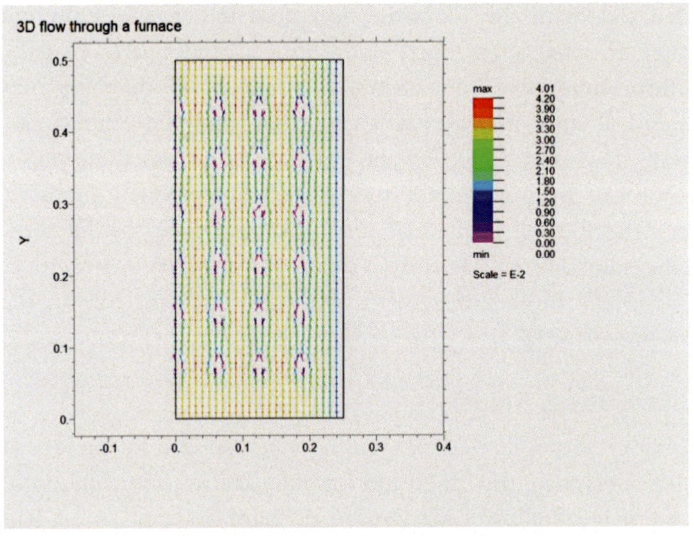

Figure 14.13 Plan view of the gas flow around carbide parts on a sinter tray during sintering (heat treatment) in a controlled gas atmosphere. *(With the permission of Element Six.)*

14.3.5 Stress distribution model

The difference between thermal expansion coefficients of WC grains and cobalt binder matrix is one of the main factors affecting the formation of internal stresses in cemented carbide articles [20]. Different phenomena, such as the formation of thermal gradients during heating or cooling processes of carbide articles cause macrostresses in the articles. Assembling carbide and steel articles (brazing, press fitting, shrink fitting, or clamping) can significantly affect the stress state in the carbide articles (see Chapter 8).

In relation to the WC grain size and metal binder content, either tensile or compressive stresses affect the WC phase and binder matrix. Multiple approaches to modeling the internal micro- and macrostresses are employed to establish the optimum pressing and sintering processes. Obtaining models as close as possible to the real application conditions requires fundamental knowledge about the material-related physical, mechanical, and metallurgical properties. Numerous trails are needed to simulate binder systems not containing WC grains, which is extremely challenging, as it is difficult to achieve the state of Co-based alloys with a certain carbon and tungsten content. It is hardly possible to make defect-free samples of the binder phase for measuring its thermal, physical, and mechanical properties. Also, different material properties like compressive strength, elastic modulus, creep resistance, Poisson's ratio, thermal expansion coefficient, thermal capacity, etc., are needed to obtain the FE simulation model. Fig. 14.14 illustrates the macroscopic thermal stress distribution in a carbide part created during cooling from sintering temperatures. It is predicted that the maximum thermal residual stress increases when moving from the outer zone to the core region. This is indicated by the color change from the green to blue region. One can assume that the model does not precisely predict the internal stress distribution in the cemented carbide articles, as different effects such as the solubility of W and C in the binder phase, structural transformations in the binder phase, etc., play an important role in the stress formation process [21].

14.3.6 Braze stress model

The brazing process followed by rapid cooling can lead to the formation of high macrostresses in the steel and carbide articles. Brazing with high-melting point braze alloys is designated as "hard brazing" (see Chapter 8). Differences in thermal expansion coefficients between the tough steel

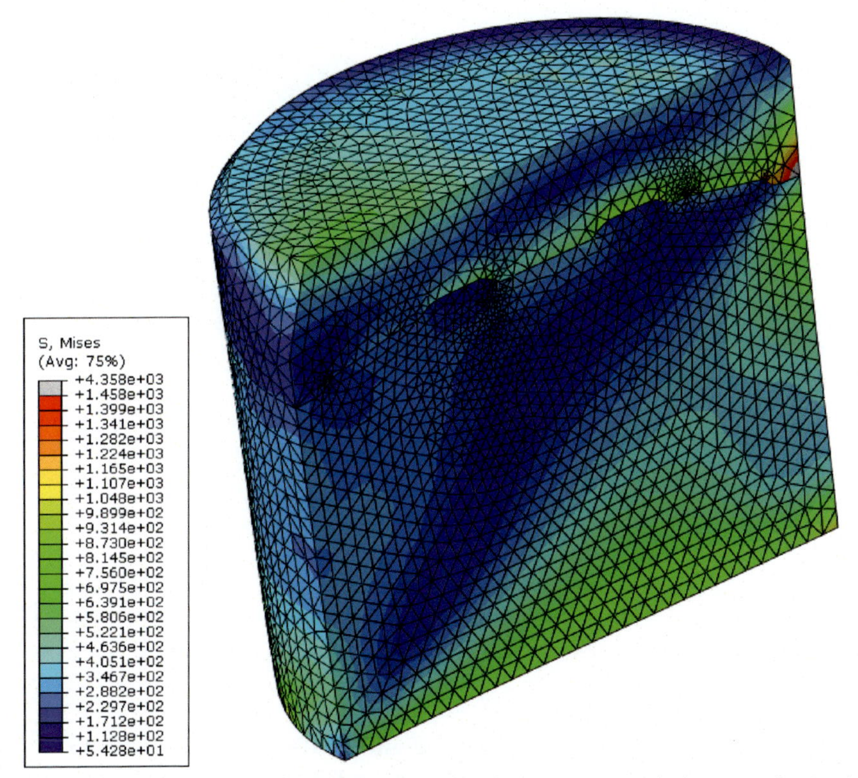

Figure 14.14 Macroscopic thermal stress distribution in a carbide part obtained during cooling from sintering temperatures [17]. *(With the permission of Element Six.)*

carrier, ductile braze alloy, and hard cemented carbide parts significantly affect the macrostress state within the whole steel/carbide composite article. Design features like sharp edges, tapers, low-thickness walls, and large brazing areas combined with small brazing gaps are additional stress raisers. Advanced FE modeling of stress states within the carbide/braze/steel composite produced by high-temperature brazing dramatically reduces a number of premature failures of such composites during operation. The stress distribution simulated in a composite carbide/braze/steel article, which is shown in Fig. 14.15, ensures fast and optimal design of the composite article. The red and green regions in Fig. 14.15 indicate high tensile stress zones within the cemented carbide tip introduced in the composite article during its cooling from the brazing temperature down to room temperature.

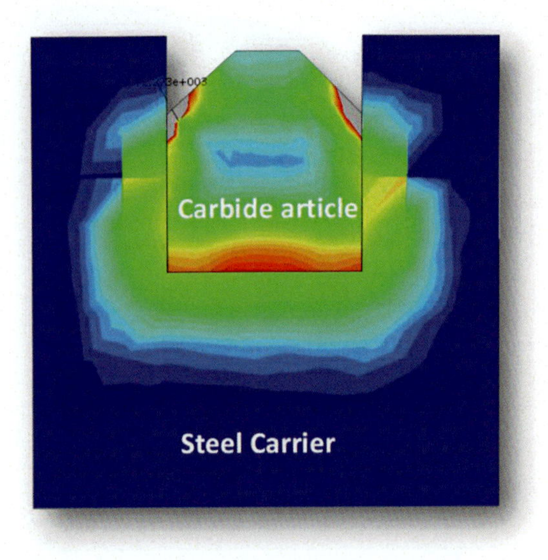

Figure 14.15 Macroscopic stress distribution in a composite carbide/braze/steel article after hard brazing at about 950°C followed by cooling down to room temperature. *(With the permission of Element Six.)*

14.3.7 Impact model

Cutting picks are tools used in mining and road-planing applications (see Chapter 12). The cutting performance of road cutting machines, continues miners, shearers, etc., are directly affected by performance and tool lifetime of such carbide tools. In the past years, a number of researchers studied theoretical aspects of the coal, rock, and asphalt cutting process [21]. Premature failures of carbide inserts and tips occurring during operation must be minimized, as this is a costly and timely factor leading to significant machines' down times. It is therefore essential to understand the influence of loads subjected to the tools during operation on the resulting stress distribution in the tools operating under dynamic application conditions. The dynamic simulations of the stress distribution can support and speed up the design optimization of reliable cutting tools. To reduce the development time and costs compared to the conventional method of "trial and error," the whole process can be studied and optimized by computer simulations. FE simulations of the impact loads in combination with choosing the appropriate material properties allow quantitative predictions of possible tool failures, which is illustrated in Fig. 14.16. The prediction of crack formation and its avoidance is a complicated task. The example of simulations of the stress distribution in a carbide tip of road-planing picks shown in Fig. 14.16, can give an idea of the von Mises stress distribution in

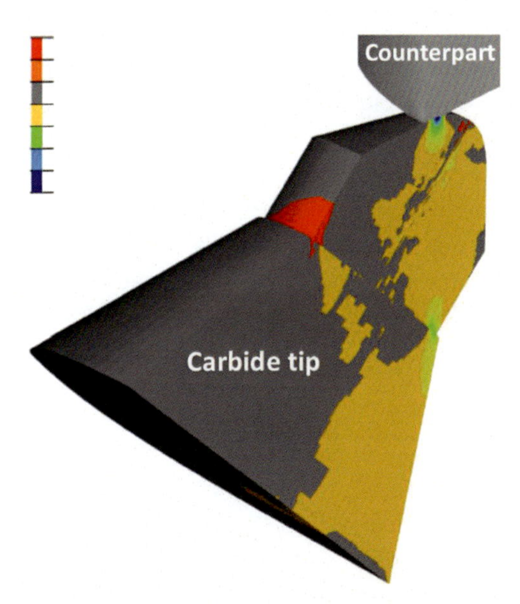

Figure 14.16 Stress distribution in a cemented carbide/braze/steel composite (carbide tip) after a single dynamic impact by the counterpart. *(With the permission of Element Six.)*

the tool operating under high impact loads. The blue and green areas at the zone subjected to high impact loads indicate compressive stresses, whereas the red area indicates the presence of tensile stresses (Fig. 14.16). An optimized stress management in relation to the cemented carbide tip shape and brazing interface can be simulated and optimized. As expected, greater stresses are generated at higher impact loads. The stress distribution was obtained by the numerical simulation of a dynamic single impact.

Modeling cyclic dynamic impacts ensures establishing defects forming due to any single impact. The elastic properties of cemented carbides are known, nevertheless, no reliable data on their plastic deformation are available; it is not possible to take into account the preexisting defects when modeling the cyclic impact processes. Consequently, the results of single impact modeling are much more realistic in comparison to those obtained on the basis of the cyclic impact models.

14.3.8 Stress distribution in high-pressure high-temperature components

High-pressure high-temperature components for diamond and cubic boron nitride synthesis include cubic anvil systems consisting of six symmetrical cemented carbide anvils (see Chapter 12). The cemented carbide anvils are press fitted into steel fixtures connected to six hydraulic pistons applying high pressures. Cubic anvils failures are characterized by similar breakage patterns. Because of high stresses within the anvil at high temperatures and high loads, a plastic zone forms underneath the nose of the anvil. At the interface between the plastic and elastic zone, there are high tensile stresses, which may lead to the crack formation. Cracks in the cubic anvils are usually initiated at the surface of one of the anvils' faces. Once cracks are visible at the surface of the anvil, the risk of failing the whole anvil in the following run is quite high. FE modeling allows the tool and process optimization and helps understand the stress distribution within the anvils. As it is shown in Fig. 14.17, the von Mises stress distribution in the double bevel area is

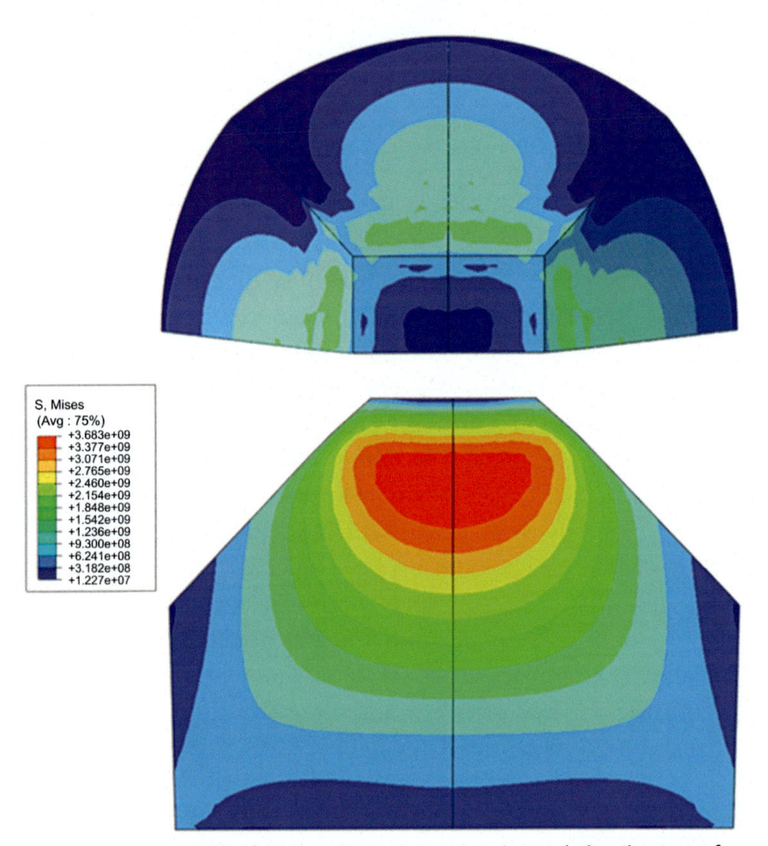

Figure 14.17 Contour plot of the von Mises stresses (spatial distribution of stresses within the anvil operating at high temperatures and ultrahigh pressures): above – top view and below – cross-section in the middle of the anvil. *(With the permission of Element Six.)*

shown in the form of a contour plot. As it can be seen in Fig. 14.17, the highest values of the von Mises stress (red and yellow areas) are located near the anvil tip. In the maximum stress regions, the calculated values of the von Mises stress are lower than the calculated failure stress values, so that it is unlikely that the anvil would fail during operation according to the von Mises stress criterion [22].

References

[1] S.A.E. Johansson, G. Wahnström, A computational study of thin cubic carbide films in WC/Co interfaces, Acta Mater. 59 (2011) 171–181.
[2] A.E. Johansson, Wahnström Theory of ultrathin films at metal-ceramic interfaces, Phil. Mag. Lett. 8 (1010) (1990) 599–609.
[3] S.A.E. Johansson, G. Wahnström, First-principles study of an interfacial phase diagram in the V-doped WC-Co system, Phys. Rev. B 86 (2012) 035403.
[4] S.A.E. Johansson, G. Wahnström, First-principle derived complexion diagrams for phase boundaries in doped cemented carbides, Curr. Opin. Solid State Mater. Sci. 20 (2016) 299–307.
[5] M.A. Gren, E. Fransson, G. Wahnström, A computational study of the temperature dependence of interface and surface energies in WC–Co cemented carbides, Int. J. Refract. Met. Hard Mater. 87 (2020) 105114.
[6] M.A. Gren, G. Wahnström, Wetting of surfaces and grain boundaries in cemented carbides and the effect from local chemistry, Materialia 8 (2019) 100470.
[7] I. Konyashin, A.A. Zaitsev, D. Sidorenko, et al., Wettability of tungsten carbide by liquid binders in WC–Co cemented carbides: is it complete for all carbon contents? Int. J. Refract. Met. Hard Mater. 62 (2017) 134–148.
[8] B.B. Straumal, I. Konyashin, B. Ries, K.I. Kolesnikova, A.A. Mazilkin, A.B. Straumal, A.M. Gusak, B. Baretzky, Pseudopartial wetting of WC/WC grain boundaries in cemented carbides, Mater. Lett. 147 (2015) 105–108.
[9] G.P. Nikishkov, Introduction into the Finite Element Method, Lecture Notes, University of Aizu, Japan, Aizu Wakamatsu, 2004.
[10] M. Magin, J.-S. Gerard, Microstructural modelling of hardmetals using innovative numeric analysis tools, in: Proc. 17th, Plansee Seminar, Reutte, Austria, vol. 3, 2009.
[11] A. Laukkkanen, T. Andersson, T. Pinomaa, et al., Crystal Plastic Modelling of Fatigue and Scale Dependencies in WC-Co Microstructure, VIT Materails & Manufacturing, Finland, 2018. Internet site Introduction to the Finite Element Method (wisc.edu), dated April 2021.
[12] A.A. Zaitsev, I. Konyashin, P. Loginov, et al., Radiation-enhanced high-temperature cobalt diffusion at grain boundaries of nanostructured hardmetal, Mater. Lett. 294 (2021) 129746.
[13] E. Jiménez-Piqué, M. Turon-Vinasa, H. Chen, et al., Focused ion beam tomography of WC-Co cemented carbides, Int. J. Refract. Metals Hard Mater. 67 (2017) 9–17.
[14] M. Gee, K. Mingard, J. Nunn, B. Roebuck, A. Gant, In-situ scratch testing and abrasion simulation of WC/Co, Int. J. Refract. Metals Hard Mater. 62 (2017) 192–201.
[15] A.J. Gant, J.W. Nunn, M.G. Gee, D. Gorman, D.D. Gohil, L.P. Orkney, New perspectives in hardmetal abrasion simulation, Wear 376–377 (2017) 2–14.

[16] K.P. Mingard, H.G. Jones, M.G. Gee, B. Roebuck, J.W. Nunn, In-situ observation of crack growth in a WC-Co hardmetal and characterization of crack growth morphology by EBSD, Int. J. Refract. Metals Hard Mater. 35 (2013) 136–142.
[17] U.A. Özden, K.P. Mingard, M. Zivcec, A. Betzold, C. Broeckmann, Mesoscopical finite element simulation of fatigue crack propagation in WC/Co-hardmetal, Int. J. Refract. Metals Hard Mater. 40 (2015) 261–267.
[18] Technical University Berlin, Structural Simulation (FEM), 2021. www.beuth-hochschule.de.
[19] T. Kraft, Optimizing press tool shapes by numerical simulation of compaction and sintering-application to a hard metal cutting insert, Model. Simulat. Mater. Sci. Eng. 11 (3) (2003) 381–400.
[20] H.K. Park, Three-dimensional microstructure modeling of particulate composites using statistical synthetic structure and its thermo-mechanical finite element analysis, Comput. Mater. Sci. 126 (2017) 265–271.
[21] S.T. Huang, L. Zhou, J. Li, Finite Element Analysis of stress distribution in the diamond picks, Key Eng. Mater. 487 (2011) 184–188.
[22] Q.-G. Han, X.-P. Jia, J.-M. Qin, R. Li, C. Zhang, Z.-C. Li, Y. Tian, H. Ma, FEM study on a double-bevel anvil and its application to synthetic diamond, High Press. Res. 29 (3) (2009) 449–456.

CHAPTER 15

Major trends in the research and development of cemented carbides

15.1 Grain boundary design

It is well known that the state and composition of WC/Co and WC/WC grain boundaries in cemented carbides play an important role with respect to obtaining a high level or mechanical and performance properties, as cracks usually propagate through such grain boundaries when failing carbide tools and articles. Once the characteristic size of materials reaches nano- or atomic scale, the mechanical properties may change drastically, and classical mechanisms of materials' failure may completely change. Nevertheless, the presence or absence of thin interlayers of several atomic monolayers, or so-called "complexions," at WC/Co interfaces depending on the carbide composition, carbon content, sintering and cooling conditions, etc., are presently not understood [1–3].

The high-resolution transmission electron microscopic (HRTEM) image of the cemented carbide bulk specimen subjected to quenching after sintering is shown in Fig. 15.1 [2]. The complexion with partially ordered structure can be seen in Fig. 15.1 at the boundary of WC and Co, which is clearly different from the structures of the both phases. By indexing the Fast Fourier Transformation (FFT) patterns and combining them with the HRTEM analysis, the crystallographic features of the complexion were obtained, as indicated in Fig. 15.1B–D. The complexion comprises two to four atomic layers, as shown in Fig. 15.1C and D, which mainly distribute at $WC_{(10\text{-}10)}$/Co and $WC_{(0001)}$/Co interfaces with coherent relationship with the WC grains. The complexion has a cubic crystal structure with the lattice parameter of 4.226Å. When taking into account results of Ref. [3], the complexion formed in the quenched cemented carbide sample is believed to consist fcc WC_x. It was reported that the phase of WC_x with the fcc crystal structure and atomic ratio of W:C lying in a range of between 1.7:1 and 1:1 normally forms and exists at temperatures above 2800°K [4].

Cemented Carbides
ISBN 978-0-12-822820-3
https://doi.org/10.1016/B978-0-12-822820-3.00014-8

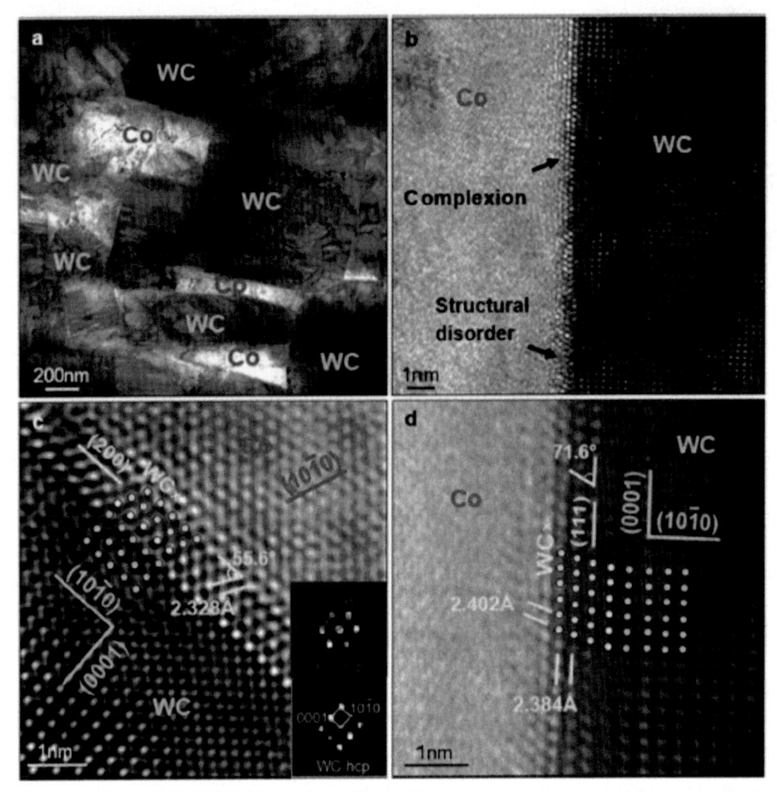

Figure 15.1 TEM and HRTEM images of the WC–Co cemented carbide quenched after sintering at 1390°C: (A) TEM image indicating the distribution of WC and Co phases; (B) complexion at WC/Co interface; (C) fcc WC_x complexion on $WC_{(10\text{-}10)}$ plane; and (D) fcc WC_x complexion on $WC_{(0001)}$ plane. *(Redrawn from Ref. [2].)*

Due to a significant difference in the solubility and diffusion rates of W and C in the liquid phase during sintering, the inhomogeneous distribution of composition occurs at the surface of WC grains, which may lead to the formation of the fcc WC_x phase at WC/Co interfaces in the process of liquid-phase sintering. Thus, the fcc WC_x formed at high temperatures can remain at WC/Co grain boundaries as a complexion after cooling down to room temperature as a result of quenching.

As it was shown in Ref. [5], a complexion of nearly two atomic monolayers shown in Fig. 15.2 can form at WC/Co grain boundaries of a cemented carbide sample with a low carbon content after conventional cooling from a sintering temperature down to room temperature (without quenching). This chemically ordered layer, complexion, is found only on straight facets of a WC grain. No chemically ordered layers were found on

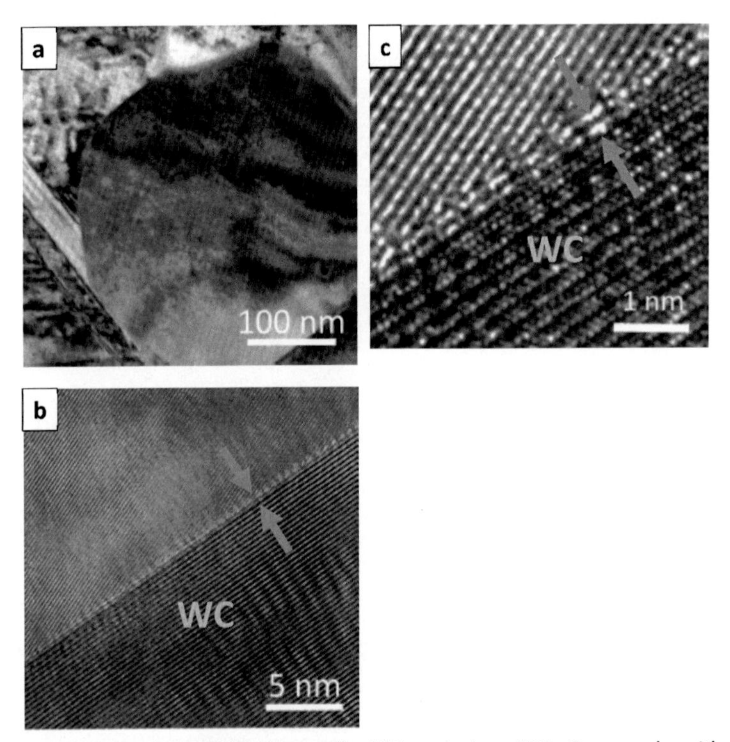

Figure 15.2 TEM and HRTEM images of a WC grain in a WC–Co sample with a low carbon content: (A) - bright field TEM micrograph of the rounded WC grain embedded in the Co-based matrix; (B) and (C) - HRTEM images of the interface between the WC grain shown in (A) and the Co-based matrix at different magnification (the complexion at the WC/Co interface is indicated by *arrows*). *(Redrawn from Ref. [5].)*

curved segments of the same WC grain. The complexion appears to form only at certain facets of the interface, where the crystallographic orientation between the WC and Co crystal lattices results in high compatibility stresses across the interface. The complexion in form of an intermediate phase is thought to be present at such interfaces to accommodate the long-range stresses and to facilitate the lattice matching.

In contrast to the results of ref. [5], no complexion is seen in Fig. 15.3 at WC/Co interfaces of cemented carbides with medium and high carbon contents after slow cooling from a sintering temperature down to room temperature.

Thus, the possibility of understanding the reasons for and mechanisms of the formation of complexions at WC/Co interfaces and obtaining such

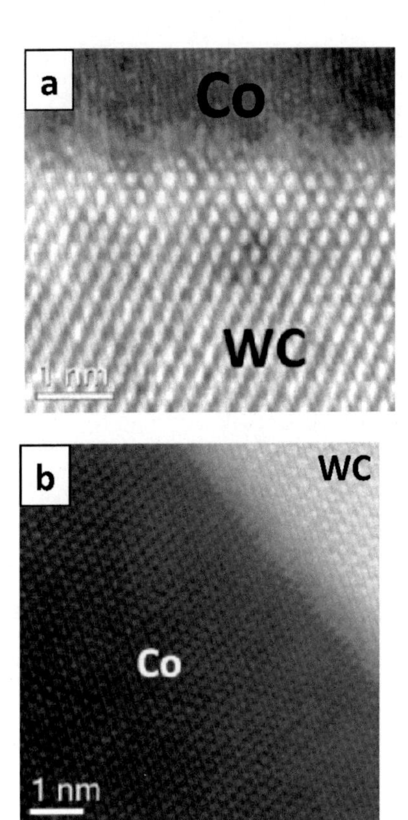

Figure 15.3 HRTEM images of WC/Co interfaces in cemented carbides with (A) medium carbon content and (B) high carbon content after slow cooling from a sintering temperature down to room temperature. *(With the permission of Element Six.)*

complexions with tailored structure and properties can potentially allow the grain boundary design of cemented carbides, which is thought to result in their significantly improved mechanical and performance properties.

15.2 *In situ* examinations of deformation and thermal processes in cemented carbides in transmission electron microscopes

As it is pointed out in Chapter 5, a relatively high rate of plastic deformation before failure of WC under loads is one the major characteristics allowing its employment in numerous applications characterized by high impact loads, severe fatigue, and intensive abrasive wear. The plastic deformation of WC grains can be enhanced due to dislocation gliding processes. The potential slip systems, involved in these processes, include the following prismatic systems: {10−10} ⟨−12−13⟩, {10−10} ⟨0001⟩, and {10−10} ⟨11−20⟩, and pyramidal systems {0−111} ⟨11−23⟩, {0−111} ⟨0−110⟩, and {11−22} ⟨11−23⟩ according to Ref. [6].

The direct *in situ* observation of the formation, movement, and annihilation of dislocations, stacking faults, and other crystal lattice defects in both the WC and binder phase in carbide samples subjected to loading in transmission electron microscopes (TEMs) has become now a powerful tool for understanding the nature of deformation processes in cemented carbides [7,8].

The process of crystallographic defects' formation and movement in a focused ion beam (FIB) lamella of WC−6%Co cemented carbide under bending loads applied by a picoindenter was examined *in situ* in a TEM (Fig. 15.4) and reported in Ref. [7]. The appearance of a small area of the FIB lamella shown in Fig. 15.4B at different stages of applying the loads is shown in Fig. 15.5. Processes of formation, movement, and annihilation of dislocations in WC can be clearly seen in Fig. 15.5. The deformation is found to result in the formation of different crystal lattice defects in the Co-based binder and WC grains. Mainly dislocations form in the tungsten carbide grains. The dislocations start moving when increasing the applied bending load. The deformation of the binder phase is found to result in the formation of mainly twins and stacking faults in the Co crystal lattice. As a result of the formation, interaction, and movement of the crystal lattice defects in the WC phase and binder phase, a significant rate of plastic deformation of the cemented carbide lamella was achieved under bending loads without its breakage.

The major objective of ref. [8] was to examine deformation processes in WC−6%Co cemented carbide in the form of a special thinned carbide sample as a result of applying tensile loads *in situ* directly in a TEM with the aid of the push-to-pull method described in Refs. [9−11]. Applying tensile loads to the sample is found to result in the plastic deformation of the Co-based binder phase leading to the formation of different crystal lattice defects in the binder. Force-time and displacement-time curves recorded when loading the samples and maintaining the loads provide evidence for continuous processes of the formation and movement of crystal lattice defects, mainly dislocations, in the WC phase and Co-based binder leading

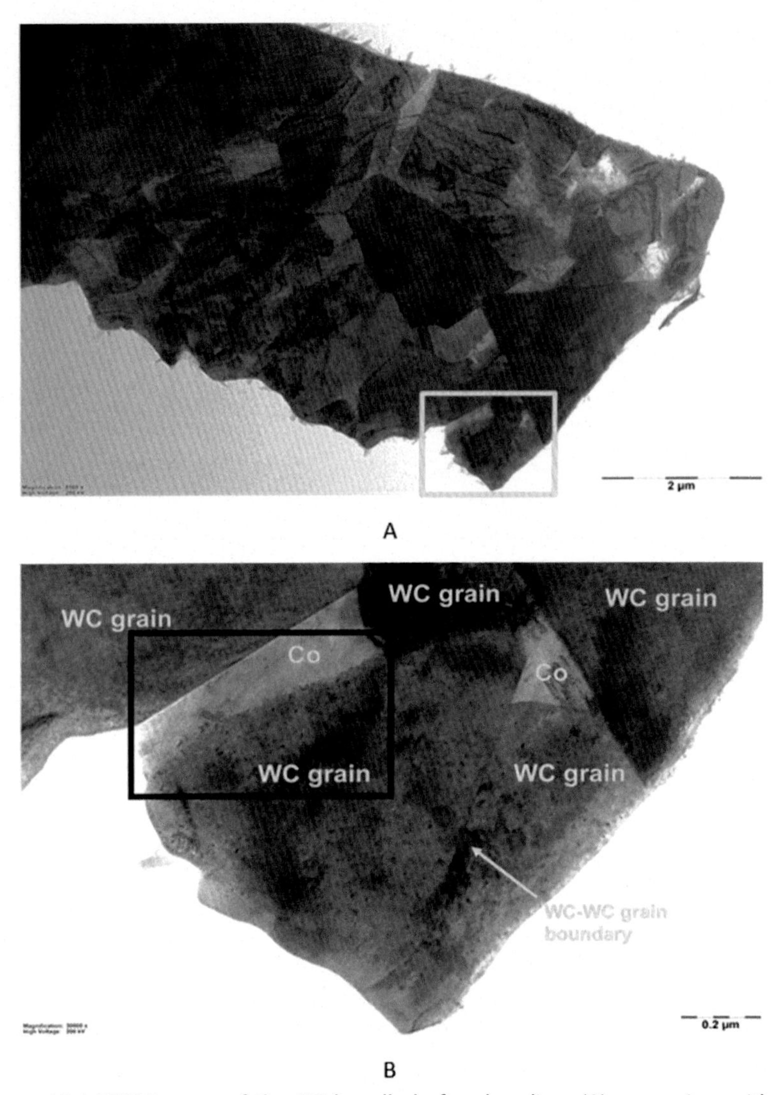

Figure 15.4 TEM images of the FIB lamella before bending: (A) - overview with the marked region examined at a higher magnification, and (B) - the region marked in (A) at the higher magnification with different WC grains and grain boundaries; the area examined in the bending experiment is marked by a *rectangle*. *(Redrawn from Ref. [7].)*

to a high rate of the binder plastic deformation. After increasing the tensile loads up to a certain level leading to the severe plastic deformation of the binder phase, the sample suddenly failed as a result of the crack initiation and propagation at WC—Co interfaces. Surprisingly, some amount of

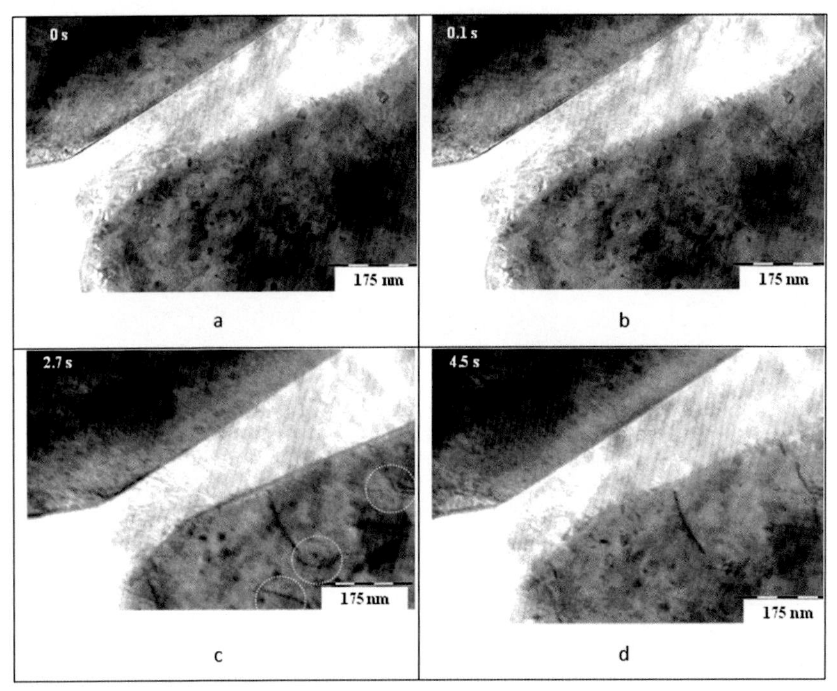

Figure 15.5 TEM images of the FIB lamella before the start of the bending test (A) and when increasing the bending load/displacement (B—D). Dislocations, stacking faults, and other crystallographic defects are indicated by *dashed circles. (Redrawn from Ref. [7].)*

cobalt was found to be present on the WC surface after breaking the sample at the WC—Co interface proving evidence that, in fact, the crack propagation path lies not directly at the WC—Co interface but in the binder region adjacent to the interface (Fig. 15.6). After performing the push-to-pull test on tensile loading, the HRTEM examination of the Co region near the crack was performed (Fig. 15.6D) indicating a number of stacking faults and other crystal lattice defects (dislocations), which presumably formed directly before failure.

By use of the *in situ* technique, it is also possible to examine diffusional processes and processes of the phase and crystallographic defects transformation in cemented carbides at elevated temperatures directly in a TEM [12], thus providing a powerful tool for *in situ* investigating such processes in TEMs.

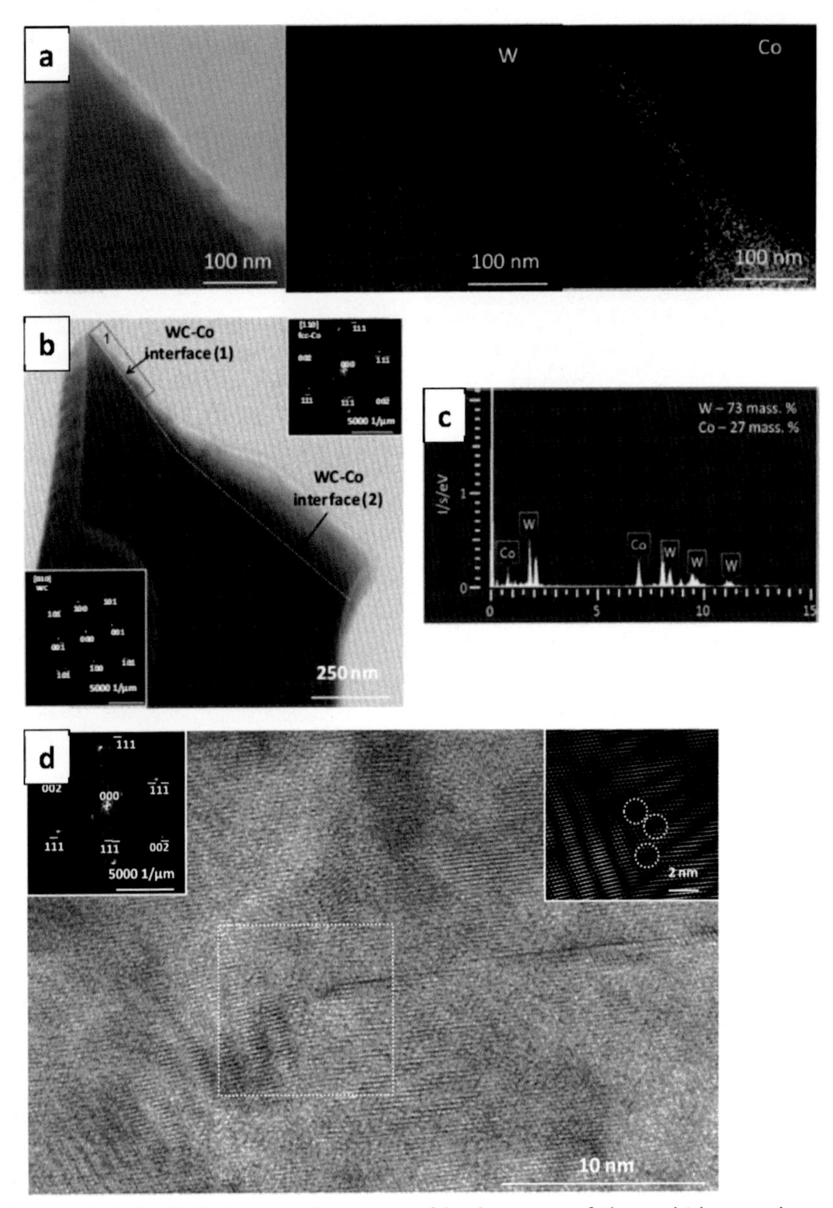

Figure 15.6 (A) STEM image of the rest of broken part of the carbide sample and corresponding EDX maps of its near-surface region at the WC—Co interface (tungsten is marked by *red color* (dark gray in print) and cobalt is marked by *green color* (light gray in print)), (B) STEM image of the broken part A at a low magnification indicating the regions of WC—Co and Co—Co interfaces with two insets showing electron diffraction patterns from WC and the Co binder, and (C) EDX spectrum taken from the near-surface area of the WC grain at the broken WC—Co interface marked by a *red frame* (dark gray in print) (B); (D) HRTEM image of an area of the Co pool in the broken part and two insets showing an electron diffraction pattern (FFT) from this area (left) and a filtered FFT image (right) taken from the region marked by a *dashed square. (Redrawn from Ref. [8].)*

15.3 Micromechanical testing of cemented carbides

In recent times, micromechanical testing of cemented carbides has become very popular for the examination of different features of WC–Co materials. Recent results on micromechanical testing of WC–Co cemented carbides and similar hard materials are summarized in the review paper [13].

The study of mechanical response of materials at small length scales has gained importance due to the recent advances in micro- and nano-fabrication as well as testing systems. A number of recent publications report results of examinations of cemented carbides by the use of carbide micropillars of diameters ranging from 1 to 4 μm, in which the specimen dimensions are comparable to the WC grain size, obtained by means of FIB milling. In this sense, it is well known that intrinsic properties of crystalline materials such as yield stress and strength are highly influenced by extrinsic factors such as the surface/volume ratio. *In situ* uniaxial compression of the micropillars and subsequent scanning electron microscopy examinations provide valuable information on the mechanical behavior of cemented carbides subjected to compressive stresses and exhibit deformation/failure mechanisms observed for WC single crystals as well as WC/Co and WC/WC grain boundaries [13–17].

In general, carbide or carbide–binder interfaces are found to be preferential sites for the nucleation of critical damage events [13].

The size of micropillars significantly affects the established values of mechanical characteristics of cemented carbides, resulting in (1) an increase of tensile strength from 2.5 to 6 GPa when varying the sample size from the centimeter to micrometer regime [18]; (2) a variation of the elastic modulus and fracture strength with the specimen size and Co area fraction within the sample [19]; and (3) a transition of the deformation/failure mechanisms from those observed for WC single crystals toward those of the bulk-like material, as the diameter of the compression tested micropillars increases from 1 to 4 μm [14].

In ref. [20], micropillars of three grades of WC–10 wt.% Co cemented carbide and different WC mean grain sizes—from fine to coarse—having a diameter of 2 μm (Fig. 15.7) were prepared and examined in uniaxial compression tests. Afterward, stress–strain curves were obtained when increasing the loads on the micropillars of each grade until their failure.

Two stress ranges corresponding to different mechanisms of deformations were established based on the characteristics of the stress–strain curves.

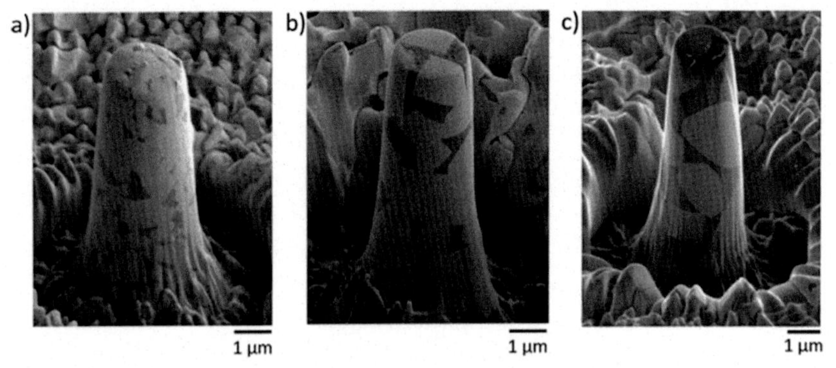

Figure 15.7 SEM images of micropillars of around 2 μm in diameter milled by means of FIB from (A) fine, (B) medium, and (C) coarse WC–Co grades.

Within the first stress range (roughly between 1 and 4 GPa), the plasticity of the micropillars is believed to be directly related to the deformation of the metallic binder. In this case, the higher strength established by examining the fine-grain micropillars is more pronounced constraining the metallic binder. Nevertheless, evidence for plastic events at stress values as low as 1.3 GPa in one medium-grain micropillar indicated the local heterogeneity of binder thickness in the micropillars examined. Although such a value is still significantly higher than the stress range where the unconstrained Co binder would be expected to deform plastically, i.e., between 0.4 and 0.8 GPa, it is also markedly lower than the stress range where the plasticity is expected to occur in bulk-like samples (i.e., above 2 GPa). Further analysis of plastic deformation of the Co binder can be done in the relatively large metallic pools found in the medium-grain grade. Deformation phenomena in the Co binder comprise shape change due to the dislocation movement and evolution of slip systems.

Within the second stress range (nearly between 4 and 6 GPa), strain bursts identified should be related to the deformation occurring either at WC/Co interface–related regions or within WC grains. Although it is reported that the carbide grains exhibit intrinsic yield stress values varying between 6 and 7 GPa, their plastic behavior is clearly evidenced. The presence of thermal residual stresses in the WC phase (compressive on average, but tensile in some regions) is assumed to result in the plasticity being developed at effective stresses lower than those measured in isolated WC single crystals.

Besides irreversible deformation events, microcracking is also established in the tested materials. In the medium-grain grade, microcracks occurred at WC/WC grain boundaries, and to a lesser extend at WC/Co interfaces; in the fine-grain alloy, they are found exclusively at WC/WC grain boundaries.

Thus, micromechanical testing of cemented carbides appears to be a very promising technique for examining mechanisms of deformation and failure of different carbide grades.

15.4 Additive manufacturing of cemented carbide articles

Additive manufacturing (AM) is a powerful tool for rapid prototyping and fabricating metal articles having a complicated geometry. However, this method is known to be used almost solely for the manufacture of articles consisting of pure metals and alloys, so that the fabrication of such metal−ceramic composites as WC−Co cemented carbides by AM is challenging. Nevertheless, the possibility of producing carbide parts with complicated geometry, that cannot be fabricated by conventional technology, e.g., U-shaped articles or cutters and end mills with cooling channels inside, makes AM a promising method, which could potentially be employed in the cemented carbide industry.

In the review paper [21], advantages and disadvantages of different AM processes used for producing WC−Co parts, including selective laser melting (SLM), selective electron beam melting (SEBM), binder jet additive manufacturing (BJAM), 3D gel-printing (3DGP), and fused filament fabrication (FFF), are discussed. The studies of microstructures, defects, and mechanical properties of WC−Co parts manufactured by different AM processes are reviewed with an emphasis on the presently existing challenges.

As it was shown in Ref. [22], porous-free or nearly porous-free cemented carbide articles can be produced at the optimized combination of parameters of the SEBM process. Nevertheless, as a result of the uneven energy distribution, the microstructure of the WC−Co samples obtained by the SLM and SEBM techniques is inhomogeneous. Because of the layer-by-layer printing process, the carbide microstructure is characterized by a layered structure shown in Fig. 15.8 [22].

Micrographs shown in Fig. 15.8 indicate that the microstructures comprise layers with medium-coarse WC grains and large or abnormally large WC grains. The abnormally coarse-grain layers form due to highly localized heating of the WC−Co granules leading to the very intensive

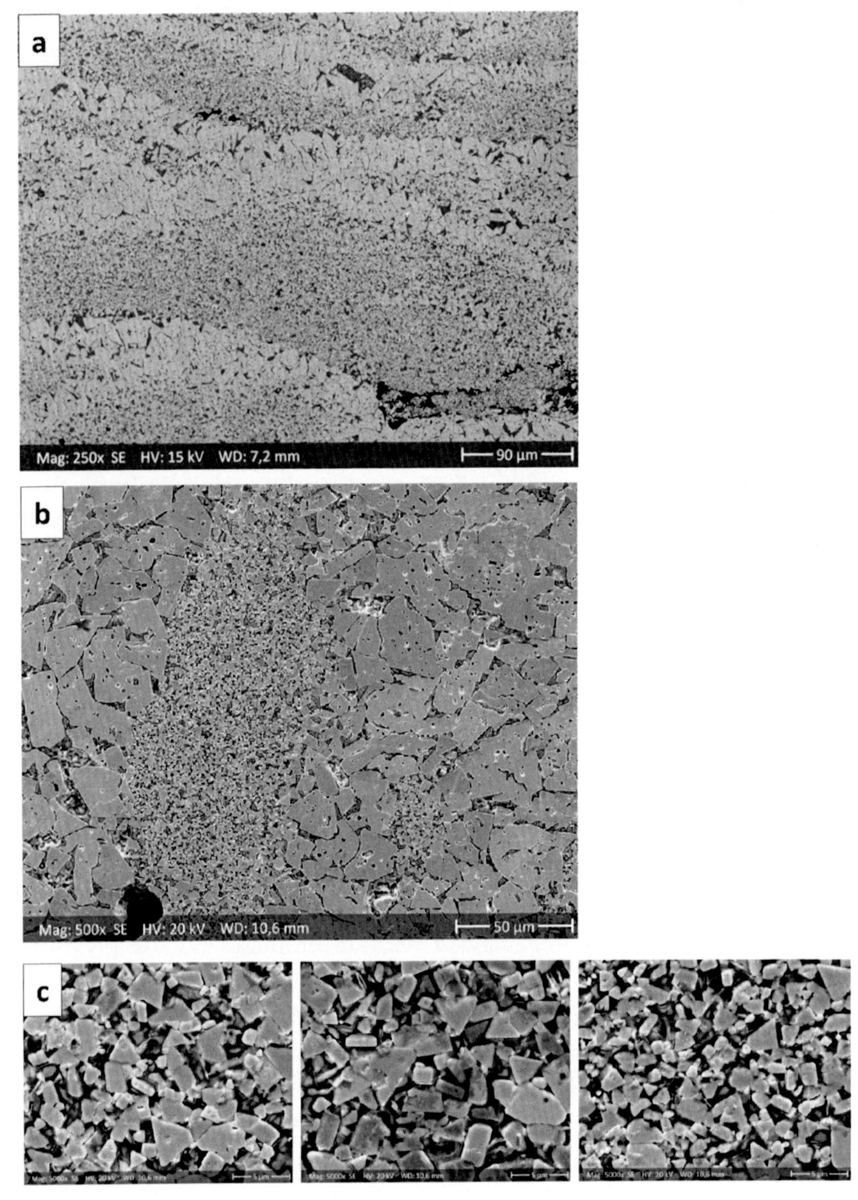

Figure 15.8 Typical microstructures of cemented carbide articles obtained by additive manufacturing (SEBM): (A,B) and (C) - HRSEM images of samples obtained at different parameters of SEBM, (D) - HRSEM images of the coarse-grain layer, (E) - optical microscopy image after etching in the Murakami reagent, (F) - SEM Image of an FIB lamella, (G) - STEM image of the FIB lamella with an *arrow* indicating the region, where an elemental map was recorded, and (H) - elemental map of the region indicated by the *arrow* in (G) comprising a Co inclusion (W is marked with *green* and Co is marked by *violet*). *(Redrawn from Ref. [23].)*

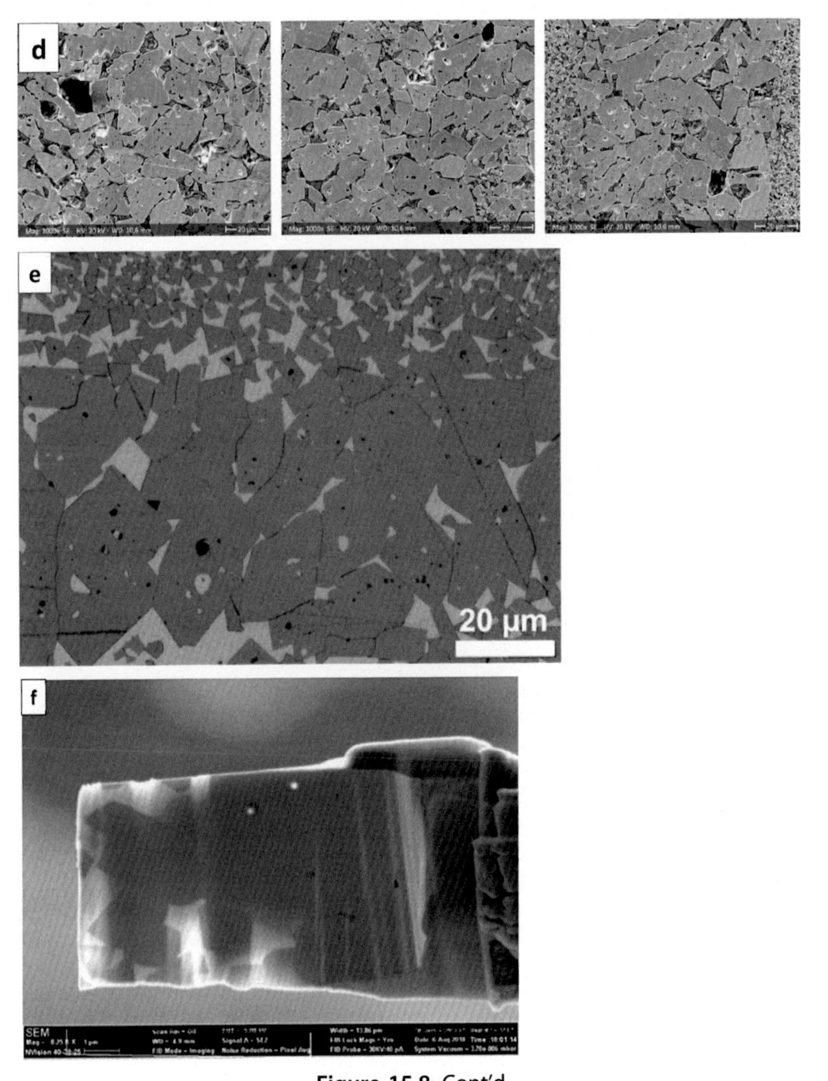

Figure 15.8 Cont'd

WC grain growth in a thin surface layer, where the electron beam is in a direct contact with WC–Co. The thickness values of the fine-grain and coarse-grain layers depend on the combinations of the beam current and scan rate. The average WC grain size in the fine-grain layers is roughly 0.9–1.0 μm, whereas that in the coarse-grain layers is about 7 μm. Usually, the Co distribution in the carbide samples is quite uniform, so that there are almost no Co pools in the microstructure; some separate Co pools

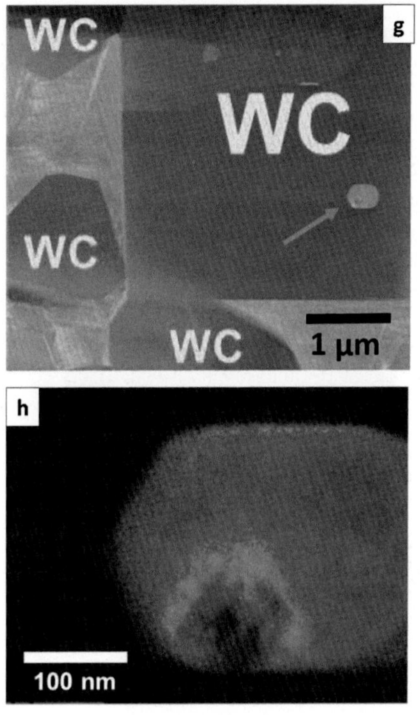

Figure 15.8 Cont'd

sometimes visible on the surface of metallurgical cross-sections are a matter of exception. As it can be seen in Fig. 15.8E, the microstructure of the carbide sample does not contain inclusions of the η-phase. Fig. 15.8E provides evidence that the abnormally large WC grains comprise inclusions of Co and black spots that can be either microporosity or traces of Co inclusions pulled out from the carbide surface during the preparation of metallurgical cross-sections. It is likely that the black spots are a result of grinding and polishing the metallurgical cross-sections, as there are no such black spots on the surface of a lamella obtained by FIB milling (Fig. 15.8F). Results of STEM/EDX studies of part of such abnormally large WC grains confirm that the bright-grey inclusions in the WC grains consist of cobalt (Fig. 15.8G and H). The phenomenon of embedding such Co inclusions into the abnormally large WC grains is related to an extremely high growth rate of the WC grains, which can reach values of several microns per millisecond, as the electron beam moves on the surface of the WC−Co granules very fast. For example, if the scan rate of the electron beam is equal to 1000 mm/s, which corresponds to the value of 1 mm per millisecond, the growth rate of the abnormally large WC grains is of the order of roughly 5—10 microns per millisecond.

Fig. 15.9 shows high-resolution TEM images of the cemented carbide samples obtained by SEBM, particularly, an image of a WC crystal lattice with atomic resolution (Fig. 15.9A), and an HRTEM image of a WC/Co grain boundary (Fig. 15.9B). As it can be seen in Fig. 15.9A, the WC crystal structure in a core region of a coarse WC grain appears to be almost defect free. In general, the WC/Co grain boundaries of the coarse WC grains are characterized by the absence of complexions or inclusions of additional phases (Fig. 15.9B), in spite of the fact that the heating and cooling rates as well as grain growth rates were very high during the AM process.

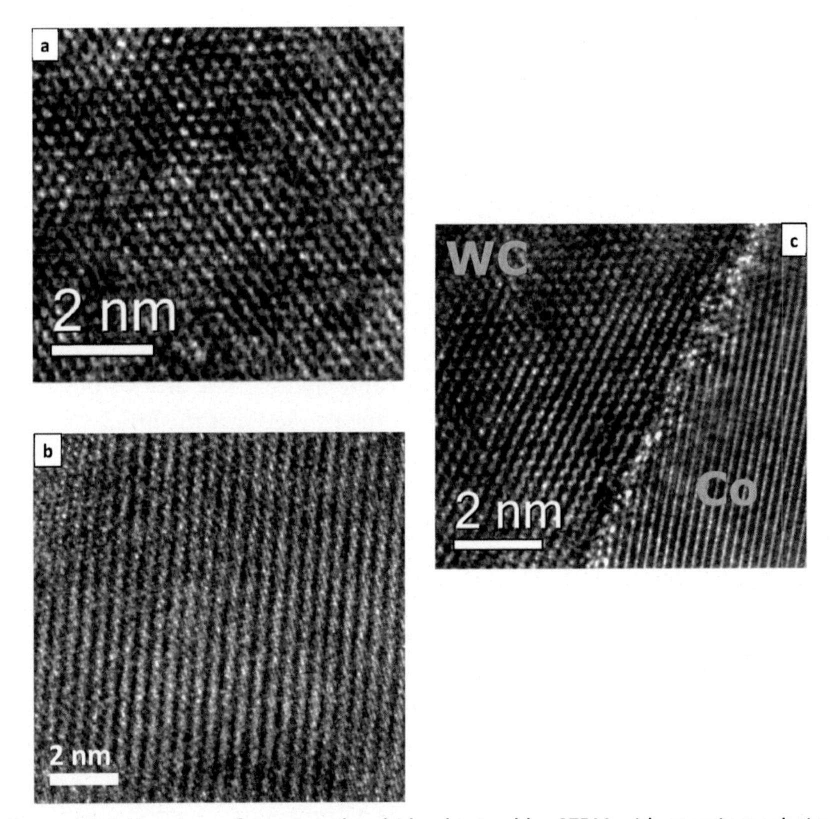

Figure 15.9 Structure of cemented carbide obtained by SEBM with atomic resolution: (A) - HRTEM image of a coarse WC grain in a core region, (B) - HRTEM image of a medium-fine WC grain in a core region, and (C) - HRTEM image of a WC/Co interface.

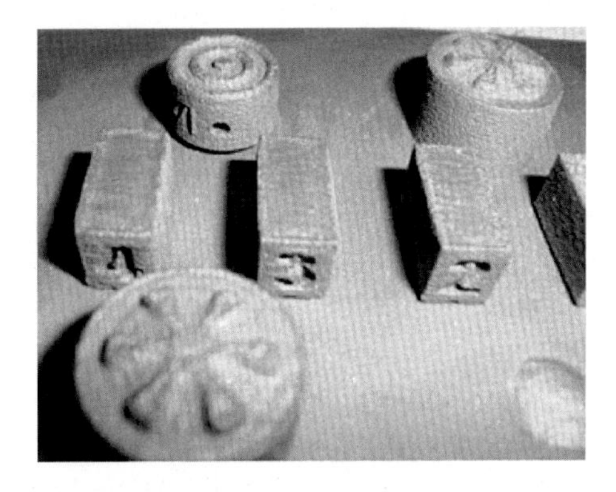

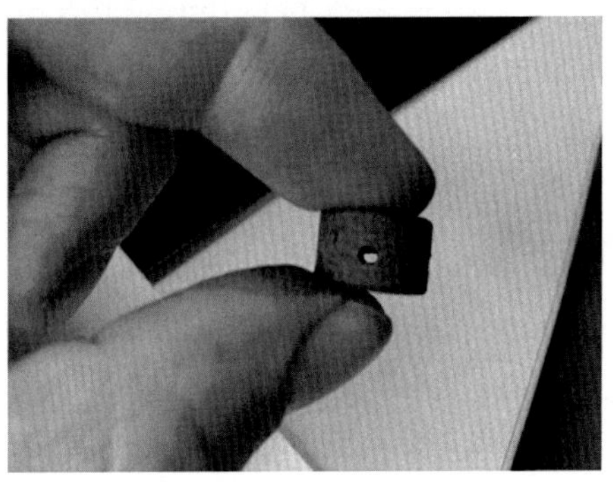

Figure 15.10 Appearance of cemented carbide articles obtained by SEBM at the optimal combination of fabrication parameters.

Fig. 15.10 shows the appearance of cemented carbide articles obtained by AM based on the SEBM technique. One can see that the articles with a very complicated geometry, which cannot be obtained by conventional powder metallurgy techniques, were produced by the AM technology. However, the carbide articles appear to be characterized by a high surface roughness. Indeed, the surface roughness of the carbide articles was found to be high with the top surface characterized by the roughness of Ra = 4.4 μm and Rz = 22.8 μm and the side surface having a roughness of Ra = 24.5 μm and Rz = 142.3 μm. For the majority of applications, such a high value of surface roughness of carbide articles is unacceptable, so the development of special surface treatment techniques of the carbide articles with complicated geometry obtained by AM is needed.

Similar to the SEBM process, in the SLM process, the very high-energy density causes the Co evaporation and carbon loss, making the samples brittle [23,25]. Cracks formation and propagation occur under the effect of high thermal stresses. When the energy density is low, the melting time of Co is insufficient to fill the pores, and the porosity of the samples is high [26].

As a result of conducting the BJAM technique, the WC–Co samples shrink at a high rate. The Co content in the samples decreases with increasing height. Lower Co contents and some decarburization when performing the AM processes lead to the formation of η-phase [27–29].

For the BJAM and 3DGP techniques, it is much easier to obtain fully dense and high-quality WC–Co parts compared to the SLM and SEBM techniques, although the printing accuracy is lower and the processes are more complicated [21]. A big challenge for the BJAM processes is that the Co content in the samples decreases with increasing the height and varying Co content leading to inconsistent properties. Therefore, further research is needed to improve the BJAM processes in order to achieve homogeneous WC–Co parts. Both the 3DGP and FFF processes allow obtaining samples with uniform microstructure and high density, and both have low requirements for powder preparation techniques. The 3DGP and FFF processes have no requirements for the powder flowability, but the WC–Co samples manufactured by these two techniques have a great surface roughness [30,31], which is unacceptable for the majority of applications. Reducing the surface roughness of WC–Co articles is the major challenge for the AM techniques.

Thus, although the AM manufacture of carbide articles shows a promise, there are numerous challenges with respect to obtaining fully dense and high-quality carbide articles of complicated geometry by the existing AM techniques. Also, due to high production costs and complications of the presently existing AM techniques, they are supposed have a limited importance in the cemented carbide industry in future and will be employed for fabricating only restricted types of carbide articles that cannot be produced by conventional powder metallurgy techniques.

15.5 Diamond-enhanced cemented carbides

In the last decade, diamond-containing cemented carbides or diamond-enhanced cemented carbides produced at ultrahigh pressure were a topic of intensive research work. One of the examples of such materials is the so-called "cellular diamond grade" developed and reported in Ref. [32]. This material is composed of "... repeating units of honeycomb composite material of polycrystalline diamond/PCD and WC–Co material ..." and said to combine the high wear resistance of PCD with high fracture toughness of WC–Co. This material can be fabricated only at ultrahigh pressures, at which diamond is thermodynamically stable in contact with WC–Co, which would presumably lead to its high production costs.

In the last years, much attention was paid to the possibility of producing diamond-enhanced carbides not employing ultrahigh pressures (at pressures below 1 GPa), at which diamond is thermodynamically unstable [24]. The production of diamond-enhanced carbides is a great challenge; however, if such materials can be produced on a large scale, this is expected to dramatically increase hardness and improve the wear resistance of WC–Co by at least one order of magnitude without sacrificing the fracture toughness. There are two major approaches to the development of diamond-enhanced carbides: (1) the use of binders not containing Fe group metals in combination with uncoated diamond and (2) the employment of Co or other Fe group metals as a binder in combination with diamond grains coated by barrier layers to prevent the diamond graphitization.

The first approach is based on sintering of diamond and carbides with binders based on either noncarbide-forming chemical elements, such as Cu, Ag, or Al, see e.g., Ref. [33]. Carbide-forming chemical elements, such as Ti or Ti with additions of Cu and Ag to reduce its melting point, see e.g., Ref. [34], can also potentially be used. For the diamond-containing cemented carbides of the first type, the major problem is the very poor wettability of diamond and other carbides by noncarbide-forming metals, which can presumably be improved by adding some carbide-forming elements such as Zr, B, Si, etc. The major problem of the diamond-containing cemented carbides of the second type is the very intensive and fast interaction of diamond with any carbide-forming element during sintering leading to the formation of high-melting point carbides, which suppresses the shrinkage during sintering.

The second approach was developed, patented, and reported in Refs. [35,36]. According to this approach, a wear-resistance material consisting of 3%–60% diamond grains having a coating of least 1 μm of carbide, nitride, or carbonitride of refractory metals can be sintered with Co, Ni, and/or Fe in the solid state under pressure of nearly 33 MPa without diamond graphitization. According to Ref. [36], a material consisting of cemented carbide plus diamond coated with SiC can be sintered at 1300°C at 41 MPa for 3 min by use of pulsed-electric current sintering. No graphitization of diamond in such a material was found and it has a 50% larger fracture toughness and 10 times better wear resistance compared to conventional cemented carbides. A new low-melting point binder of the system, Fe group metal—chromium—silicon—carbon with a melting point of nearly 1140°C for fabrication of diamond-enhanced carbides was developed and patented (Ref. [37]). The diamond-enhanced carbides produced from diamond grains having a TiC coating of 0.5 μm, WC, Co, Cr_3C_2, and Si were sintered at 1160°C with the presence of liquid phase, which allowed obtaining its full density (Fig. 15.11). The wear resistance of this material in a grinding test against a diamond grinding wheel was found to be nearly 100 times higher than that of conventional WC—Co cemented carbides.

Nevertheless, even if carbide—diamond composites are produced at ultrahigh pressures allowing the complete elimination of problems related to the diamond graphitization, their significantly improved wear resistance can be achieved only in tests similar to the grinding test reported in Ref. [37]. As it was established in Ref. [38], additions of a limited number of large diamond grains to the WC—Co material do not allow obtaining significant improvements in wear resistance in the ASTM B611 test and laboratory performance tests on percussive drilling of quartzite. As it can be seen in Fig. 15.12, the wear pattern of such composites after performing the ASTM B611 test is characterized by the presence of heavily worn WC—Co regions among the diamond grains. These regions form due to the impact of extremely fine abrasive alumina particles during the test, which leads to the significant wear of the WC—Co matrix resulting in the fact that the diamond grains become unsupported and can be easily damaged and

Figure 15.11 Diamond-enhanced carbide produced by the use of a binder with a low melting point. *(Redrawn from Ref. [24].)*

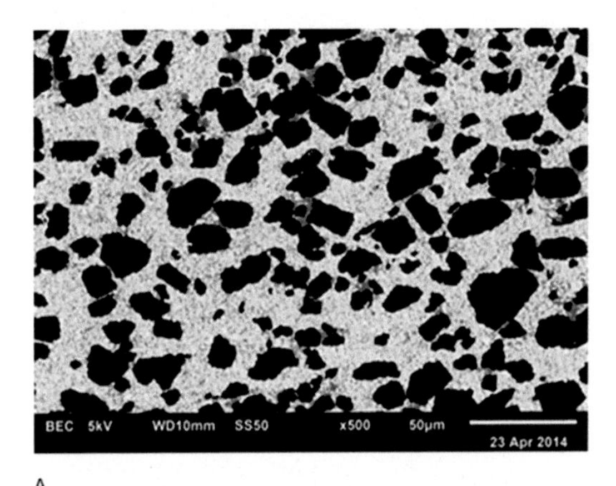

A

B

Figure 15.12 Microstructure of the WC–Co cemented carbide containing 42 wt.% diamond visible as black grains (above) and its wear pattern after performing the ASTM B611 test (below); the damaged diamond grains or holes formed as a result of their detachment from the surface when performing the test are indicated by *arrows*. *(Redrawn from Ref.* [38].*)*

detached during the test (Fig. 15.12). A similar phenomenon occurs in percussive drilling leading to the detachment of numerous diamond grains, which become unsupported as a result of fast wear of the WC–Co matrix. When taking into account the abovementioned, the carbide–diamond composites must contain very fine and densely packed diamond grains forming a structure similar to that of PCD (see Chapter 11) in order that they will have a significantly improved combination of hardness, fracture toughness, and wear resistance in comparison with conventional WC–Co materials. In other words, the thickness of the WC–Co interlayers among the diamond grains must be low (of the order of 0.1–0.3 μm) to prevent the selective wear of such interlayers by fine abrasive particles in mining and construction applications.

It should be noted that in spite of intensive research and numerous patents in the field of diamond-enhanced carbides with a metal binder obtained at pressures lower than 1 GPa, there are still no such materials on the market. This is presumably associated to difficulties related to producing dense diamond-containing cemented carbides with binders based on noncarbide-forming metals. In case of diamond-enhanced cemented carbides with binders on the basis of Fe group metals, the major difficulty appears to be the fabrication of thick, continuous, and defect-free barrier coatings on diamond, which must completely prevent diamond graphitization during liquid-phase sintering.

References

[1] B.B. Straumal, A.A. Mazilkin, B. Baretzky, Grain boundary complexions and pseudopartial wetting, Curr. Opin. Solid State Mater. Sci. 20 (5) (2016) 247–256.
[2] X. Liu, X. Song, H. Wang, X. Liu, F. Tang, H. Lu, Complexions in WC-Co cemented carbides, Acta Mater. 149 (2018) 161–178.
[3] R.H. Willens, E. Buehler, B.T. Matthias, Superconductivity of the transition metal carbides, Phys. Rev. 159 (1967) 327–330.
[4] A.S. Kurlov, A.I. Gusev, Tungsten carbides and WC phase diagram, Inorg. Mater. 42 (2006) 156–163.
[5] I. Konyashin, A. Sologubenko, T. Weirich, B. Ries, Complexion at WC-Co grain boundaries of cemented carbides, Mater. Lett. 187 (2017) 7–10.
[6] X. Liu, J. Zhang, C. Hou, H. Wang, X. Song, Z. Nie, Mechanisms of WC plastic deformation in cemented carbide, Mater. Des. 150 (2018) 154–164.
[7] P. Loginov, A. Zaitsev, I. Konyashin, et al., *In-situ* observation of hardmetal deformation processes by transmission electron microscopy. Part I: deformation caused by bending loads, Int. J. Refract. Metals Hard Mater. 84 (2019) 104997.
[8] P. Loginov, A. Zaitsev, I. Konyashin, et al., *In-situ* observation of hardmetal deformation processes by transmission electron microscopy. Part II: deformation caused by tensile loads, Int. J. Refract. Metals Hard Mater. 84 (2019) 105017.
[9] H. Guo, K. Wang, Y. Deng, et al., Nanomechanical actuation from phase transitions in individual VO2 micro-beams, Appl. Phys. Lett. 102 (2013) 231909–231914.
[10] J.W. Ma, W.J. Lee, J.M. Bae, et al., Carrier mobility enhancement of tensile strained Si and SiGe nanowires via surface defect engineering, Nano Lett. 15 (11) (2015) 7204–7210.
[11] A. Kobler, T. Beuth, T. Klöffel, et al., Nanotwinned silver nanowires: structure and mechanical properties, Acta Mater. 92 (2015) 299–308.
[12] A.A. Zaitsev, I. Konyashin, et al., Radiation-enhanced high-temperature cobalt diffusion at grain boundaries of nanostructured hardmetal, Mater. Lett. 294 (2021) 129746.

[13] A. Naughton-Duszová, T. Csanádi, R. Sedlák, P. Hvizdoš, J. Dusza, Small-scale mechanical testing of cemented carbides from the micro- to the nano-level: a review, Metals 9 (5) (2019) 502, https://doi.org/10.3390/met9050502.
[14] D.A. Sandoval, A. Rinaldi, J.M. Tarragó, J.J. Roa, J. Fair, L. Llanes, Scale effect in mechanical characterization of WC-Co composites, Int. J. Refract. Metals Hard Mater. 72 (2018) 157–162, https://doi.org/10.1016/j.ijrmhm.2017.12.029.
[15] D.A. Sandoval, J.J. Roa, O. Ther, E. Tarrés, L. Llanes, Micromechanical properties of WC-(W,Ti,Ta,Nb)C-Co composites, J. Alloys Compd. 777 (2019) 593–601.
[16] D.A. Sandoval, A. Rinaldi, A. Notargiacomo, O. Ther, J.J. Roa, L. Llanes, WC-base cemented carbides with partial and total substitution of Co as binder: evaluation of mechanical response by means of uniaxial compression of micropillars, Int. J. Refract. Metals Hard Mater. 84 (2019) 105027.
[17] I. El Azhari, J. García, M. Zamanzade, F. Soldera, C. Pauly, C. Motz, L. Llanes, F. Mücklich, Micromechanical investigations of CVD coated WC-Co cemented carbide by micropillar compression, Mater. Des. 186 (2020) 108283.
[18] T. Klünsner, S. Wurster, P. Supancic, R. Ebner, M. Jenko, J. Glätzle, A. Püschel, R. Pippan, Effect of specimen size on the tensile strength of WC–Co hard metal, Acta Mater. 59 (2011) 4244–4252.
[19] T. Namazu, T. Morikaku, H. Akamine, T. Fujii, Mechanical reliability of FIB-fabricated WC–Co cemented carbide nanowires evaluated by MEMS tensile testing, Eng. Fract. Mech. 150 (2015) 126–134, https://doi.org/10.1016/j.engfracmech.2015.
[20] D.A. Sandovala, A. Rinaldic, A. Notargiacomod, O. There, E. Tarrése, J.J. Roaa, L. Llanesa, Influence of specimen size and microstructure on uniaxial compression of WC-Co micropillars, Ceram. Int. 45 (2019) 1593–1594.
[21] Y. Yang, C. Zhang, D. Wang, et al., Additive manufacturing of WC-Co hardmetals: a review, Int. J. Adv. Manuf. Technol. 108 (2020) 1653–1673, https://doi.org/10.1007/s00170-020-05389-5.
[22] I. Konyashin, H. Hinners, B. Ries, et al., Additive manufacturing of WC-13%Co by selective electron beam melting: achievements and challenges, Int. J. Refract. Metals Hard Mater. 84 (2019) 105028.
[23] R.S. Khmyrov, V.A. Safronov, A.V. Gusarov, Obtaining crackfree WC-Co alloys by selective laser melting, Phys. Proc. 83 (2016) 874–881, https://doi.org/10.1016/j.phpro.2016.08.091.
[24] I. Konyashin, in: V. Sarin (Ed.), Cemented Carbides for Mining, Construction and Wear Parts, Comprehensive Hard Materials, Elsevier Science and Technology, 2014, pp. 425–451.
[25] A. Fortunato, G. Valli, E. Liverani, A. Ascari, Additive manufacturing of WC-Co cutting tools for gear production, Lasers Manuf. Mater. Proc. 6 (2019) 247–262, https://doi.org/10.1007/s40516-019-00092-0.
[26] S. Kumar, A. Czekanski, Optimization of parameters for SLS of WC-Co, Rapid Prototyp. J. 23 (2017) 1202–1211, https://doi.org/10.1108/rpj-10-2016-0168.
[27] C.L. Cramer, N.R. Wieber, T.G. Aguirre, R.A. Lowden, A.M. Elliott, Shape retention and infiltration height in complex WC-Co parts made via binder jet of WC with subsequent Co melt infiltration, Addit. Manuf. 29 (2019) 100828, https://doi.org/10.1016/j.addma.2019.100828.
[28] R.K. Enneti, K.C. Prough, T.A. Wolfe, A. Klein, N. Studley, J.L. Trasorras, Sintering of WC-12%Co processed by binder jet 3D printing (BJ3DP) technology, Int. J. Refract. Metals Hard Mater. 71 (2018) 28–35, https://doi.org/10.1016/j.ijrmhm.2017.10.023.
[29] R.K. Enneti, K.C. Prough, Wear properties of sintered WC12%Co processed via binder Jet 3D printing (BJ3DP), Int. J. Refract. Metals Hard Mater. 78 (2019) 228–232, https://doi.org/10.1016/j.ijrmhm.2018.10.003.

[30] X. Zhang, Z. Guo, C. Chen, W. Yang, Additive manufacturing of WC-20Co components by 3D gel-printing, Int. J. Refract. Metals Hard Mater. 70 (2018) 215–223.
[31] W. Lengauer, I. Duretek, M. Fürst, V. Schwarz, J. Gonzalez-Gutierrez, S. Schuschnigg, C. Kukla, M. Kitzmantel, E. Neubauer, C. Lieberwirth, Fabrication and properties of extrusion based 3D-printed hardmetal and cermet components, Int. J. Refract. Metals Hard Mater. 82 (2019) 141–149.
[32] A. Griffo, R. Brown, K. Keshvan, Oil and gas drilling materials, Adv. Mater. Process. (2003) 59–60.
[33] R. Kösters, A. Lüdtke, Wearing Part Consisting of a Diamondiferous Composite, 2005. PCD patent application WO2005/118901A1.
[34] T. Ikawa, K. Miyashi, Fitting Diamond in Metal, 1972. Japan patent JP47044134.
[35] H. Brandrup-Wognsen, S. Ederyb, Diamond-impregnated Hard Material, 1996. US patent 5,723,177.
[36] Y. Miyamoto, H. Moriguchi, A. Ikegaya, Ultrahard Composite Members Comprising Diamond Particles Dispersed in Cemented Carbide or Cermets Matrix Having High Hardness and Wear-Resistance and Their Preparation, 1997. Japan patent JP10310838.
[37] I. Konyashin, S. Hlawatschek, B. Ries, F. Lachmann, A Superhard Element, a Tool Comprising Same and Methods for Make Such Superhard Element, 2010. PCT Patent Application WO2010/103418A1.
[38] A.J. Gant, I. Konyashin, B. Ries, A. McKie, R.W.N. Nilen, J. Pickles, Wear mechanisms of diamond-containing hardmetals in comparison with diamond-based materials, Int. J. Refract. Metals Hard Mater. 71 (2018) 106–114.

Index

Note: 'Page numbers followed by "*f*" indicate figures and "*t*" indicate tables.'

Printed in the United States
by Baker & Taylor Publisher Services